STUDIES OF TEMPORAL TRENDS OF POLLUTION IN SELECTED COASTAL AREAS BY THE APPLICATION OF ISOTOPIC AND NUCLEAR TECHNIQUES

The following States are Members of the International Atomic Energy Agency:

AFGHANISTAN
ALBANIA
ALGERIA
ANGOLA
ANTIGUA AND BARBUDA
ARGENTINA
ARMENIA
AUSTRALIA
AUSTRIA
AZERBAIJAN
BAHAMAS
BAHRAIN
BANGLADESH
BARBADOS
BELARUS
BELGIUM
BELIZE
BENIN
BOLIVIA, PLURINATIONAL STATE OF
BOSNIA AND HERZEGOVINA
BOTSWANA
BRAZIL
BRUNEI DARUSSALAM
BULGARIA
BURKINA FASO
BURUNDI
CABO VERDE
CAMBODIA
CAMEROON
CANADA
CENTRAL AFRICAN REPUBLIC
CHAD
CHILE
CHINA
COLOMBIA
COMOROS
CONGO
COOK ISLANDS
COSTA RICA
CÔTE D'IVOIRE
CROATIA
CUBA
CYPRUS
CZECH REPUBLIC
DEMOCRATIC REPUBLIC OF THE CONGO
DENMARK
DJIBOUTI
DOMINICA
DOMINICAN REPUBLIC
ECUADOR
EGYPT
EL SALVADOR
ERITREA
ESTONIA
ESWATINI
ETHIOPIA
FIJI
FINLAND
FRANCE
GABON
GAMBIA, THE
GEORGIA
GERMANY
GHANA
GREECE
GRENADA
GUATEMALA
GUINEA
GUYANA
HAITI
HOLY SEE
HONDURAS
HUNGARY
ICELAND
INDIA
INDONESIA
IRAN, ISLAMIC REPUBLIC OF
IRAQ
IRELAND
ISRAEL
ITALY
JAMAICA
JAPAN
JORDAN
KAZAKHSTAN
KENYA
KOREA, REPUBLIC OF
KUWAIT
KYRGYZSTAN
LAO PEOPLE'S DEMOCRATIC REPUBLIC
LATVIA
LEBANON
LESOTHO
LIBERIA
LIBYA
LIECHTENSTEIN
LITHUANIA
LUXEMBOURG
MADAGASCAR
MALAWI
MALAYSIA
MALI
MALTA
MARSHALL ISLANDS
MAURITANIA
MAURITIUS
MEXICO
MONACO
MONGOLIA
MONTENEGRO
MOROCCO
MOZAMBIQUE
MYANMAR
NAMIBIA
NEPAL
NETHERLANDS, KINGDOM OF THE
NEW ZEALAND
NICARAGUA
NIGER
NIGERIA
NORTH MACEDONIA
NORWAY
OMAN
PAKISTAN
PALAU
PANAMA
PAPUA NEW GUINEA
PARAGUAY
PERU
PHILIPPINES
POLAND
PORTUGAL
QATAR
REPUBLIC OF MOLDOVA
ROMANIA
RUSSIAN FEDERATION
RWANDA
SAINT KITTS AND NEVIS
SAINT LUCIA
SAINT VINCENT AND THE GRENADINES
SAMOA
SAN MARINO
SAUDI ARABIA
SENEGAL
SERBIA
SEYCHELLES
SIERRA LEONE
SINGAPORE
SLOVAKIA
SLOVENIA
SOMALIA
SOUTH AFRICA
SPAIN
SRI LANKA
SUDAN
SWEDEN
SWITZERLAND
SYRIAN ARAB REPUBLIC
TAJIKISTAN
THAILAND
TOGO
TONGA
TRINIDAD AND TOBAGO
TUNISIA
TÜRKİYE
TURKMENISTAN
UGANDA
UKRAINE
UNITED ARAB EMIRATES
UNITED KINGDOM OF GREAT BRITAIN AND NORTHERN IRELAND
UNITED REPUBLIC OF TANZANIA
UNITED STATES OF AMERICA
URUGUAY
UZBEKISTAN
VANUATU
VENEZUELA, BOLIVARIAN REPUBLIC OF
VIET NAM
YEMEN
ZAMBIA
ZIMBABWE

The Agency's Statute was approved on 23 October 1956 by the Conference on the Statute of the IAEA held at United Nations Headquarters, New York; it entered into force on 29 July 1957. The Headquarters of the Agency are situated in Vienna. Its principal objective is "to accelerate and enlarge the contribution of atomic energy to peace, health and prosperity throughout the world".

IAEA-TECDOC-2070

STUDIES OF TEMPORAL TRENDS OF POLLUTION IN SELECTED COASTAL AREAS BY THE APPLICATION OF ISOTOPIC AND NUCLEAR TECHNIQUES

REPORT OF A COORDINATED RESEARCH PROJECT

INTERNATIONAL ATOMIC ENERGY AGENCY
VIENNA, 2025

COPYRIGHT NOTICE

Publishing Section
International Atomic Energy Agency
Vienna International Centre
PO Box 100
1400 Vienna, Austria
tel.: +43 1 2600 22529 or 22530
email: sales.publications@iaea.org
www.iaea.org/publications

For further information on this publication, please contact:

Radiometrics Laboratory
Marine Environment Laboratories
International Atomic Energy Agency
4a Quai Antoine 1er, 98000
Principality of Monaco
Email: Official.Mail@iaea.org

Printed by the IAEA in Austria
March 2025
https://doi.org/10.61092/iaea.q1tx-6r1g

IAEA Library Cataloguing in Publication Data

Names: International Atomic Energy Agency.
Title: Studies of temporal trends of pollution in selected coastal areas by the application of isotopic and nuclear techniques / International Atomic Energy Agency.
Description: Vienna : International Atomic Energy Agency, 2025. | Series: IAEA TECDOC series, ISSN 1011-4289 ; no. 2070 | Includes bibliographical references.
Identifiers: IAEAL 24-01711 | ISBN 978-92-0-131024-8 (paperback : alk. paper) | ISBN 978-92-0-131124-5 (pdf)
Subjects: LCSH: Marine pollution — Analysis. | Water — Pollution. | Radioactive dating. | Sedimentation analysis.

FOREWORD

Monitoring and assessing the impact of radioactive and non-radioactive contamination on the environment is a growing need that can be addressed using nuclear techniques. The IAEA supports Member States in using nuclear applications to assess the status and trends of marine pollution with the purpose of making environmental management decisions based on scientific information. One such technique is the quantification of unsupported ^{210}Pb concentrations in sediments to track the movement of sediment and contaminants in the environment, to determine sedimentation rates and dates of sediment deposition and to reconstruct historical contamination trends. Several projects have been carried out in the context of the collaboration between the IAEA and the Food and Agriculture Organization of the United Nations (FAO) and, since the end of the twentieth century, the use of isotopic techniques has been promoted to facilitate soil conservation, for which ^{210}Pb concentrations have been determined to study soil redistribution, especially in the face of anthropogenic practices and climate change impacts.

The IAEA has also encouraged the use of ^{210}Pb dating of sediment to estimate the sediment deposition in coastal areas and assess the effects of anthropogenic activities on marine environmental health. Within the scope of technical cooperation projects, ^{210}Pb dating was used to track growing sedimentation in the wider Caribbean region and its effects on marine biota.

Analytical difficulties highlighted by all these projects have encouraged proficiency tests to identify sources of error and common problems in ^{210}Pb quantification. To further address common issues in applying ^{210}Pb dating to environmental studies, a coordinated research project entitled Study of Temporal Trends of Pollution in Selected Coastal Areas by the Application of Isotopic and Nuclear Tools was implemented with the objective of producing a comprehensive status and gap analysis of the application of nuclear techniques, in particular ^{210}Pb sediment dating, in coastal pollution studies.

The IAEA is grateful to the participants of the coordinated research project. The IAEA officers responsible for this publication were M. Rožmarić, K. Telfeyan and I. Osvath of the IAEA Marine Environment Laboratories.

CONTENTS

1. INTRODUCTION

1.1. BACKGROUND

Coastal zones, and the ocean as a whole, provide critical environmental functions, such as climate regulation, biogeochemical cycling hotspots, storm attenuation, and habitats for biodiversity [1.1]. About 40% of the global population lives near the coast (within 100 kilometers) and relies on coastal ecosystems for their livelihoods, including food, transportation, shipping, industry, recreation, and tourism. In recent decades, the rapid urbanization and industrialization of the coasts has led to unprecedented levels of contamination by microplastics, persistent organic pollutants, heavy metals, radionuclides, excess nutrients, etc. [1.2, 1.3]. The growing extent and degree of marine pollution has prompted policy makers to enact strategies to minimize the detrimental effects of anthropogenic activities that stress coastal resources, and UNEP [1.4] has laid out guidelines for coastal governance strategies. Yet, determining the efficacy of these measures remains challenging owing to the complex interactions of biogeochemical processes and limitations on analytical capabilities. Furthermore, identifying the source, transport routes, contamination hotspots, and background values requires a spatially comprehensive long-term time-series. Where historical data do exist, comparison of contaminant concentrations before and after large releases can be made to determine the extent of contamination, zones of accumulation and partitioning between water, sediment and biota. which are often unavailable. However, in many places of the world such records are unavailable. In the absence of long-term information, sediment dating techniques provide one such tool to study contamination trends in coastal systems to inform coastal management policy. The evaluation of contaminant concentrations in sediment cores dated with the ^{210}Pb method allows reconstructing historical trends of contamination and establishing background levels to serve as a baseline to evaluate the contamination degree and setting remediation targets.

The specific tasks carried out during the CRP K41016 consisted of discussing coastal contamination studies in selected areas around the world and the use of ^{210}Pb-dated sediment cores for the reconstruction of contamination trends. It included an overview on the ^{210}Pb dating method, examples of its applications in specific case studies, conducting a critical review of pitfalls and challenges in the application of the method, preparing a harmonized methodology and guidelines to reconstruct temporal trends of contamination; and preparing two technical documents (IAEA-TECDOCs), one on sediment dating protocols, and a second, on coastal contamination case studies from participating member states.

The tasks were accomplished through expert meetings and interlaboratory comparisons (ILC). The first ILC involved a ^{210}Pb, ^{137}Cs, and ^{226}Ra radioisotope dating exercise of two sediment cores; and the second, labs determined activities of radionuclides and concentrations of heavy metals in sediments and produced an age-depth model of a sediment core. The ILC results highlighted common mistakes or erroneous assumptions that lead to misuse of the ^{210}Pb dating method. Two peer-reviewed publications (Barsanti et al., 2020; García-Tenorio et al., 2020) evaluated the results of the ICLs to identify the possible problems and limitations of the ^{210}Pb dating method. One publication was produced to provide guidelines for sampling, radiometric analysis and ^{210}Pb-dating of sediment cores; and this publication provides the compilation of coastal pollution studies conducted by participating members of the CRP.

1.2. OBJECTIVE

The overarching objective of this publication is to provide, by way of 10 case studies, an overview of current research on sources, distribution, and trends of contaminants in marine environments, including the application of radiometric dating of sediments. This publication was developed by experts in radiometric dating and contamination assessment in the marine environment. It is designed to inform researchers and Member States about the monitoring and evaluation of temporal trends of contamination in the marine realm around the world, that could serve the readers as up-to-date examples of studies addressed to assist in environmental coastal management. It also presents examples on the latest and most accurate reconstruction of age-depth models and its application to the planning of sustainable coastal resource management which, in a general sense, is a considerable challenge facing many Member States.

1.3. SCOPE

This publication provides an overview of local and regional case studies on radioactive and nonradioactive contamination around the globe. Case studies were selected to represent a wide range of geographic locations, including coasts in Africa, Asia, Europe, North and South America. The studies represent a range of geologically different coasts and contaminants in order to be relevant to as many readers across the globe as possible. On the temporal trends of concentration and inventory of artificial radionuclides, Strumińska-Parulska et al., identify the Chernobyl release as the greatest contributor of Pu isotopes to certain regions across the Baltic Sea. Kusakabe and Kambayashi demonstrate the different residence times of Cs in seawater and in sediments and the mixing within a sediment column that can occur, providing information that is valuable for understanding post-depositional migration of ^{137}Cs, after the Fukushima Daiichi Nuclear Power Plant accident in March 2011. Eriksson et al. compare actinide concentrations in a dated sediment core with historical record to analyze releases from the Studsvik nuclear facility.

Some case studies in this publication demonstrate the strengths and weaknesses of the ^{210}Pb dating method for the retrospective reconstruction of temporal variations of excess contamination or, the effects of corrective measure to mitigate or eliminate the contamination in coastal areas. For example, Laissaoui et al. used the ^{210}Pb method to date sediment cores in coastal Moroccan lagoons and identified maxima of trace metal concentrations corresponding to the operation of industrial plants and decreasing concentrations corresponding to the enactment of environmental management policies. Similarly, Al-Rousan et al. used ^{210}Pb age models to quantify increases in sedimentation rates resulting from rapid urbanization in the port of Aqaba. Ruiz Fernández et al. highlight the utility of ^{210}Pb dating for establishing a baseline of contaminant concentration prior to industrialization, providing critical information for assessing remediation efforts and separating natural background levels from pollution. They demonstrate that the load of certain trace metal concentrations, which have naturally enriched background concentrations, had increased further due to a combination of factors related to the oil industry, but they also note that environmental regulations have successfully led to some decrease in contamination in recent decades.

The case studies also highlight that care needs to be taken when applying $^{210}Pb_{ex}$ age-depth models. Variations in sediment accumulation rates and post-depositional physical disturbances can render invalid the assumptions inherent in the age models. Yii et al. used the Constant Flux-Constant Sedimentation (CF:CS) model, which proved challenging due to complicated $^{210}Pb_{ex}$ profiles. The lack of confidence in their age model is attributed to sedimentary processes at the study site, including variations sources, mixing and accumulation rates of the sediment, as well as grain size distribution. Such sedimentary processes are common confounding factors in ^{210}Pb dating and highlight the importance of site reconnaissance prior to sampling. Nevertheless, they were able to use this data to estimate preliminary sediment accumulation rates and assess trace metal enrichment. Dung et al. attempt to use a variety of age-depth models and conclude that the Constant Initial Concentration (CIC) and the CF:CS models do not apply to the whole cores, but can apply to segments of the cores, and that a composite Constant Rate of Supply model can be applied with caution and with ^{137}Cs validation.

The use of ^{137}Cs as a validation tool for the ^{210}Pb age dating method is the most common, but its use is limited in the Southern Hemisphere and worldwide with time as ^{137}Cs decays, triggering the need for alternative validation methods. Godoy et al. present alternative methods that rely on site-specific knowledge. For example, they observed the PCB-based Askarel in sediment cores and constrained the sediment dating to its use over a specific time interval from its introduction to ban.

1.4.STRUCTURE

This chapter provides a brief background on the objectives of the CRP that supported this publication. It introduces the purpose, utility, and challenges of ^{210}Pb dating, which are elaborated on in subsequent sections. The remainder of this publication is organized geographically. Chapters 2 and 3 examine contamination trends in Africa (Morocco and Namibia), Chapters 4-7 in Asia (Japan, Jordan, Malaysia, Vietnam), Chapters 8 and 9 in Europe (Poland and Sweden), Chapter 10 in South America (Brazil), and Chapter 11 in Central America (Mexico). Each case study examines historical trends in contamination with the help of the ^{210}Pb dating technique. The authors highlight the benefits and the difficulties encountered during applications of the ^{210}Pb dating model. Chapter 12 provides general conclusions and implications for coastal zone management.

1.5.REFERENCES

[1.1] IOC/UNESCO, IMO, FAO, UNDP, A Blueprint For Ocean and Coastal Sustainability, IOC/UNESCO, Paris (2015).

[1.2] LU, Y., YUAN, J., LU, X., SU, C., ZHANG, Y., WANG, C., CAO, X., LI, Q., SU, J., ITTEKKOT, V., BARBUTT, R. A., BUSH, S., FLETCHER, S., WAGEY, T., KACHUR, A., SWEIJD, N., Major threats of pollution and climate change to global coastal ecosystems and enhanced management for sustainability, Environ. Pollut. **239** (2018) 670-680.

[1.3] WILHELMSSON, D., THOMPSON, R. C., HOLMSTROM, K., LINDEN, O., ERIKSSON-HAGG, H., "Marine pollution," Managing Ocean Environments in a Changing Climate: Sustainability and Economic Perspectives, Elsevier, Amsterdam, The Netherlands (2013).

[1.4] FLETCHER, S., LU, Y., ALVAREZ, P., MCOWEN, C., BANINLA, Y., FET, A. M., HE, G., HELLEVIK, C., KLIMMEK, H., MARTIN, J., MENDOZA ALFARO, R., PHILIS, G., RABALAIS, N., RODRIGUEZ ESTRADA, U., WASTELL, J., WINTON, S., YUAN, J. A., Governing Coastal Resources: Implications for a Sustainable Blue Economy, Report of the International Resource Panel, United Nations Environmental Programme, Nairobi, Kenya (2021).

2. RADIOMETRIC DATING OF SEDIMENT CORES FROM OUALIDIA AND SIDI MOUSSA COASTAL LAGOONS (WESTERN MOROCCO): IMPLICATIONS FOR ESTABLISHING RECENT POLLUTION RECORDS

A. LAISSAOUI, A. BENKDAD, A. EL YAHYAOUI, H. AIT BOUH
National Centre for Nuclear Energy, Science and Technology
Rabat, Morocco
E-mail: laissaoui@cnesten.org.ma

A. BENMHAMMED, N. ZIAD, E. CHAKIR
University Ibn Tofail
Kenitra, Morocco

N. MEJJAD, O. EL HAMMOUMI, A. FEKRI
University Hassan II of Casablanca
Casablanca, Morocco

Abstract

The Oualidia and Sidi Moussa lagoons are two coastal ecosystems located on the Atlantic coast of Morocco and constitute important areas of large potential for economic development. They are affected by various anthropogenic activities such as urban, tourism, and agriculture development, and increasing industry having NORM as a by-product. The paper addresses the assessment of present and past environmental conditions in the study area by investigating sediment cores as environmental archives for recent pollution history. Five sediment cores were collected from the lagoons and analyzed by gamma spectrometry, ICP-MS and Instrumental NAA for radionuclides and trace metal determination. Substantial increases of ^{210}Pb activities were detected at specific depths in the Oualidia lagoon cores, attributable to high atmospheric ^{222}Rn concentration in the study area. A modified Constant Rate of Supply model, taking account of the change in ^{210}Pb flux, was applied for establishing the relations age/depth and sediment accumulation rates.^{137}Cs was used as a time-marker for validation of the obtained chronologies, although the profiles were affected by downward post-depositional migration, especially in cores with low sedimentation rates. Activities of ^{238}U and ^{210}Pb in the upmost layers of the cores collected from the Sidi Moussa lagoon were relatively high, reaching values as high as 350 Bq/kg and 2500 Bq/kg, respectively. In addition, Cd, Pb, Cr, Cu and Zn concentrations were higher than those reported for the Upper Continental Crust. Their profiles presented a visible peak of maximum concentration corresponding to the period of 1980s according to the obtained chronology. Enrichment Factors were calculated to assess the degree of sediment contamination throughout the years. Maximal concentrations were registered in Cu, Co, As and Ni profiles at a depth fitting the commissioning of the plants of phosphate processing. Decreasing concentrations in recently deposited sediment is indicative of the positive impact of the environmental policy implemented since early 2000s by the phosphate industrial operator.

2.1. INTRODUCTION

The Oualidia and Sidi Moussa lagoons are two seawater bodies located on the Atlantic littoral of Morocco and belonging to a wetland of ecological and biological importance protected by the Ramsar Convention (site N° 1474). The lagoonal complex is composed of two coastal lagoons and some wetlands separated from the ocean by elongated dunes. Anthropogenic activities such as oyster farming, agriculture, fishing, tourism and industry involving NORM are the main environmental stressors in the area.

The Oualidia lagoon is abundant with natural resources and, therefore, offers a space with a large potential for economic activities as well as for biodiversity enrichment of both faunal and floral species such as wading birds, marine organisms and seagrasses. The lagoon hosts several kinds of migratory birds and turtle species. Fishing and oyster farming have been providing, since 1970, a source of income to the local population, among other activities such as livestock and crop production, which introduce large amounts of manufactured fertilizer to the environment. Many research studies were carried out during the last decades in the Oualidia lagoon with the aim to study its environmental conditions by using multidisciplinary approaches [2.1, 2.2, 2.3].

The lagoon of Sidi Moussa is extended around 4 km^2 area, and it is protected from ocean currents and waves by sandy combined dunes. The lagoon is connected to the Atlantic Ocean by a narrow pass and composed of a main waterway from which several small channels branch off. Bathymetry is characterized by a maximum depth of about 5 m, decreasing gradually towards the internal part of the lagoon. It is affected by anthropogenic activities, of which the phosphate industry is the most important among others. The biggest phosphoric acid plant of Jorf Lasfar, located at 15 km from the lagoon, started production in 1986. Many environmental researches were carried out in Sidi Moussa [2.4, 2.5, 2.6, 2.7] to evaluate levels contamination by toxic metals in sediment and organisms commonly used as ecosystem quality indicators. In a recent study including five lagoons along the Atlantic littoral of Morocco, the Sidi Moussa lagoon was found to be the most contaminated by trace metals. The primary source of the abnormally high concentrations, in particular those of cadmium, has been attributed to the effluents discharged from the fertilizer plants rich enough of toxic elements [2.9]. In addition, the North-Western African upwelling was also reported to be a natural source of Cd to the lagoons [2.10]. It is worth noting that there are, based on available knowledge, no data on radionuclide activities in the scientific literature in the Sidi Moussa lagoon and, therefore, the results produced in this research could be considered as reference values for forthcoming monitoring work. High activities of ^{210}Pb in the Oualidia lagoon sediments were reported in recent studies [2.2, 2.11, 2.12], attributed mainly to atmospheric deposition.

Dating of sediment deposition in natural aquatic systems using ^{210}Pb as a chronometer is widely used in environmental studies involving age-depth associations over the last 120 years. Classic models, such as Constant Initial Concentration (CIC), Constant Flux-Constant Sedimentation (CF:CS), and Constant Rate of Supply (CRS) are commonly used for this purpose. A compilation of the potential and limitations of these models can be found in Kirchner et al [2.13]. The CRS model usually provides better results than the CIC or CF:CS models, when validation of the obtained chronologies is carried out using ^{137}Cs or other independent tracers. However, some assumptions are to be fulfilled for the appropriate application of the mentioned models; otherwise, the computed ages could be deemed as not reliable. Constant delivery of ^{210}Pb to sediment and no redistribution of particles after deposition are the most sensitive assumptions.

Processes such as compaction, variable sedimentation rate and/or initial concentration and sediment reworking were included in other complex dating models [2.14, 2.15, 2.16]. Conventional models always assume a constant ^{210}Pb flux over several decades, which is not often the case. Indeed, as ^{210}Pb is a disintegration product of the gaseous ^{222}Rn, high concentrations of ^{222}Rn in the atmosphere could produce a large amount of ^{210}Pb, which is transferred to the water column via dry and wet deposition and, consequently, to the sediment [2.17, 2.18]. Other sources of ^{210}Pb in coastal sediments include that resulting from in-situ decay of ^{222}Rn coming itself from ^{226}Ra decay and leaching from adjacent continental areas. Nevertheless, significant variations at different timescales of ^{210}Pb Flux have been reported in some sites worldwide [2.19]. In such cases, significant fluctuations in unsupported ^{210}Pb delivery to the sediment-water interface were deduced from vertical profiles in sediment columns.

The paper addresses the potential use of radiometric dating of sediment cores, which are usually perceived as natural archives, to reconstruct the recent history of contaminant inputs to Oualidia and Sidi Moussa lagoons. Such an approach allows establishing the depth-age relationships and sediment accumulation rates through the appropriate use of ^{210}Pb and ^{137}Cs profiles as a chronometer and independent time marker, respectively. The sediment deposition chronologies obtained will be used to assess the degree of sediment enrichment by metals throughout the last decades by computing the Enrichment Factors throughout the cores.

2.2. MATERIAL AND METHODS

2.2.1. Sampling and samples pretreatment

Five sediment cores were retrieved from the bank main channels of both lagoons at low tide using a Uwitec corer fitted with PVC tubes with a 10-cm internal diameter. Two of them, CO-1 and CO-2 were collected from Oualidia lagoon (32°45'12" N, 009°00'52" W, 52 cm and 32°46'24'' N; 008°59'07" W, 35 cm, respectively) and three, CSM-1, CSM-2 and CSM-3, for Sidi Moussa lagoon (32°59'44.1" N; 8°44'7.4" W, 55 cm, 32°59'42.2" N; 8°44'3.1" W, 34 cm and 32°59'16" N; 8°44'9" W, 44 cm, respectively). Figure 2.1 shows the coring sites location, which were selected on the basis of currentology considerations [2.20]. Weak maximal currents, less than 0.40 m s^{-1}, prevail in the sampling sites, which should promote the deposition of particles. In addition to the cores, three samples of 5 L of water were collected in polyethylene bottles along the Oualidia lagoon. The samples were acidified with concentrated hydrochloric acid to pH 2 to prevent adsorption of radionuclides onto the internal walls.

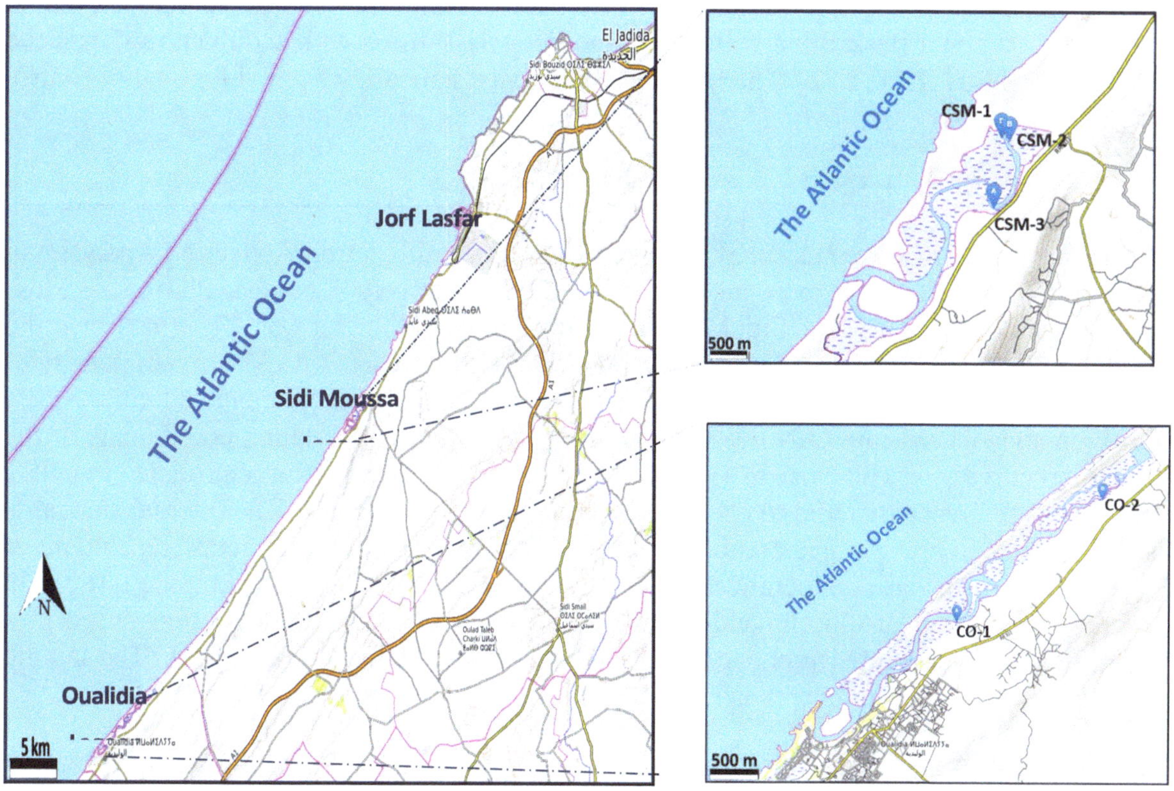

FIG. 2.1. Map of the study area showing the two lagoons with core sampling locations.

Immediately after the collection, sediment cores were sliced at 2-cm intervals to impede any particle redistribution throughout the cores. Samples were labelled and transferred to the laboratory and properly stored before radiometric and elemental analyses. Sediment sections were dried at 80°C in an oven to constant weight, gently ground with a mortar and pestle and homogenised. Bulk density was determined as the ratio of dry mass and volume of each section. Water samples were filtered using 0.45-µm Millipore® filters under vacuum and then evaporated on a hot plate to a final volume of around 200 ml. Concentrated solutions were transferred quantitatively to polyethylene flasks by adding 2M hydrochloric acid to dissolve solids. The flasks were sealed hermetically to allow equilibrium between ^{226}Ra and its progenies and kept for one month prior to gamma counting.

2.2.2. Radionuclide measurements

For the analysis of ^{226}Ra, ^{210}Pb, ^{137}Cs and ^{238}U, an ORTEC gamma spectrometer of low-background and consisting of a broad energy (from 1 KeV to 10 MeV) germanium detector, housed in 10-cm-thick lead shield of high purity and a copper layer was used. The detector has a relative efficiency of 50% and a resolution in the region of the 1332 keV 60Co γ-peak of 1.8 keV. The Maestro® software was used for spectra visualisation and data acquisition.

Known masses of sediment samples were weighed in Nalgene flasks and sealed to avoid radon escape and stored for three weeks or more to reach equilibrium among ^{226}Ra and its decay products. An Amersham liquid multi-gamma source was used for efficiency and energy calibration in the working geometry (500 ml for water and 50 ml for sediment). Detector efficiency, count rate, gamma intensity and sample weight were the input parameters for activity calculation (in Bq kg^{-1}). Supported ^{210}Pb was assumed to be in equilibrium with ^{226}Ra, so that excess ^{210}Pb was obtained as the total ^{210}Pb minus ^{226}Ra activity at each layer. Correction of ^{210}Pb activities for self-absorption was carried out by counting a punctual source of ^{210}Pb on an empty flask, on a flask filled with water and on a flask filled with sediment to obtain the attenuation factors [2.21, 2.22]. The IAEA-327 reference standard was used for quality control check.

2.2.3. Elemental analyses

Metal concentrations were measured by INAA (Instrumental Neutron Activation Analysis) in samples from Oualidia lagoon and by ICP-MS (Inductively Coupled Plasma – Mass Spectrometry) in samples from Sidi Moussa lagoon. The spectrometer was a Thermo Scientific XSERIES 2, adjusted to deliver minimal ratios of Ba^{2+}/Ba^{+} and CeO^{+}/Ce^{+} optimal intensity of the chemical species. External calibration was achieved by running monoelement liquids SPEX CertiPrep and Astasol-mix (AN 9088 (MN)) for lanthanoids). Correlation coefficients of the calibration curves were all close to 1. SRM 694 'Western Rock Phosphate' provided by NIST was used as reference material to check the accuracy of the procedure. The internal standards used were Rh and Ir at a concentration of 0.01 mg l^{-1}. For the analysis of metals, an amount of 50 mg of sediment was demineralized in a microwave oven with 3 ml of ultrapure nitric acid 60 % and 5 ml of hydrofluoric acid 40 %. The solution was quantitatively transferred to a Teflon flask and evaporated after adding boric acid to eliminate any remaining HF. The sample was then diluted with ultrapure water to the working volume and measured by ICP-MS.

For INAA analysis, samples were irradiated with thermal neutrons in the Moroccan research reactor, Triga Mark II, at fluxes of 5.0 10^{12} n cm^{-2} s^{-1} and 2.0 10^{13} n cm^{-2} s^{-1} at the pneumatic transfer system (PTS) and at the rotary specimen rack, respectively. For long period elements, samples were irradiated for four hours, while for those of short periods, samples were irradiated for 36 to 60 seconds along with 500 mg of sulfur sublimed powder used as flux monitor. The samples and flux monitor, packed in polyethylene rabbits, are sent by means of a pneumatic transfer system to the reactor core to be successively irradiated. A hyper-pur germanium detector linked to a PC running Maestro software was used in the counting of gamma emitters in the activated samples. The K0_IAEA software was used for computing element concentrations.

2.3. RESULTS AND DISCUSSION

2.3.1. The Oualidia lagoon

2.3.1.1. Radionuclides and sediment deposition chronologies

Total ^{210}Pb, ^{226}Ra and excess ^{210}Pb activities throughout the two cores are plotted in Fig. 2.2. ^{226}Ra activities were quite uniform in the cores, with activities varying from 20 to 50 Bq kg^{-1}, being within the typical values for non-contaminated coastal sediments. In contrast, activities of total ^{210}Pb increase upcore to values in the sediment-water interface much higher than typical activities measured in other Moroccan coastal ecosystems [2.23]. Maanan et al [2.24] reported ^{210}Pb activities in surface sediment of the same area as high as 1400 Bq kg^{-1}.

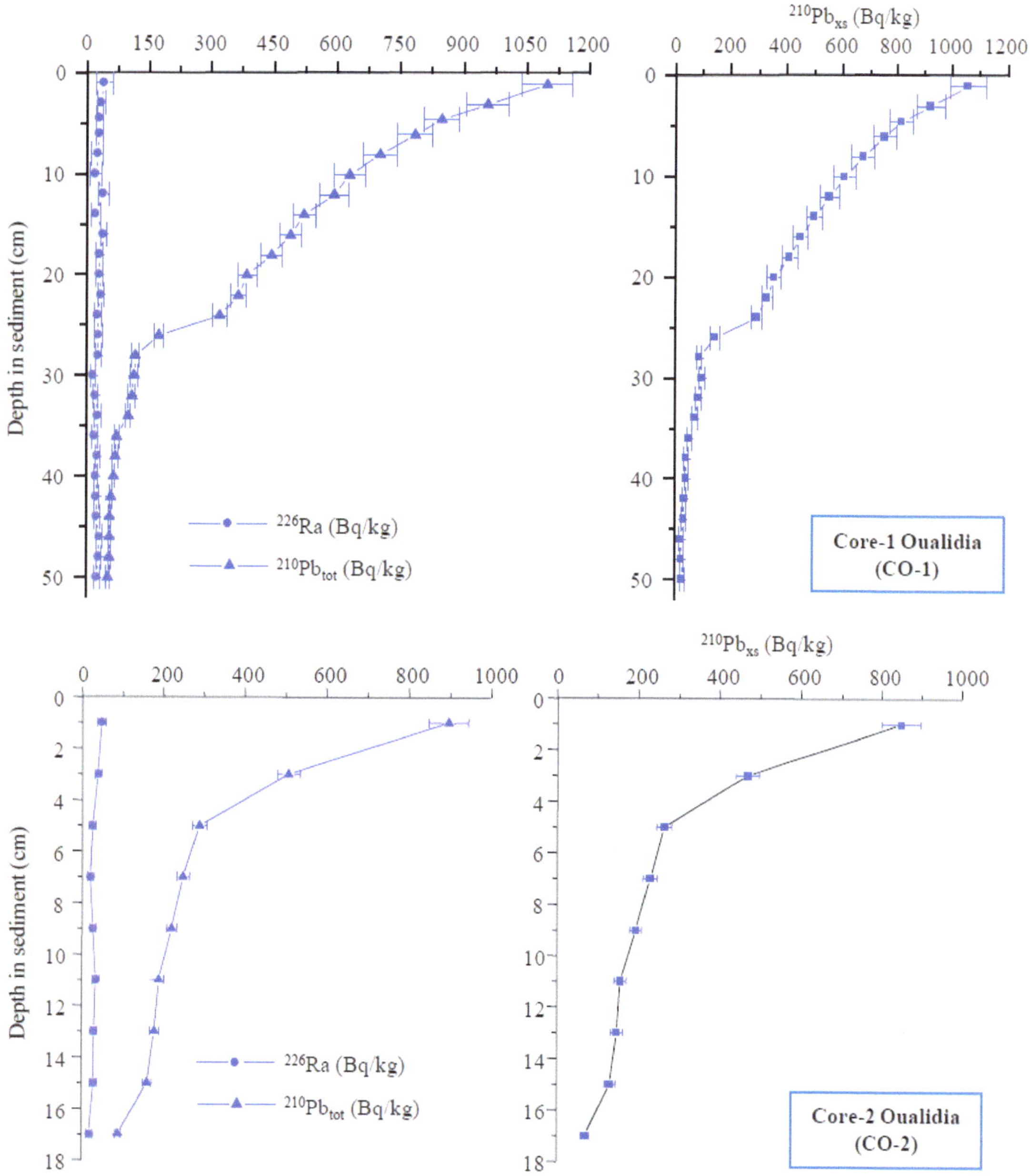

FIG. 2.2. Depth profiles of ^{226}Ra, $^{210}Pb_{tot}$ and $^{210}Pb_{xs}$ measured in CO-1 and CO-2 "(produced from Ref. [2.11] with permission courtesy of [Springer])".

Such abnormally high activities of ^{210}Pb in the Oualidia lagoon sediment are evidently due to human activities in the surrounding area (e.g., industry, agriculture). ^{210}Pb is incorporated to the lagoon sediment by two possible ways: (1) with marine waters rich in dissolved and/or particulate ^{210}Pb. Such assumption if questionable due to the fact that the studied sediment is not enriched in ^{226}Ra, whose activities were in the range of those of uncontaminated sediment. This could be indicative of a substantial predominance of the unsupported ^{210}Pb; (2) deposition of atmospheric ^{210}Pb originating from ^{222}Rn radioactive decay (half-life 3.8 d). It is well established that radon escapes from rocks, soils and anthropogenic facilities and can be removed from the atmosphere by processes such as with precipitations and/or wet deposition. This hypothesis is the most appropriate to explain the high ^{210}Pb activities but needs additional studies to confirm it. In this context, with the aim to check whether dissolved radionuclide concentrations were remarkably high, the three water samples collected along the Oualidia lagoon were analyzed by gamma spectrometry. ^{226}Ra and ^{210}Pb activities were all below the detection limits (<0.4 Bq L^{-1} for ^{210}Pb and <0.1 Bq L^{-1} for ^{226}Ra for 24 h counting time of 2-L sample volume), which excluded the assumption that disintegration of dissolved ^{226}Ra could be behind high ^{210}Pb activities in recently deposited sediment. Indeed, only a small amount of ^{210}Pb is produced in the water column from in-situ decay of ^{226}Ra.

^{210}Pb profiles of Fig. 2.2 display fast increases at 24 cm depth in the sediment core CO-1 and 4 cm in CO-2, indicative of a substantial increase of ^{210}Pb delivery to the sediment starting at the respective deposition years. This change altered the expected typical exponential decrease in ^{210}Pb activities in the studied cores. It has been supposed that such abrupt enhancement matches the beginning of industrial development in 1986 and intensive agricultural activities. Thus, a change in the flux of ^{210}Pb to the sediment has been evidenced in the profiles and, therefore, conventional dating models could not be used to establish the chronologies, as the assumption of constant flux seems to be invalid. The CRS (Constant Rate of Supply) model was adapted to take account of the change in the flux of unsupported ^{210}Pb to the sediment [2.11]. A key parameter for implementing this approach is the exact knowledge of the year in which the substantial increase of the flux occurred. Total excess ^{210}Pb inventory is calculated, in this modified CRS model, as the sum of partial inventories covering the whole core; 1) (Σ_1) covers the part from the surface sediment layer to that where the sudden increase was detected, and 2) (Σ_2) covers the rest of downcore layers in the core until the last section where excess ^{210}Pb is measurable. In this way, the total inventory (Σ_{tot}) can be written as follows:

$$\Sigma_{\text{tot}} = \int_0^{t1} F_1 \cdot e^{-\lambda \cdot t} dt + \int_{t1}^{\infty} F_2 \cdot e^{-\lambda \cdot t}$$

$$= \frac{F_1}{\lambda} \cdot \left(1 - e^{-\lambda \cdot t1}\right) + \frac{F_2}{\lambda} \cdot e^{-\lambda \cdot t1} \quad (2.1)$$

λ : the ^{210}Pb radioactive decay constant.
t1 : the elapsed time between the sampling year (2014) and that in which ^{210}Pb flux to the sediment changed (1986). *t1* is 28 years in this case.
F_1 and F_2 : the fluxes of unsupported ^{210}Pb corresponding, respectively, to the intervals covering the period before and after the substantial increase started. F_1 and F_2 were determined using the expressions as follows:

$$F_1 = \frac{\Sigma_1 \cdot \lambda}{\left(1 - e^{-\lambda \cdot t1}\right)} \quad and \quad F_2 = \frac{\Sigma_2 \cdot \lambda}{e^{-\lambda \cdot t1}} \quad (2.2)$$

The calculated partial and total inventories are given in Table 1 of Ref. 2.11 along with the fluxes sediment accumulation rates. Σ_1 in CO-1 is much higher (about three fold) than that of CO-2 due to the different prevailing hydrodynamic conditions at each coring site; which results in quite different rates of sediment accumulation (SAR). Indeed, computed depth-averaged values of SAR were 0.1 g cm^{-2} y^{-1} and (0.5 g cm^{-2} y^{-1} in CO- and CO-2, respectively. ^{210}Pb partial inventories, Σ_1 and Σ_2, and fluxes, F_1 and F_2, exhibited different variation patterns in the two sediment cores. Inventories in the upper and lower parts in CO-2 were quite comparable, while the flux in the upper part was nearly two-fold lower than that corresponding to the downcore part. On the other hand, the partial inventory and flux of the upper interval of CO-1 were larger by about four-fold than those calculated for the downcore part.

From the aforementioned equations, the ages can be written as follows:

$$t(z) = -\frac{1}{\lambda} \cdot Ln\left[(\Sigma(z) - \Sigma_2) \cdot \frac{\lambda}{F_1} + e^{-\lambda \cdot t1}\right] \quad t > t_1$$

$$t(z) = -\frac{1}{\lambda} \cdot Ln\left[\frac{\Sigma(z) \cdot \lambda}{F_2}\right] \quad t < t_1 \tag{2.3}$$

Σ(z) is ^{210}Pb inventory below the sediment layer at depth *z*. The obtained dating results (SARs and ages) for both cores are plotted in Fig. 2.3 along with ^{137}Cs activity profiles. ^{137}Cs is a man-made radionuclide whose presence in the environment is due to nuclear weapons testing. It is commonly used in recent sediment geochronology as a time marker for validation of ages. ^{137}Cs activities in Fig. 2.3 were in the range of values reported in sediments in Moroccan coastal systems [2.23].

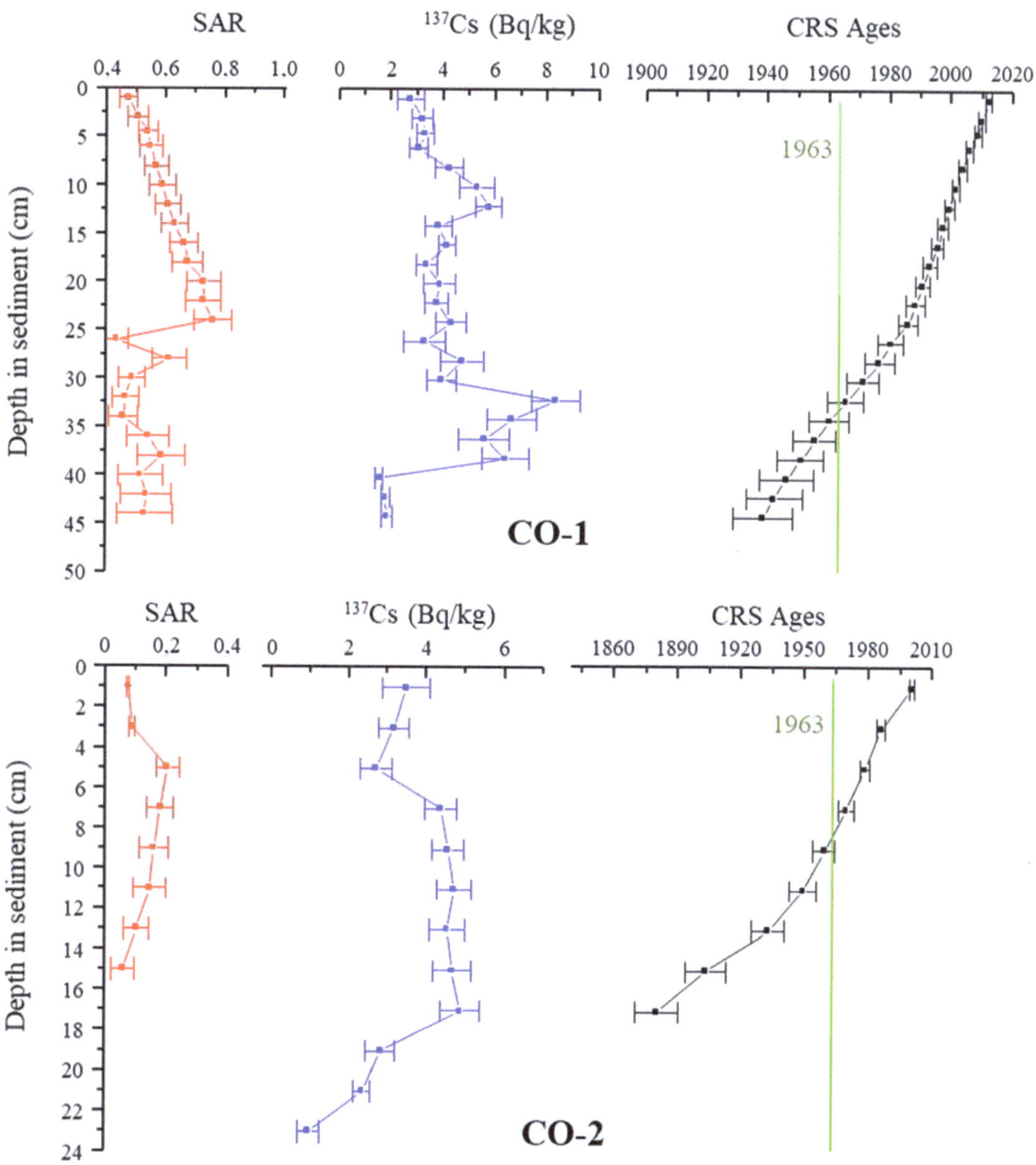

FIG. 2.3. Vertical profiles of ^{137}Cs, computed Sediment Accumulation Rates (SAR in g cm^{-2} y^{-1}) and ages in the two cores from Oualidia lagoon "(produced from Ref. [2.11] with permission courtesy of [Springer])".

A peak of maximum ^{137}Cs activity was recorded in the CO-1 profile, but the peak is not clear enough, and vertical migration produced a "broadening effect" both up and down the peak. The ^{137}Cs maximum activity matches the date of maximum atmospheric fallout (1963) given by the modified CRS model. The detectable activities of ^{137}Cs in layers deposited before nuclear tests are a consequence of downward diffusion after sediment deposition. Although diffusion occurs, the depth of ^{137}Cs maximum concentration remains preserved [2.26, 2.27], but it could be a constraint on its use as a time marker in some cases. Diffusion could also affect ^{210}Pb so that the assumption of no migration or diffusion becomes not valid and, therefore, could result in ^{210}Pb profile alteration. Nevertheless, downward migration could always be considered as minimal, depending mainly on the porosity of the sediment column, when no evidence is visible in ^{210}Pb profiles such as peaks and/or inflexions.

No visible peak of ^{137}Cs activity was apparent in the profile of CO-2 core, possibly because of the low sedimentation rate, which was around 0.11 g cm^{-2} y^{-1}. Thus, the depth of maximum ^{137}Cs fallout was impossible to distinguish throughout the profile. According to the established chronology using ^{210}Pb, this depth should be located at 8-10 cm in the sediment column. When low sedimentation rates predominate, vertical diffusion would broaden considerably the ^{137}Cs peak [2.28], making it difficult to identify. In addition, considering the low core resolution (2-cm) and weak SAR, the activity in each layer should be understood as the mean value of activities in several thin sediment layers formed during a period of a few years.

Further validation of the modified CRS chronologies can be obtained from ^{238}U profiles plotted in Fig. 2.4. Indeed, both profiles reveal increases in ^{238}U activities at depths matching the period around 1986. The activities were fairly uniform from downcore to around 1980. ^{238}U activities in these parts of the profiles were around 40 Bq kg^{-1} being within the range of values reported for non-contaminated sediment. Disequilibrium in U^{238} decay chain is visible in Fig. 2.4, especially in the upper part of the CO-1 core, indicating a substantial enhancement of uranium in recently deposited sediment due to anthropogenic activities around the study site.

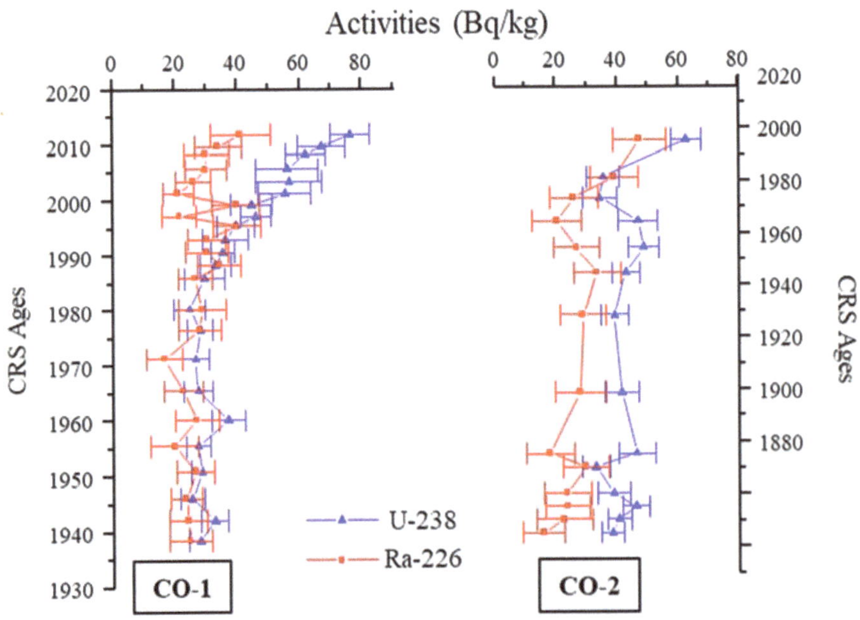

FIG. 2.4. Vertical distributions of ^{238}U and ^{226}Ra versus CRS ages in the cores from the Oualidia lagoon.

2.3.1.2. Vertical distribution of metals

Figure 2.5 displays the vertical distributions in CO-1 and CO-2 cores of Ni, Cu, Zn, Cd, Pb, V, Co, As, Th, U and Fe versus the ages given by the modified CRS model. Almost all elements displayed a similar pattern of variation throughout the cores. Indeed, in core CO-1, high concentrations in the bottom, a decrease during the two decades is observed followed by an increase to reach maximal concentrations in the topmost layers. In core CO-2, a slight increase in concentrations in the deep layers before reaching the first maximum in the 1930s, similar in amplitude and temporal location, in most cases, to the peaks recorded in the deepest section of the core CO-1. Thus, for metals such as Ni, Cu, Zn, Pb, V and Th, similar distributions were observed for the period extended from 1930 to the present. The increase of metal concentrations in the Oualidia lagoon seems to be due to human activities, such as agriculture, aquaculture and urban area development, which intensified since the 1960s [2.24, 2.25]. In addition, the relatively high levels of metal contents in recently deposited sediment could be due to the industrial development in the area, in particular the phosphate industry, commissioned in 1986.

The concentrations in downcore layers of CO-2, deposited before the year 1900, could be considered as local background levels. In view of such similarity between vertical distributions in both sediment cores, the modified CRS model, which was applied to ^{210}Pb profiles, provided consistent results despite the differences in terms of hydrodynamic conditions and sedimentation patterns recorded in the two coring sites.

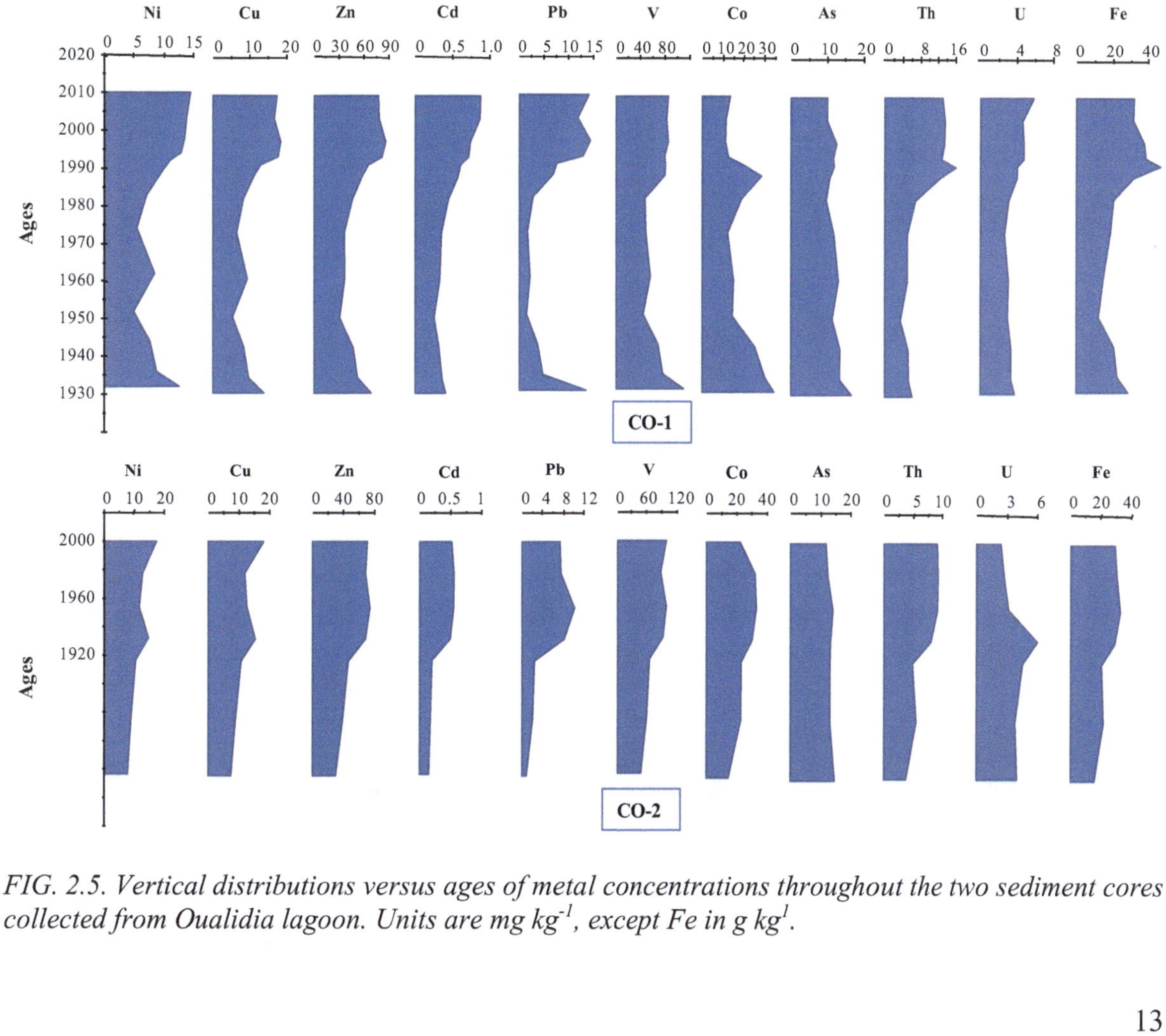

FIG. 2.5. Vertical distributions versus ages of metal concentrations throughout the two sediment cores collected from Oualidia lagoon. Units are mg kg^{-1}, except Fe in g kg^{1}.

The Enrichment Factor (LOI has been commonly used as an indicator of metal enhancement in sediment and, therefore, allows sorting the studied sediment in terms of the degree of enrichment, ranging from "no enrichment" to "extremely severe enrichment", compared to the local geological background [2.29].

The EFs of Cu, Cd, Co, Pb As and U were computed using the following expression:

$$EF = \frac{(X/_{Fe})_{sample}}{(X/_{Fe})_{background}} \quad (2.4)$$

where X is the Fe-normalized element concentration. Element concentrations in the upper continental crust were taken as background levels. Iron was used as a reference element owing to its geochemistry which is similar to almost all trace metals, and its natural content tends to be uniform [2.30]. Besides, Fe is largely employed as a normalizing element for river and marine sediment [2.31, 2.32, 2.33]. EF can be classified in five degrees: EF < 2 no enrichment, 2 ≤ EF < 5 moderate enrichment, 5 ≤ EF < 20 significant enrichment, 20 ≤ EF < 40 very high enrichment and EF > 40 Extremely high enrichment. The time evolutions of EFs throughout the two cores are depicted in Fig. 2.6.

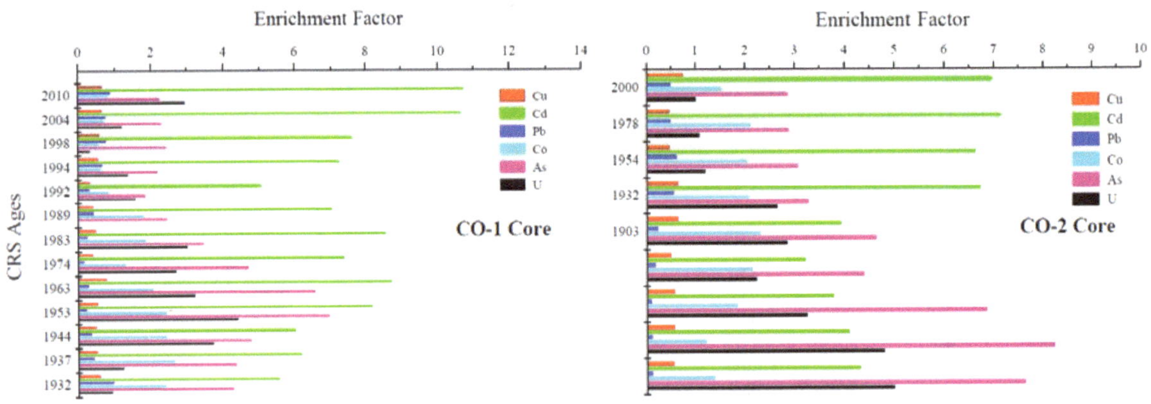

FIG. 2.6. Time evolution of enrichment factors of U, Cd, Pb, Co, Cu and As in both sediment core from Oualidia lagoon.

Pb and Cu showed no enrichment along the cores suggesting no anthropogenic input into the Oualidia lagoon sediments during the last decades. In contrast, Co and U displayed minor and moderate enrichment showing that an important fraction of these metals were issued from non-crustal materials and, therefore, a detectable source has been feeding the lagoonal sediment. As and Cd indicate moderate and significant enrichment revealing an anthropogenic source. Significant enrichment was observed in the period extending from 1950 to the sampling date (2014). This period is well connected to the history of human activities growth in the studied area [2.12]. Arsenic showed substantial enrichment in the deeper layers; this is due to its non-conservative behaviour owing to the co-precipitation with iron oxides or diffusion processes from sediment pore water [2.34, 2.35]. These results suggest that the Oualidia lagoon sediment cores are enriched by As and Cd at different extents and horizons, thus indicating a clear contribution of anthropogenic sources.

2.3.2. The Sidi Moussa lagoon

2.3.2.1. Radionuclides and sediment deposition chronology

Figure 2.7 shows the vertical profiles of total ^{210}Pb, excess ^{210}Pb and ^{226}Ra in the cores from the Sidi Moussa lagoon. Topmost layers of the three studied cores displayed total ^{210}Pb activities (1946 Bq kg^{-1}, 1980 Bq kg^{-1} and 2588 Bq kg^{-1} in CSM-1, CSM-2 and CSM-3, respectively) higher by more than twofold those found in surface layers of the Oualidia cores [2.2, 2.11]. These quite elevated activities of unsupported ^{210}Pb in sediment are attributed to a combination of the atmospheric deposition of ^{210}Pb and inputs with marine currents. The lagoon is located at about 15 km from the phosphate industrial complex and surrounded by agricultural lands and, therefore, high radionuclide and metal concentrations were predictable. The profile of total ^{210}Pb in CSM-1 core displayed a mixing zone exceeding 20 cm depth and then an exponential decay of activities to levels somewhat greater than those of supported ^{210}Pb. This core was discarded from radiometric dating because of the large mixing depth, likely affected by episodes of fast sedimentation and/or mechanical reworking.

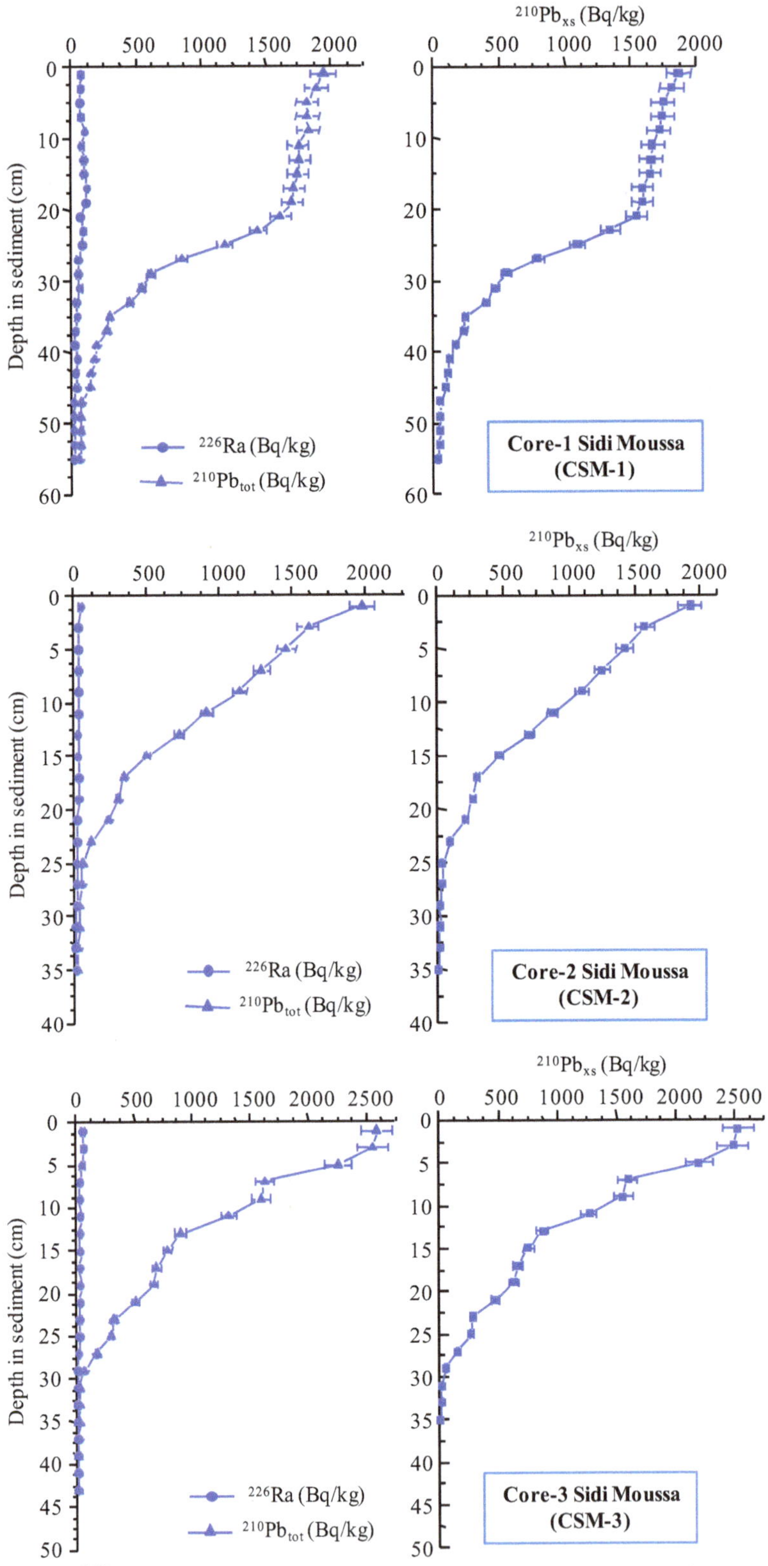

FIG. 2.7. Excess ^{210}Pb profiles in the three cores collected from the Sidi Moussa lagoon.

^{226}Ra exhibited, in general, a smooth increase from downcore to surface layers with the exception of the presence of a maximum in CSM-1 core. Activities of ^{226}Ra around 20 Bq kg^{-1} were found in bottom layers, being typical of unperturbed coastal sediments [2.23]; thus they could be considered as local background activities. It was expected to observe ^{226}Ra activities much higher than those of Fig. 2.7, as it was the case for ^{210}Pb, because phosphogypsum is enriched in ^{210}Pb and ^{226}Ra (1638 and 1566 Bq kg^{-1}, respectively [2.36]), thus containing 80% of the initial content of these radioelements in the phosphate rock. In saline waters, radium has a weak affinity for sediment particles [2.37] and, therefore, any ^{226}Ra discharged in the ocean at few kilometers from the study site is likely to be mainly in the dissolved form. Unsupported ^{210}Pb in different layers of the cores was obtained by subtracting ^{226}Ra, assumed to be in equilibrium with supported ^{210}Pb, from total ^{210}Pb at each layer because ^{226}Ra presented significant vertical variability (18 - 126 Bq kg^{-1} in CSM-1 core, 14 - 55 Bq kg^{-1} in CSM-2 core and 16 - 71 Bq kg^{-1} in CSM-3 core). Excess ^{210}Pb profiles for CSM-2 and CSM-3 cores shown in Fig. 2.7 were appropriate to carry out radiometric dating. The CRS model assumes a constant atmospheric ^{210}Pb flux to sediment over the last decades. Nevertheless, the rate of supply of ^{210}Pb has possibly been increased since the start of the fertilizer production plants. The excess ^{210}Pb profiles plotted in Fig. 2.7 showed an exponential decay with depth to ^{210}Pb supported activities without indications of flux changes throughout the years. Thus, the classical CRS dating model could be applied to CSM-2, and CSM-3 profiles as the necessary assumptions would be valid. This model uses total and partial excess ^{210}Pb inventories to assign an age to a given layer of the sediment core, and then computes sediment accumulation rate for each two adjacent layers.

Inventories are a function of sediment accumulation rates and radionuclide concentrations and, therefore, they are quantitative indicators of contaminants delivery in sediment [2.38]. In CSM-3 core, total ^{210}Pb inventory and flux were slightly higher than those in SM-3 core due only to the difference in activities in both cores as sedimentation rates are similar. Although the activities in CSM-1 core are similar to those in the other cores, the inventory and flux are twofold higher, thus indicating high sedimentation rates in CSM-1 coring site, especially in recent years.

The values of calculated inventories and delivery rates were higher than those estimated in the previous section in the Oualidia lagoon and significantly greater than those reported for Moroccan coastal systems. Anthropogenic activities involving NORM and extensive agriculture are behind increasing ^{210}Pb inputs to sediment of the Sidi Moussa lagoon. Depth-averaged accumulation/sedimentation rates were comparable in both cores. SARs vertical profiles are plotted in Fig. 2.8, showing an increasing trend from bottom to top in both cores, due to the surrounding land development, since several decades, for industrial, agricultural and fishing activities, which have produced a gradual increase of the amount of eroded material in the lagoon, similar to the Oualidia lagoon [2.39].

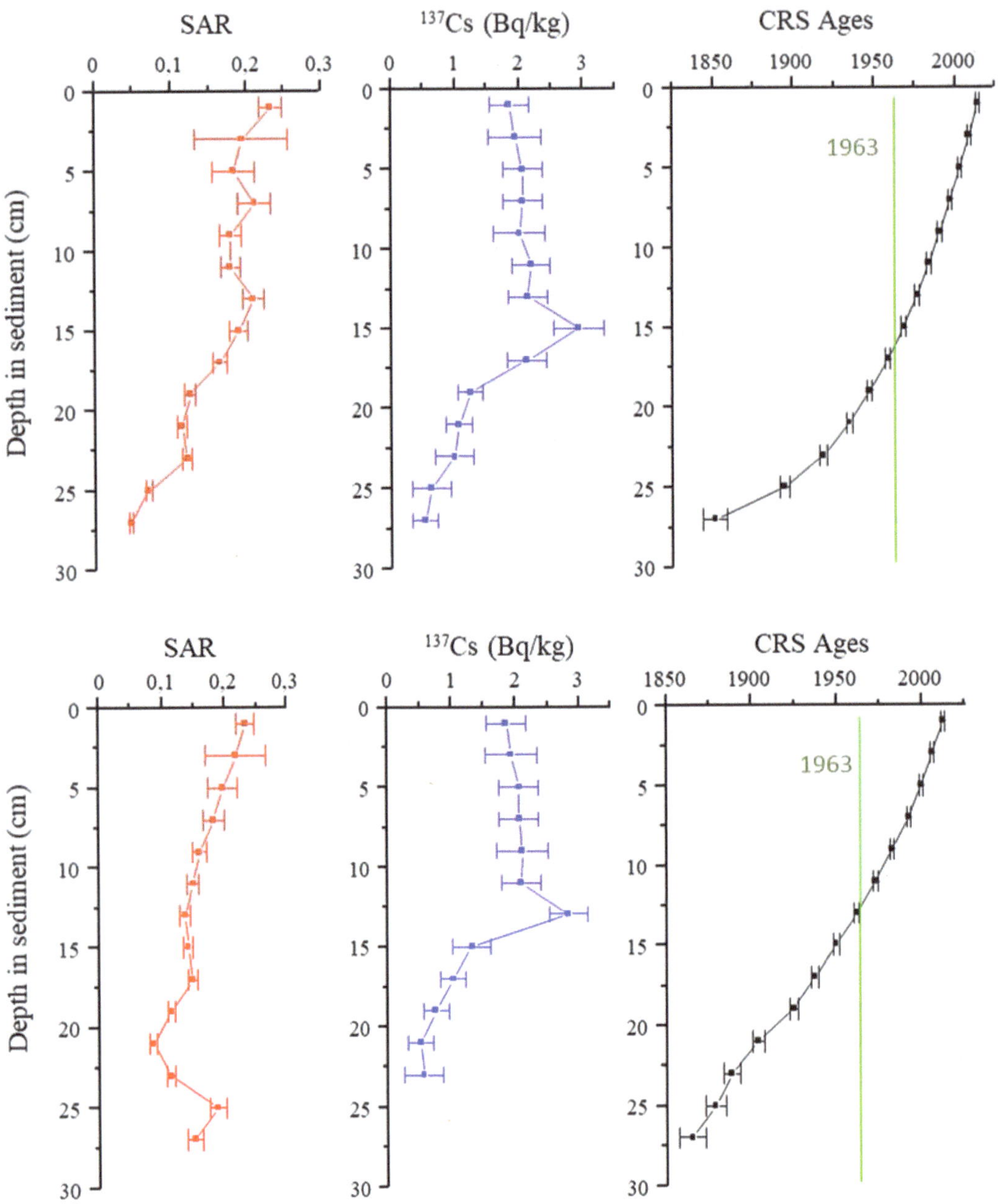

FIG. 2.8. Vertical profiles of SARs, ^{137}Cs activities and CRS ages in the cores from the Sidi Moussa lagoon.

Vertical distributions of ^{137}Cs activities in CSM-2 and CSM-3 cores are also shown in Fig. 2.8, along with the ages calculated using excess ^{210}Pb profiles and the CRS model. In both profiles, ^{137}Cs activities displayed a subsurface maximum at 13 cm in CSM-2 and 15 cm in CSM-3, matching the year of maximum fallout (1963). Activities in upcore layers deposited from 1975 to present were almost uniform, then decreased below the peak layers to 0.5 Bq kg^{-1} in bottom layers. According to the CRS ages, deposition of downcore layers occurred before the early 1950s, marked by the beginning of nuclear atmospheric tests, so that the presence of ^{137}Cs in detectable levels, as is in the Oualidia lagoon [2.11], is considered contradictory. Such a behaviour has been previously reported in several studies [2.40], and is often attributed to Cs downward diffusion in the sediment column. It is worth noting that ^{210}Pb is unlikely to be as well affected by diffusion because of its high distribution coefficient ($K_d \sim 10^5$) and, consequently, remains mostly attached to particles of sediment.

Inventories of ^{137}Cs in the two studied cores were respectively 277 ± 15 Bq m^{-2} and 318 ± 16 Bq m^{-2}, similar to those estimated in the Alboran basin [2.41]. There are no data on ^{137}Cs in the study area, which impedes comparison with results from previous studies. ^{137}Cs profiles were used to estimate mean accumulation and sedimentation rates through corresponding the peaks in the profiles to the year 1963. The calculated SARs were 0.156 and 0.163 g cm^{-2} y^{-1} and sedimentations rates were 0.24 and 0.29 cm y^{-1}, in CSM-2 and CSM-3, respectively. The values are in good agreement with those provided by the CRS model and reported in Table 1 of Ref. 2.47.

It can be observed in ^{238}U vertical distributions presented in Fig. 2.9 that activities in surface sediment (250 Bq kg^{-1} and 350 Bq kg^{-1} in CSM-2 and CSM-3, respectively) are up to 6 times higher than those in bottom layers. All activities, except those recorded in bottom sections of CSM-2 core, exceed by far the worldwide mean activity reported by UNSCEAR [2.17], 35 Bq kg^{-1}. It is well established that dissolved uranium is incorporated to the sediment during transport processes across the sediment-water interface [2.42]. It was reported that ^{238}U content in the phosphate rock is about 1600 Bq kg^{-1}, and around 10 % is removed with the solid sub-product (phosphogypsum) in the phosphoric acid production process [2.37]. According to the temporal evolution of activities plotted in Fig. 2.9, the enhancement of ^{238}U in Sidi Moussa lagoon sediment had started much before 1986, the year when the phosphate industry initiated production. The presence of ^{238}U in deep sections of the cores is most possibly due to its high mobility prompted by downward diffusion of uranyl ions in porewater to deep sections, usually characterized by reducing conditions that accelerate UO_2 precipitation. On the other hand, the profiles showed similar patterns in terms of $^{238}U/^{226}Ra$ disequilibrium throughout the cores. Disequilibrium became larger in the upcore layers reaching the maximum at the sediment-water interface. This is attributable to the different geochemical behaviour of each radionuclide.

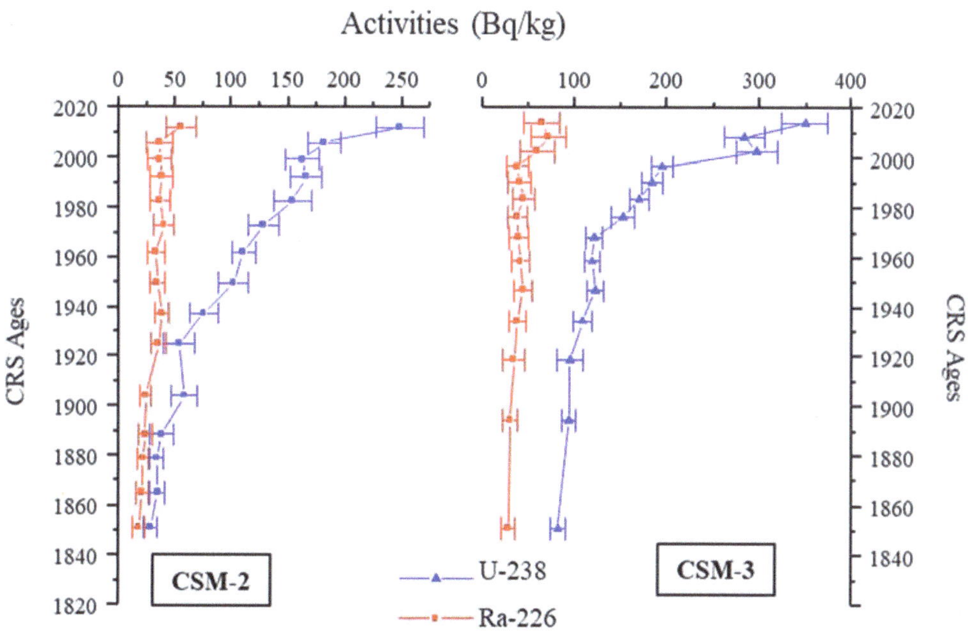

FIG. 2.9. Vertical distributions of ^{238}U versus CRS ages in the cores retrieved from Sidi Moussa lagoon.

2.3.2.2. Vertical profiles of metals in the sediment cores

Metal concentration profiles in the two cores are presented in Fig. 2.10. Concentrations of all metals, except As, were higher than those presented in the previous section for the Oualidia lagoon, distant from the Sidi Moussa lagoon of about 50 km south. Most metal profiles displayed significant variability and the presence of one or more maxima and minima throughout the cores, while Fe showed a quite constant distribution around a mean averaged value of 12.28 g kg^{-1} (SD = 1.02). It can be observed that downcore layers contents in metals were similar to those reported for the Upper Continental Crust (UCC) [2.43], and high concentrations were registered in a depth between 8 and 12 cm, which decreased towards the upper layers. However, U concentrations exhibited a smooth increase from bottom to top layers reaching a maximum greater than the UCC level at the topmost layer. Such high U concentration could be due to the high water content and its high solubility [2.44]. In contrast, two maxima can be seen in Th profiles, both lesser than the reference UCC level.

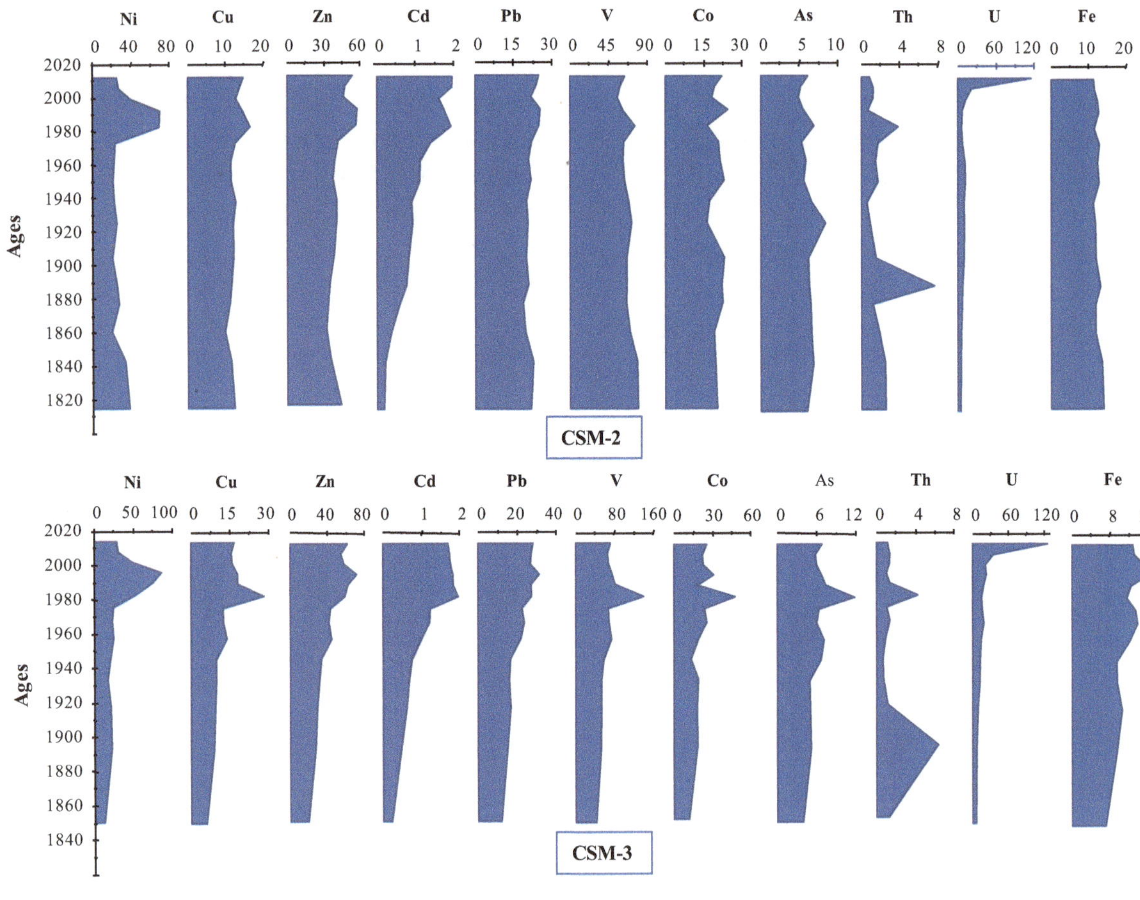

FIG. 2.10. Vertical distributions versus ages of metal concentrations throughout the two sediment cores from Sidi Moussa lagoon. Units are mg kg^{-1}, except Fe in g kg^{1}.

It can be seen in the trace metal profiles that most elements, such as Co, Cu, AS and Ni, displayed in both cores a maximum at a depth corresponding to the years when the phosphate plants started. The same profiles showed in the layers above the maxima a decreasing trend in concentrations, suggesting the positive impact of the environmental policy undertaken since the early 2000s by the phosphate company [2.45]. Previous studies held in the area [2.4, 2.6, 2.8, 2.46] had reported high and positively correlated concentrations of these metals in sediment, revealing the same origin, discharges from the phosphoric acid facilities and intensive agriculture. Nevertheless, almost all metals were correlated significantly one to another in the CSM-3 sediment core but weakly in the CSM-2 core [2.47]. Such peculiar difference can be explained by the positive correlation of Fe with all metals in CSM-3 core and not correlated to any element. Therefore, this behavior suggests that all elements are related to oxyhydroxydes of iron in one core and not in the other.

U, Cd, Pb, Co, Cu, and As enrichment factors were calculated using the expression (4) and Fe as normalizer and background concentrations as those reported in Taylor and McLennan [2.43]. The concentrations measured in bottom layers were close to background concentrations in most cases. Vertical profiles of EFs are plotted in Fig. 2.11. Enrichment factors of cadmium displayed continuous increase throughout the two cores, reaching EFs greater than 9, indicative of moderately severe enrichment, during the 1980's and in the surface layer. In bottom layers, no Cd enrichment (EF < 1) was noticed prior to the year 1850. Weak enrichment (EF < 3) and moderate enrichment (3 < EF < 5) were recorded in the middle of the cores. On the other hand, uranium seemed to be enriched as well in the sediment cores, with strong enrichment in the surface layers and no enrichment in the bottom layers. No enrichment of Pb, Co, Cu and As was noticeable in the two cores.

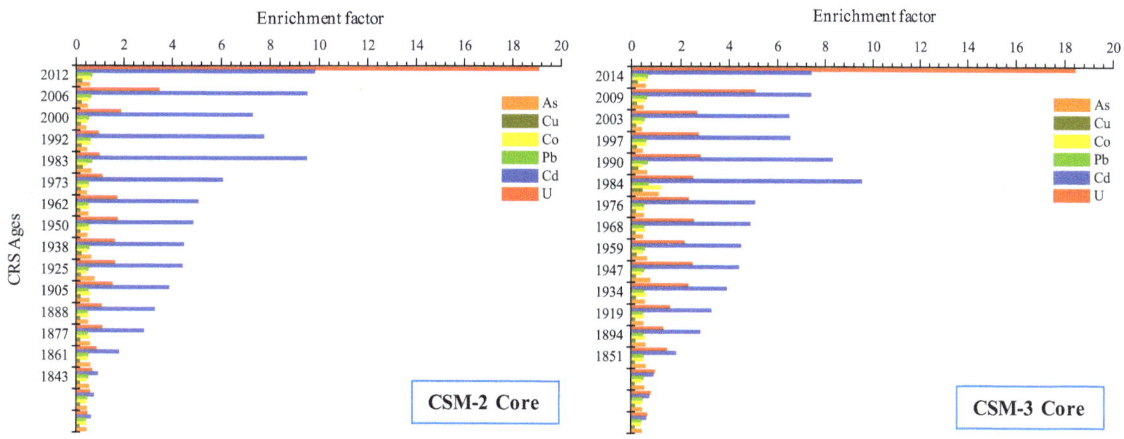

FIG. 2.11. Time evolution of enrichment factors of U, Cd, Pb, Co, Cu and As in both sediment core from Sidi Moussa lagoon.

Such enrichment in Cd and U in Sidi Moussa lagoon sediment can be likely produced by the negative impact of the effluents discharging from the phosphate industrial facilities at Jorf Lasfar, as has been pointed out in previous studies [2.6, 2.7] and lately by Boutahar et al [2.8]. It needs to be emphasized that given the high toxicity of Cd [2.48] and the need to comply with international standards regarding Cd content in the phosphate products, processes have been developed in Morocco to reduce its content in the phosphoric acid [2.49]. However, two effects could be behind the presence of Cd and U enriched in layers formed much before the launch of industrial activities in the area: i) their geochemical behavior characterized by low distribution coefficients [2.50], 10^3 and 10^2, respectively, responsible for their post-deposition downward diffusion with porewater, and ii) the rising of deep ocean waters with the so-called "North-West African upwelling" [2.10], as a source of Cd, among other elements, affecting the Moroccan littoral in the south until the study site. Beside cadmium and uranium, Co and As presented weak enrichment in CSM-3 core, but only in the layer deposited between 1984 and 1990.

2.4. CONCLUSIONS

The Oualidia Sidi Moussa lagoon complex is of particular ecological importance but also a key in regional socio-economic development. The environmental issues arisen as a consequence of the increasing human activities since several decades make the lagoons an authentic natural laboratory to highlight the potential use of sediment deposition chronology in environmental reconstructions. Lead-210 profiles have provided evidence for a recent increase atmospheric ^{210}Pb delivery to the Oualidia lagoon sediment, which makes inappropriate the application of conventional dating models. In such conditions, a modified version of the CRS model has been developed and applied to the studied sediment cores. Nevertheless, ^{210}Pb in Sidi Moussa cores showed typical exponential decay profiles but with enhanced activity concentrations, exceeding typical values reported in non-perturbed environments. The CRS dating model was applied and the chronologies were consistent with the peaks of maximal fallout in ^{137}Cs profiles, used as a time marker, but the depth of ^{137}Cs penetration reached the pre-fallout horizons, indicating that downward diffusion has occurred. It is worth noting that the increase of ^{210}Pb activities suggests a non-constant ^{210}Pb flux to the sediment and, therefore, constitutes a contraindication to the use of conventional models. Despite this, the model was applied as a first approximation to determine the age-depth relationship, but further studies are needed to test the reliability of ^{210}Pb chronologies taking account also of possible changes in sedimentary conditions in the study area.

Results obtained regarding historical inputs of toxic metals bound to sediment particles, in particular their low concentrations in topmost layers, showed the positive effect of the environmental strategy implemented by the phosphate industrial facility since the early 2000s. In addition, the high concentrations previously reported for Cd in sediments of the study area was not found to be due entirely to the industrial effluents from the phosphate processing plants, as always pointed out. Another natural source, the most plausible being the North-West African upwelling, have contributed to the enhancement of Cd in sediment in accordance with the established historical inputs inferred from radiometric dating of sediment cores.

^{137}Cs, used as a time-marker for validation of the chronologies obtained, has shown limitations, in particular in environments characterized by low sedimentation rates. Indeed, post-depositional transport in the sediment produces a broadening of the 1963 peak and diffusion of ^{137}Cs to pre-fallout sediment layers, making it sometimes impossible to identify the horizon of maximum fallout.

The above findings demonstrate that radiometric dating of recent sediment is a powerful tool in environmental studies requiring age-depth relationships, especially in establishing the time evolution of coastal quality conditions. However, some points still need to be addressed for a better understanding of the applicability of conventional dating models especially if ^{137}Cs profiles are impacted by diffusion within the sediment column.

2.5. ACKNOWLEDGEMENTS

This work was supported by the International Atomic Energy Agency under the Contract Research Project CRP K41016: "*Study of temporal trends of pollution in selected coastal areas by the application of isotopic and nuclear tools.*" The authors acknowledge the use of sampling and laboratory equipment provided within the framework of the IAEA Technical Cooperation Project RAF7/015.

2.6. REFERENCES

[2.1] MAANAN, M., BARJY, M. E., HASSOU, N., ZIDANE, H., ZOURARAH, B., MAANAN, M., Origin and Potential Ecological Risk Assessment of Trace Elements in the Watershed Topsoil and Coastal Sediment of the Oualidia Lagoon, Morocco, Hum. Ecol. Risk Assess. Int. J. **24** 3 (2018) 602–614.

[2.2] MEJJAD, N., LAISSAOUI, A., EL-HAMMOUMI, O., BENMANSOUR, M., BENBRAHIM, S., BOUNOUIRA, H., BENKDAD, A., BOUTHIR, F. Z., FEKRI, A., BOUNAKHLA, M., Sediment Geochronology and Geochemical Behavior of Major and Rare Earth Elements in the Oualidia Lagoon in the Western Morocco, J. Radioanal. Nucl. Chem. **309** 3 (2016) 1133–1143.

[2.3] ZOURARAH, B., MAANAN, M., CARRUESCO, C., AAJJANE, A., MEHDI, K., CONCEIÇÃO FREITAS, M., Fifty-Year Sedimentary Record of Heavy Metal Pollution in the Lagoon of Oualidia (Moroccan Atlantic Coast), Estuar, Coast. Shelf Sci. **72** 1 (2007) 359–369.

[2.4] CHEGGOUR, M., LANGSTON, W. J., CHAFIK, A., TEXIER, H., IDRISSI, H., BOUMEZZOUGH, A., Phosphate Industry Discharges and Their Impact on Metal Contamination and Intertidal Macrobenthos: Jorf Lasfar and Safi Coastlines (Morocco), Toxicol. Environ. Chem. **70** 1–2 (1999) 159–179.

[2.5] CHEGGOUR, M., CHAFIK, A., LANGSTON, W. J., BURT, G. R., BENBRAHIM, S., TEXIER, H., Metals in Sediments and the Edible Cockle Cerastoderma Edule from Two Moroccan Atlantic Lagoons: Moulay Bou Selham and Sidi Moussa, Environ. Pollut. **115** 2 (2001) 149–160.

[2.6] KAIMOUSSI, A., CHAFIK, A., MOUZDAHIR, A., BAKKAS, S., The Impact of Industrial Pollution on the Jorf Lasfar Coastal Zone (Morocco, Atlantic Ocean): The Mussel as an Indicator of Metal Contamination, Comptes Rendus Académie Sci. - Ser. IIA - Earth Planet. Sci. **333** 6 (2001) 337–341.

[2.7] MAANAN, M., ZOURARAH, B., CARRUESCO, C., AAJJANE, A., NAUD, J., The Distribution of Heavy Metals in the Sidi Moussa Lagoon Sediments (Atlantic Moroccan Coast), J. Afr. Earth Sci. **39** 3 (2004) 473–483.

[2.8] BOUTAHAR, L., MAANAN, M., BOUOUAROUR, O., RICHIR, J., POUZET, P., GOBERT, S., MAANAN, M., ZOURARAH, B., BENHOUSSA, A., BAZAIRI, H., Biomonitoring Environmental Status in Semi-Enclosed Coastal Ecosystems Using Zostera Noltei Meadows, Ecol. Indic. **104** (2019) 776-793.

[2.9] GAUDRY, A., ZEROUAL, S., GAIE-LEVREL, F., MOSKURA, M., BOUJRHAL, F.-Z., EL MOURSLI, R. C., GUESSOUS, A., MOURADI, A., GIVERNAUD, T., DELMAS, R., Heavy Metals Pollution of the Atlantic Marine Environment by the Moroccan Phosphate Industry, as Observed through Their Bioaccumulation in Ulva Lactuca, Water. Air. Soil. Pollut. **178** 1 (2007) 267–285.

[2.10] AUGER, P. A., MACHU, E., GORGUES, T., GRIMA, N., WAELES, M., Comparative Study of Potential Transfer of Natural and Anthropogenic Cadmium to Plankton Communities in the North-West African Upwelling, Sci. Total. Environ. **505** (2015) 870–888.

[2.11] LAISSAOUI, A., MEJJAD, N., ZIAD, N., AIT BOUH, H., EL HAMMOUMI, O., BENKDAD, A., FEKRI, A., Evidence for a Recent Increase in Delivery of Atmospheric 210Pb to Oualidia Lagoon, Coastal Morocco, Environ. Monit. Assess. **190** 11 (2018) 642.

[2.12] MEJJAD, N., LAISSAOUI, A., EL-HAMMOUMI, O., FEKRI, A., AMSIL, H., EL-YAHYAOUI, A., BENKDAD, A., Geochemical, Radiometric, and Environmental Approaches for the Assessment of the Intensity and Chronology of Metal

Contamination in the Sediment Cores from Oualidia Lagoon (Morocco), Environ. Sci. Pollut. Res. Int. **25** 23 (2018) 22872–22888.

[2.13] KIRCHNER, G., 210Pb as a Tool for Establishing Sediment Chronologies: Examples of Potentials and Limitations of Conventional Dating Models, J. Environ. Radioact. **102** 5 (2011) 490–494.

[2.14] APPLEBY, P. G., PILIPOSIAN, G. T., Radiometric Dating of Sediment Records from Mountain Lakes in the Tatra Mountains, Biologia **61** 18 (2006) S51–S64.

[2.15] ABRIL, J. M., A New Theoretical Treatment of Compaction and the Advective-Diffusive Processes in Sediments: A Reviewed Basis for Radiometric Dating Models, J. Paleolimnol. **30** 4 (2003) 363–370.

[2.16] ABRIL, J. M., Radiometric Dating of Recent Sediments: On the Performance of 210Pb-Based CRS Chronologies under Varying Rates of Supply, Quat. Geochronol. **51** (2018) 1-14.

[2.17] UNITED NATIONS SCIENTIFIC COMMITTEE ON EFFECTS OF ATOMIC RADIATION. Exposures from Natural Radiation Sources, UNSCEAR Report, New York (2000).

[2.18] BASKARAN, M., Progeny of Radon (210Pb) as a Tracer and Chronometer in Continents and Aqueous Systems. In: Radon: A Tracer for Geological, Geophysical and Geochemical Studies, Springer Geochemistry, Springer, Cham. (2016).

[2.19] MAI, K., PHAM, POVINEC, P.P., NIES, H., BETTI, M., Dry and wet deposition of 7 Be, 210Pb and 137Cs in Monaco air during 1998-2010: Seasonal variations of deposition fluxes, J. Environ. Radioact. **120** (2013) 45–57.

[2.20] HILMI, K., ORBI, A., LAKHDAR IDRISSI, J., Hydrodynamisme de la lagune de Oualidia (Maroc) durant l'été et l'automne 2005, Bull. Inst. Sci. **31** (2009) 29-34.

[2.21] RAMOS-LERATE, I., BARRERA, M., LIGERO, R.A., CASAS-RUIZ, M., A New Method for Gamma-efficiency Calibration of Voluminal Samples in Cylindrical Geometry, J. Environ. Radioact. **38** 1 (1998) 47–57.

[2.22] KHATER, A.E.M., EBAID, Y.Y., A simplified gamma-ray self-attenuation correction in bulk samples. Appl. Radiat. Isot. **66** 3 (2008) 407-413.

[2.23] LAISSAOUI, A., MAS, J. L., HURTADO, S., ZIAD, N., VILLA, M., BENMANSOUR, M., Radionuclide Activities and Metal Concentrations in Sediments of the Sebou Estuary, NW Morocco, Following a Flooding Event, Environ. Monit. Assess. **185** 6 (2013) 5019–5029.

[2.24] MAANAN, M., RUIZ-FERNANDEZ, A.C., MAANAN, M., FATTAL, P., ZOURARAH, B., SAHAB, M., A long-term record of land use change impacts on sediments in Oualidia lagoon, Morocco, Int. J. Sediment Res. **29** (2014) 1-10.

[2.25] AUDRY, S., SCHÄFER, J., BLANC, G., JOUANNEAU, J.M., Fifty-year sedimentary record of heavy metal pollution (Cd, Zn, Cu, Pb) in the Lot River reservoirs (France), Environ. Pollut. **132** 3 (2004) 413–426.

[2.26] OUGHTON, D.H., BØRRETZEN, P., SALBU, B., TRONSTAD, E., Mobilisation of 137Cs and 90Sr from sediments: potential sources to arctic waters, Sci. Total Environ. **202** 1 (1997) 155–165.

[2.27] FOSTER, I.D.L., MIGHALL, T.M., PROFFITT, H., WALLING, D.E., OWENS P.N., Post-depositional 137Cs mobility in the sediments of three shallow coastal lagoons, SW England, J. Paleolimn. **35** 4 (2006) 881–895.

[2.28] OMOKHEYEKE, O., SIKOKI, F. D., LAISSAOUI, A., AKPULUMA, D., ONYAGBODOR, P. O., BENKDAD, A., BENMANSOUR, M., Sediment Geochronology and Spatio-Temporal and Vertical Distributions of Radionuclides in the Upper Bonny Estuary (South Nigeria), Geochronometria **41** 4 (2014) 369–376.

[2.29] KOCH, R. W. SALOMONS, U. FÖRSTNER., Metals in the Hydrocycle, Berlin — Heidelberg — New York — Tokyo: Springer Verlag, 1984, 349 S., 149 Abb., DM 98, Acta Hydrochim. Hydrobiol. **13** 2 (1985) 267–267.

[2.30] BHUIYAN, M. A. H., SURUVI, N. I., DAMPARE, S. B., ISLAM, M. A., QURAISHI, S. B., GANYAGLO, S., SUZUKI, S., Investigation of the Possible Sources of Heavy Metal Contamination in Lagoon and Canal Water in the Tannery Industrial Area in Dhaka, Bangladesh, Environ. Monit. Assess. **175** 1 (2011) 633–649.

[2.31] SALAH, E. A. M., ZAIDAN, T. A., AL-RAWI, A. S., Assessment of Heavy Metals Pollution in the Sediments of Euphrates River, Iraq, J. Water Resour. Prot. **4** 12 (2012) 1009–1023.

[2.32] VAROL, M., Assessment of Heavy Metal Contamination in Sediments of the Tigris River (Turkey) Using Pollution Indices and Multivariate Statistical Techniques, J. Hazard. Mater. **195** (2011) 355–364.

[2.33] ABRAHIM, G. M. S., PARKER, R. J., Assessment of Heavy Metal Enrichment Factors and the Degree of Contamination in Marine Sediments from Tamaki Estuary, Auckland, New Zealand, Environ. Monit. Assess. **136** 1 (2008) 227–238.

[2.34] ANDREAE, M. O., BYRD, J. T., FROEHLICH, P. N., Arsenic, Antimony, Germanium, and Tin in the Tejo Estuary, Portugal: Modeling a Polluted Estuary, Environ. Sci. Technol. **17** 12 (1983) 731–737.

[2.35] ANDREAE, M. O., ANDREAE, T. W., Dissolved Arsenic Species in the Schelde Estuary and Watershed, Belgium, Estuar, Coast. Shelf Sci. **29** 5 (1989) 421–433.

[2.36] RENTERÍA-VILLALOBOS, M., VIOQUE, I., MANTERO, J., MANJÓN, G., Radiological, Chemical and Morphological Characterizations of Phosphate Rock and Phosphogypsum from Phosphoric Acid Factories in SW Spain, J. Hazard. Mater. **181** 1 (2010) 193–203.

[2.37] INTERNATIONAL ATOMIC ENERGY AGENCY, The Environmental Behaviour of Radium: Revised Edition. Technical Reports Series 476 Vienna (2014).

[2.38] ABRIL, J. M., Radiometric Dating of Recent Sediments: On the Performance of 210Pb-Based CRS Chronologies under Varying Rates of Supply, Quat. Geochronol. **51** (2019) 1–14.

[2.39] MEJJAD, N., LAISSAOUI, A., FEKRI, A., HASSEN, N.El., BENMHAMMED, A., EL-HAMMOUMI, O., BENKDAD, A., AMSIL, H., Tracking natural and human impact on sediment dynamics using radiometric approach in Oualidia lagoon (Morocco), Int. J. Environ. Anal. Chem. **100** (2020) 1-16.

[2.40] PFITZNER, J., BRUNSKILL, G., ZAGORSKIS, I., 137Cs and Excess 210Pb Deposition Patterns in Estuarine and Marine Sediment in the Central Region of the Great Barrier Reef Lagoon, North-Eastern Australia, J. Environ. Radioact. **76** (1–2) (2004) 81–102.

[2.41] LAISSAOUI, A., BENMANSOUR, M., ZIAD, N., MAJAH, M., ABRIL, J. M., MULSOW, S., Anthropogenic Radionuclides in the Water Column and a Sediment Core from the Alboran Sea: Application to Radiometric Dating and Reconstruction of Historical Water Column Radionuclide Concentrations, J. Paleolimnol. **40** (2008) 823–833.

[2.42] SWARZENSKI, P., BASKARAN, M., Uranium Distribution in the Coastal Waters and Pore Waters of Tampa Bay, Florida, Mar. Chem. **104** (2007) 43–57.

[2.43] TAYLOR, S. R., MCLENNAN, S. M., The Geochemical Evolution of the Continental Crust, Rev. Geophys. **33** 2 (1995) 241–265.

[2.44] LYSANDROU, M., PASHALIDIS, I., Uranium Chemistry in Stack Solutions and Leachates of Phosphogypsum Disposed at a Coastal Area in Cyprus, J. Environ. Radioact. **99** 2 (2008) 359–366.

[2.45] EL GUEDDAR, O., SAHIB EDDINE, A., RAHMOUNI, F., Aajjane, A., Governance and management system environmental in the company: case of ocp jorf lasfar, Intern. J. Innov. Appl. Stud. **10** 1 (2015) 155–165.

[2.46] CHAFIK, A., CHEGGOUR, M., COSSA, D., Quality of Moroccan Atlantic Coastal Waters: Water Monitoring and Mussel Watching, Aquat. Living Resour. **14** 4 (2001) 239–249.

[2.47] BENMHAMMED, A., LAISSAOUI, A., MEJJAD, N., ZIAD, N., CHAKIR, E., BENKDAD, A., AIT BOUH, H., EL YAHYAOUI., A. Recent pollution records in Sidi Moussa coastal lagoon (western Morocco) inferred from sediment radiometric dating, J. Environ. Radioactiv. **227** (2021) 106464.

[2.48] ZHANG, H., REYNOLDS, M. Cadmium Exposure in Living Organisms: A Short Review, Sci. Total Environ. **678** (2019) 761–767.

[2.49] SAMRANE, K., BOULIF, R., DHIBA, D., BOUHAOUSS, A., Improvements and intensification of industrial co-crystallization process for cadmium removal from wet phosphoric acid, Int. J. Eng. Sci. Technol. **7** 10 (2018) 152–163.

[2.50] INTERNATIONAL ATOMIC ENERGY AGENCY, Sediment distribution coefficients and concentration factors for biota in the marine environment, Technical reports series 422 Vienna (2004).

3. SEDIMENT PROFILES FROM ANOXIC REGIONS OFF THE NAMIBIAN COAST

D.C. LOUW
National Marine Information and Research Centre (NatMIRC), Ministry of Fisheries and Marine Resources, Swakopmund, Namibia
Email: deonlo@yahoo.com

R. GARCIA TENORIO
Department of Applied Physics II, University of Sevilla
Centro Nacional de Aceleradores, Univ. Sevilla-J. Andalucía-CSIC
Sevilla, Spain

M. ROZMARIC
Office of Radiation Protection and Environmental Monitoring
Environmental Protection Agency (EPA)
Dublin, Ireland
*Environment Laboratories, International Atomic Energy Agency (IAEA)
Monaco

P. McGINNITY
Environment Laboratories, International Atomic Energy Agency (IAEA)
Monaco

I. OSVATH
Environment Laboratories, International Atomic Energy Agency (IAEA)
Monaco

M. FUJAK
Environment Laboratories, International Atomic Energy Agency (IAEA)
Monaco

J. BARTOCCI
Environment Laboratories, International Atomic Energy Agency (IAEA)
Monaco

O. BLINOVA
Environment Laboratories, International Atomic Energy Agency (IAEA)
Monaco

M. PHAM
Environment Laboratories, International Atomic Energy Agency (IAEA)
Monaco

B. GASSER
Environment Laboratories, International Atomic Energy Agency (IAEA)
Monaco

A.K. van DER PLAS
National Marine Information and Research Centre (NatMIRC), Ministry of Fisheries and Marine Resources, Swakopmund, Namibia

B. MUDUMBI
Natural Science Research National Commission on Research, Science and Technology (NCRST)
Windhoek, Namibia

Abstract

During two scientific surveys organized in 2014 and 2015, in total 21 sediment cores were collected along the Namibian coast from 18° to 26° latitude and at different distances from the shoreline. The Namibian coastal region is part of the northern Benguela upwelling system, one of the four major eastern boundaries coastal upwelling ecosystems of the world. It is characterized by strong wind-driven upwelling conditions, extensive areas of bottom waters containing minimum amounts of oxygen (< 2 mL/L), affected by intensive sulfuric eruptions and/or dominated by marine phosphate deposits. The radiometric characterization of sediments from such a dynamic system provided necessary information on areas of net sedimentation that are ideal for collecting sediment cores for studying temporal evolution of potential anthropogenic contaminants during the last century by ^{210}Pb dating method. Obtained data suggested that sediment cores can be associated with a sedimentary area only by the analysis of the uppermost layer which consequently significantly simplifies the work, allowing to concentrate efforts on cores from which valuable information can be gained. Locations affected by anoxic conditions were identified within sedimentary areas and were further investigated. The detailed study of one sediment core affected by anoxic conditions presented in this study shows that radionuclides mostly used for ^{210}Pb dating (^{210}Pb and ^{226}Ra) are not affected by variable enhancements or depletions in the sediments due to the redox conditions. The ^{210}Pb dating technique can be thus applied with confidence in coastal sediment cores affected by anoxic conditions. The U profile shows enrichment along the core consistent with a reduced form of uranium which is present in anoxic conditions in the areas with high productivity and/or deposition of organic matter. Specific profiles of some heavy metals (manganese, cadmium, zinc and cooper) significantly influenced by anoxic conditions can be used as the ideal fingerprint of the sediment conditions in the investigated area.

3.1. INTRODUCTION

Many coastal areas, especially in the vicinity of industrial, urban, and agricultural activities have been increasingly impacted by loads of contaminants (heavy metals, organic compounds, radionuclides etc.) since the beginning of 20th century. Despite previously being considered the least polluted, the rapid industrial development, high population growth rates, consumption of massive amounts of natural resources and the subsequent production of large amounts of waste in developing countries has gradually negatively impacted coastal zones. Coastal areas threatened by continuously increasing contamination require sustainable management of marine ecosystems and resources.

Various contaminants have entered the marine environment from anthropogenic sources directly through discharge to marine waters, or indirectly via rivers and runoff from land-based activities and atmospheric deposition. Furthermore, by oceanic and atmospheric circulation, they are subject to long-range transport even to relatively pristine areas with the tendency of finally being deposited onto the seabed. Marine sediments effectively retain and preserve many of the previously mentioned contaminants [3.1, 3.2] and as such can be used to reconstruct sources and magnitudes of contaminant inputs. The use of properly dated sediments (mostly by ^{210}Pb and ^{137}Cs), as natural environmental archives, may allow the definition of temporal variability and trends of concentration levels for a variety of contaminants in the marine environment. Radiometric characterization of marine sediments thus complements conventional marine monitoring programmes and/or reconstructs past environmental conditions, even in the regions that lack long-term monitoring data. [3.3]

Namibia obtained its independence in 1993 from South Africa, and shortly after that the Ministry of Fisheries and Marine Resources was established with the responsibility and mandate to manage the marine environment. A newly extended Exclusive Economic Zones (EEZs) up to 200 nautical miles were introduced and furthermore, the Benguela Current Large Marine Ecosystem (BCLME) was established, composed of Angola, Namibia and South Africa, bordering Namibia in the north and south, respectively. The cold Benguela current is one of the main features of this Large Marine Ecosystem (LME) and extends from approximately 37° at Cape Point to 5° in northern Angola [3.4]. A unique features is that it is bound by two warm water systems, the Angola Current and Agulhas Current, in the north and south, respectively. On the landward side, the arid Namib Desert is situated. These boundaries are highly dynamic and the surrounding warmer waters directly influence the ecosystem as a whole. The LME is dominated by an intense wind-driven coastal upwelling system, the predominant centre being near Lüderitz (27°S, southern Namibia). The earlier studies in the BCLME by Shannon and O'Toole [3.5-3.7], have described the system as complex and highly variable as demonstrated by the seasonal, interannual, and decadal variability as well as periodic shifts in local fish populations in this ecosystem. The Benguela Current LME experiences a temperate climate and has an important influence on global climate and ocean processes.

The study area, the northern Benguela upwelling system (nBUS), is located in the central part of this LME, where the coastal upwelling is driven by quasi-permanent intense along-shore south-east trade winds [3.8]. This upwelling process ensures high phytoplankton productivity throughout the year especially during summer and autumn, when upwelling is reduced, at least north of the main upwelling cell. However, the decay of organic matter leads to a decrease of oxygen levels (0.5 mL/L) near the seabed. This area of anoxic water is found between Conception Bay and Cape Cross within the 100 m isobaths. The anoxic condition is flushed out of the system with the onset of winter and spring, during the upwelling. However, these local conditions are enhanced by seasonal input from oxygen-poor water from the Angola Dome region [3.8, 3.9]

Another seasonal phenomenon is the sulphur eruption along the Namibian coast, which is the production of hydrogen sulphide in the sediment due to the decay of the organic matter as mentioned, by nitrate reducing bacteria. Emeis et al. described this process as an escape, either by diffusion or eruption, of a mixture of methane and hydrogen sulphide which has accumulated in the pore water of diatomaceous mud [3.10].

A large part of the Namibian inner shelf consists of Holocene phosphorite which is from a well-established diatomaceous mud belt along the coast [3.11-3.13]. A large deposit primarily outcrops offshore between Lüderitz and Walvis Bays over an area of 18350 km^2 offshore, and a secondary deposit occurs offshore between the mouth of the Kunene River and Palgrave Point over an area of 6340 km^2 [3.14]. Coupled biostratigraphy and strontium isotope records indicated that the oldest phosphorite is from the late Miocene (7–5 Ma), but the majority of the phosphorite was formed between the Pliocene to Pleistocene (3 and 0.5 Ma). Phosphorite formation is actively occurring in shallower waters along the Namibian coast (up to about 130 m water depth). The deeper laying phosphorite deposits (around 200 m depth and deeper) off the Namibian coast were formed between 0.5 and 7 million years ago as previously described.

Due to the importance of BCLME for Namibia and lack of relevant information on possible contaminants in their marine environment, the Environment sub-division of the Ministry of Fisheries and Marine Resources in Namibia requested the IAEA Environment Laboratories to participate in a monthly oceanographic monitoring (MOM) survey of the Namibian coast on the new multidisciplinary Research Vessel (RV) Mirabilis. The primary aim of the mission was to carry out a baseline study of marine radioactivity levels in the nBUS. Furthermore, to extend the existing database by including additional information on inshore and offshore trace elements and rare earth elements (REE) and to provide assistance to Namibia to set up a future marine radioactivity monitoring programme. A large number of marine samples (28 seawater, 435 sediment and 45 biota samples) was collected and pre-treated during two MOM surveys onboard the RV Mirabilis in 2014 and 2015. A significant amount of obtained data has been presented elsewhere [3.15-3.17] or is in the publication process. This paper is a small part of a larger study of coastal sediments as a sink, and in some cases, as a source of the contamination.

Baseline information on sedimentary conditions in the investigated area along the Namibian coast obtained by the application of radiometric tools as a basis to evaluate the temporal evolution of the levels of some contaminants that potentially could affect the area will be presented in detail elsewhere. After the initial study on characterization of surface sediment layers by gamma-ray spectrometry, baseline maps of several natural radionuclides were constructed and information indicating if the sampling location can be considered as a good repository due to the accumulation and preservation of the material deposited was obtained. Areas with net sedimentation appropriate to study history of contaminants are confirmed by in-depth analysis of the profiles of several natural radionuclides in selected sediment cores from zones with different levels of dissolved oxygen.

The main objective of this paper was to explore the possibility of using sediment cores collected from the anoxic areas as natural archives appropriate to study recent history of contamination and to emphasize the importance of radiometric characterization of surface sediments to define sedimentary conditions in the region prior to in-depth studies of contamination history.

3.2. MATERIAL AND METHODS

3.2.1. Study area

The Benguela Current Large Marine Ecosystem (BCLME) is situated along the coast of south-western Africa. Its southern border extends to the east of the Cape of Good Hope, and its northern border touches the Angola (Cabinda) Front. It covers one of the four major global coastal upwelling ecosystems occuring at the eastern boundaries of the oceans. The combination of BCLME's distinctive hydrography, bathymetry, tropodynamics, and chemistry result in one of the most productive ocean zones in the world. The Benguela upwelling system (BUS) is divided by the Lüderitz upwelling cell at 27°S into the northern and southern BUS (nBUS and sBUS). The 1500 km long Namibian coast is a part of the nBUS.

Two scientific surveys organised in 2014 and 2015 covered the entire shelf inshore to offshore, between 18 °S and 27 °S latitudes (Fig. 2.1). The nearest sampling station to the coastline was located two nautical miles (nm) from the shore, followed by 5 and 10 nm and every 10 nm thereafter until 70 nm in an offshore direction.

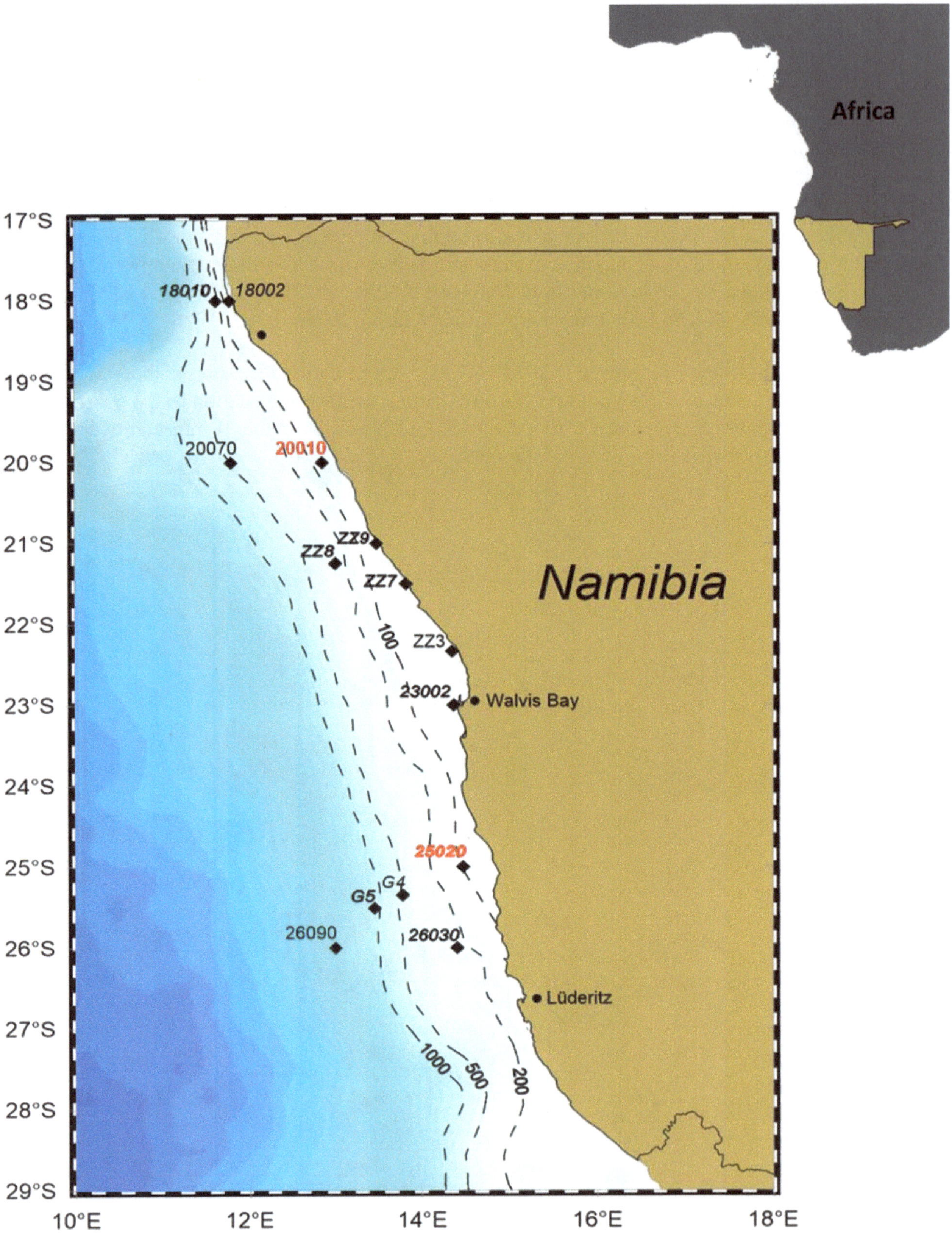

FIG. 3.1. Map of the study area indicating sampling points off Namibia (data for core sediments collected at red marked sampling points is presented in this work).

3.2.2. Sampling of sediment cores and sample pre-treatment

During the two surveys core sediment samples were collected at 21 stations in total. In 2014, sediments were collected by using MC-200-4 MiniMulti-Corer, which is the smallest corer in the MC series, carries four sample tubes 6.25 cm x 32.5 cm with an effective penetration of 20 cm. In 2015, additional core sediment samples were taken by a MC-400 Hedrick/Marrs Multi-Corer corer which is constructed of anodized aluminum and carries four individually sealing polycarbonate sample tubes. Overall sample tube length is 58 cm, with a maximum penetration of 34.5 cm (Fig 2.2). After the collection, sediment samples were sliced per 1 cm up to 10 cm and per 2 cm after 10 cm of each core. Samples were packed, labelled, frozen and transported to IAEA laboratory in Monaco for further analysis. In the laboratory, sediments were freeze dried, grinded, homogenized and properly stored before analysis.

FIG. 3.2. Deployment of box corers on the RV Mirabilis and on board slicing of core sediments.

3.2.3. Radionuclides analysis by gamma-ray and alpha-particle spectrometry

Core sediment samples were analysed for naturally-occurring and anthropogenic gamma-ray emitting radionuclides using high purity germanium (HPGe) detectors in the IAEA Environment Laboratories' low level counting facilities in Monaco. After pre-treatment, samples were prepared in small volume containers (depending on the sample volume available with diameters equal to either 40 mm or 58 mm) and sealed. They were then set aside for at least a 21-day period to allow sufficient in-growth for attaining equilibrium between ^{226}Ra and ^{222}Rn daughters. Samples were counted on a number of different detectors in the IAEA Environment Laboratories' low-level underground and surface counting facilities covered by an ISO/IEC 17025 compliant quality management system (QMS). The methods used are compliant with relevant ISO standards [3.18-3.20] and have been verified using IAEA certified reference materials (CRMs) (IAEA-385 Radionuclides in Irish Sea sediment, IAEA-410 Radionuclides in Bikini Atoll sediment and IAEA-412 Radionuclides in Pacific Ocean sediment) and regular participation in independent PTs and ILCs. All spectral analysis was performed using the Genie 2000 software [3.21]. Depending on the specific detector used, efficiency calibrations were derived either from measurements of calibration sources, prepared with reference standards traceable to national standards, or verified Monte Carlo methods [3.22, 3.23]. All necessary corrections for true coincidence summing and self-attenuation were calculated using EFFTRAN [3.24], GESPECOR [3.23] or LABSOCS [3.22].

For the analysis of uranium isotopes (^{234}U, ^{235}U and ^{238}U), each sediment sample was spiked with a known activity of ^{232}U (National Physical Laboratory), wet digested using concentrated HNO_3, HF, HCl and 30% H_2O_2. The dried residues were dissolved in 3M HNO_3 and uranium was purified by using UTEVA prepacked columns. UTEVA was preconditioned with H_2O, 1 M HNO_3 and 3 M HNO_3 followed by subsequent elution of interfering elements with 3 M HNO_3, 9 M HCL and 5 M HCl/0.05M $H_2C_2O_4$. Uranium was eluted with 1 M HCl, alpha sources were prepared by micro-precipitation using neodymium (III) carrier, $TiCl_3$ and concentrated HF [3.25] and counted by alpha-particle spectrometry with low background ion-implanted silicon detectors (Ultra-AS Ortec, counting efficiency of ~20%). The reliability of the analytical results was ensured by analysis of the IAEA CRM IAEA-384 (Radionuclides in Fangataufa Lagoon sediment) and reagent blanks.

3.2.4. Elemental analysis by ICP-MS

The concentrations of elements of interest in the different layers of selected sediment cores have been determined by ICP-MS technique using a Triple Quadrupole ICP-MS (Agilent 8800). The Method 200.8 is usually applied for determination of dissolved elements in underground waters, surface waters and drinking water [3.26]. However, this method has been also successfully applied to determine element concentrations in liquid wastes, sludges and soil samples. In this study, it has been applied to sediments after total dissolution and conditioning of the solution for the measurement. The samples were analysed directly by pneumatic nebulisation. For quantitative analyses, a minimum of three replicate subsamples were made. The mean value of these measurements was used for data reporting. A rinse blank is used to flush the system between samples. Quality check of results was done by measuring standard and blank samples after every 10 samples. In addition, the reagents used for conditioning of the samples were analysed in order to evaluate the background contribution to the sample measurements.

3.3. RESULTS AND DISCUSSION

3.3.1. Radiometric characterization of surface sediments along the Namibian coast

Radiometric characterization of surface layers (0-1 cm) of 21 sediment cores collected along the Namibian coast by analysis of naturally occurring radionuclides from uranium and thorium series, ^{40}K and ^{137}Cs with the aim to define sedimentary conditions along the Namibian coast will be presented in detail elsewhere while only relevant information to support this study was given below.

Massic activity ratios $^{238}U/^{226}Ra$ higher than one are observed in the great majority of the surface sediments analysed. The coastal zone off Namibia, as part of the nBUS, is characterized by high deposition of organic matter caused by high productivity associated with the strong upwelling and very low oxygen or anoxic conditions at the water bottom surface in most of the region. Such extreme conditions could explain relatively high levels of uranium due to its higher tendency to associate with the organic matter and to precipitate in reduced form under the predominant anoxic conditions. Further support for this conclusion is based on the fact that ^{238}U determined in seawater collected in the area showed a clear deficiency in comparison to the expected values estimated from the salinity data. According to uranium salinity correlations, expected uranium concentrations for salinities 34.5 to 35 are around 40 mBq/L whereas the obtained results in the examined area are around 30 mBq/L.

Another important fact about the region is that some defined areas off Namibia are dominated by marine phosphate deposits which are clearly evident from very high levels of radionuclides from the uranium series (^{238}U, ^{226}Ra, ^{210}Pb) in near secular equilibrium obtained in sediment core G4. In less proportions, the presence of phosphate deposits can be confirmed from the results obtained for sediment cores ZZ7 and 26090.

In-depth analysis of massic activities of ^{210}Pb and ^{226}Ra in the surface layer with the aim to define sedimentary conditions in the region indicated existence of two significantly different groups of sediments. Sediments with ^{210}Pb massic activities higher than ^{226}Ra (18002, 20010, ZZ3, WBHC, 23002, 25020, G5 and LUDHC) indicating a presence of high fraction of unsupported ^{210}Pb which could be associated with areas of net sedimentation. Such cores might be a good tool for geochronological studies focused on history of deposition of potential contaminants. On the other hand, sediments showing the absence of unsupported ^{210}Pb where difference in massic activities of ^{210}Pb and ^{226}Ra was not so evident (18010, ZZ9, ZZ7, ZZ1, G4, 26030, 26090 and SWB) could be associated with transportation bottoms without continuous sediment accumulation rates and as such are not used for the geochronological studies.

3.3.2. Vertical profiles of ^{210}Pb and ^{226}Ra in sediment cores from anoxic zones

Sediment cores 20010 and 25020 were collected in the northern and southern net sedimentation zones with minimum level of oxygen (< 2 mL/L) at the sediment-water interface (Fig 3a). Focusing on the vertical profiles of ^{210}Pb and ^{226}Ra as critical radionuclides used for the ^{210}Pb dating, it is visible that there is no significant difference in their shape by comparing it with sediments collected in oxic zones. The unsupported ^{210}Pb profile is visible in both cores (Fig 2.3b and 2.3c) although in the case of core 20010 the information is partially masked due to high limits of detection associated with the ^{210}Pb values. On the contrary, core 25020 shows a very good resolution of ^{210}Pb profile and constant values in the uppermost layers were associated with the characteristics of the material deposited over time. Obtained results confirmed the previous assumption established by analysis of surface layers that both cores were collected in sedimentation areas and could be potentially used for geochronological studies related to the evolution of potential contaminants in the area.

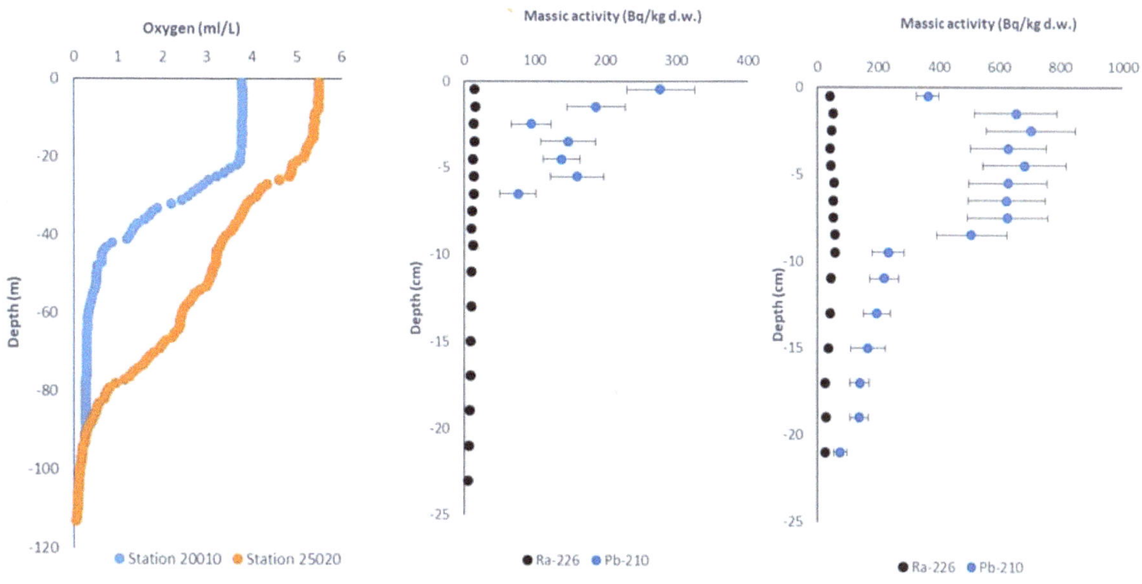

a) *b)* *c)*

FIG. 3.3. Oxygen in seawater vertical profiles a) and ^{210}Pb and ^{226}Ra vertical profiles in sediment cores 20010 b) and 25020 c) collected in anoxic zones off Namibia.

3.3.3. Vertical profiles of ^{238}U in sediment cores from anoxic zones

Comparing the ^{238}U profiles of sediment cores collected in anoxic areas it can be observed that massic activities are up to one order of magnitude higher than in cores collected from well oxygenated zones and that they are increasing by depth. Such quite unusual ^{238}U profiles in anoxic sediment cores as presented in Fig 2.4a. are rarely reported and mostly obtained in the areas with high productivity and/or deposition of organic matter [3.27].

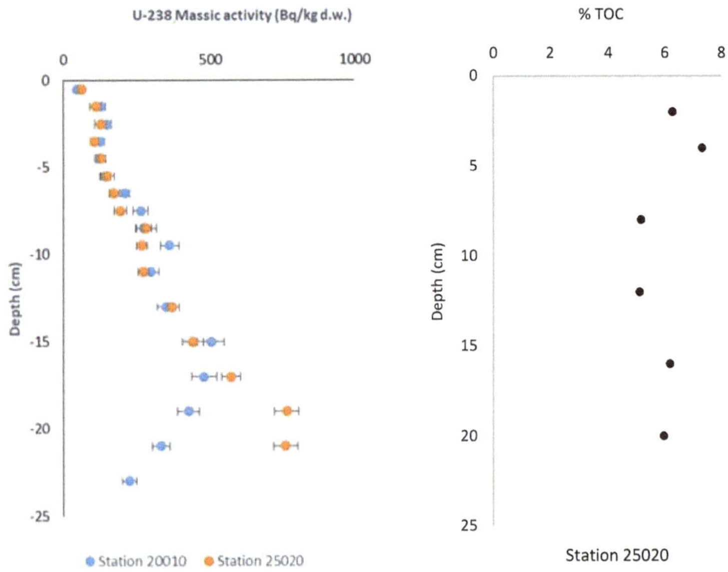

a) *b)*

FIG. 3.4. a) ^{238}U vertical profiles in sediment cores 20010 and 25020 collected in anoxic zones off Namibia, and b) total organic carbon (TOC) in the 25020.

Such peculiar profiles of uranium can be explained by two effects:

a) Under the predominate anoxic conditions, uranium is present in its reduced form with a precipitation tendency of U(IV) oxides (hydroxides) contrary to its highly oxidized U(VI) form that exists in solution under oxic conditions.

b) Uranium exhibits a tendency of more efficient absorption by organic matter than by clay minerals and amorphous Fe and Mn-oxyhydroxides [3.27]. Therefore, the very high deposition of organic matter in the area, as expected in an upwelling zone with extremely high productivity and evidenced by the high TOC levels measured in the layers of the cores (Fig 2.4b), contributes to the fixation of the U precipitated in the different core layers and further explains the U obtained profiles.

The reliability of data was confirmed by determination of ^{238}U massic activities by two additional techniques (ICP-MS and alpha-particle spectrometry). In Fig 2.5 comparison of results obtained by gamma-ray spectrometry and ICP-MS is shown. Data is also confirmed by determination of ^{238}U by alpha-particle spectrometry in selected layers of this core. A good correlation between data was obtained by using three independent techniques with a slope equal to one, and with a value for the ordinate in the origin equal to zero.

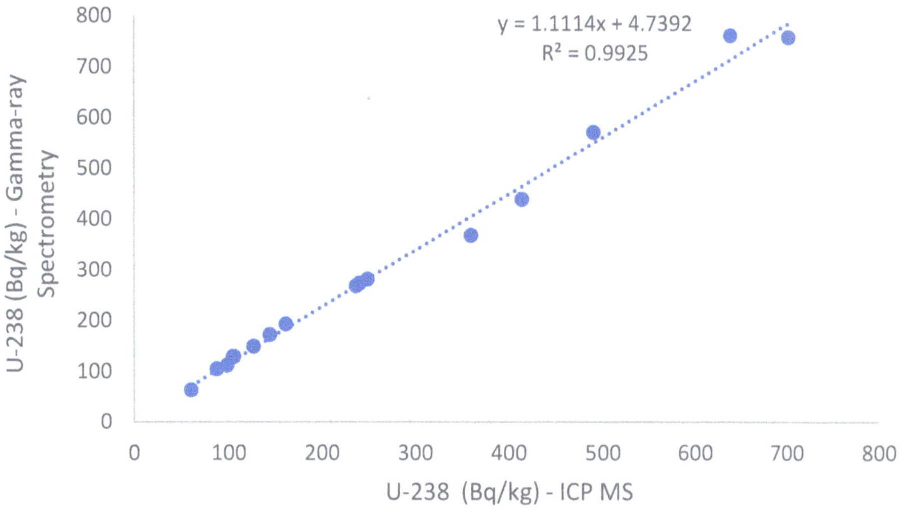

FIG. 3.5. Comparison of results for ^{238}U massic activities (Bq/kg) in sediment core 25020 obtained by gamma-ray spectrometry and ICP-MS.

Results of $^{234}U/^{238}U$ activity ratios obtained by alpha-particle spectrometry (Table 2.1) are in accordance with ratios obtained in seawater from the same locations which is expected due to the main origin of uranium in sediments. Due to anoxic conditions uranium in reduced form is precipitated from the water column without fractionation between radioisotopes and therefore the same ratio is observed in different layers of the sediment.

TABLE 3.1. $^{234}U/^{238}U$ ACTIVITY RATIOS OBTAINED BY ALPHA-PARTICLE SPECTROMETRY IN SELECTED LAYERS OF THE CORE 25020

Sediment layer (cm)	U-234/U-238 activity ratio
8-10	1.11 ± 0.03
4-16	1.13 ± 0.03
18-20	1.12 ± 0.03

3.3.4. Heavy metals as fingerprints of anoxic conditions

As shown from previously given results, uranium behaviour can clearly indicate anoxic conditions. However, behaviour of specific heavy metals is also significantly influenced by anoxic conditions and consequently they can be used as the ideal fingerprint of the conditions of the sediments in the respective area [3.28, 3.29].

In general, the elements belong to two groups: elements whose valence state is associated to the prevailing redox potential conditions and elements which don’t change their valence but form highly insoluble sulphides under anoxic conditions.

Beside uranium, manganese also exists in multiple valence states. While uranium is converted to reactive or insoluble species of lower valence under anoxic conditions, manganese behaviour is the opposite. It forms a highly insoluble oxide under prevailing oxic conditions, whereas under reducing conditions, it is predominantly soluble. Therefore, low concentrations of manganese in comparison with the found ones in oxic areas are expected in sediment cores collected from areas under the anoxic conditions as it mostly reflects lithogenic manganese associated to the silicate fraction. The expected manganese profile is confirmed by results obtained for core 25020 (Fig 2.6).

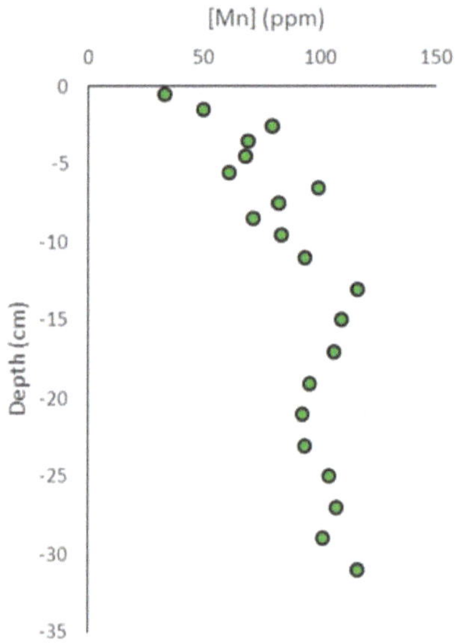

FIG. 3.6. Vertical profile of Mn in sediment core 25020 collected in anoxic zones off Namibia.

Elements such as cadmium, zinc and cooper are usually removed from the solution in the presence of H_2S, thus increasing their precipitation and presence in sediment under anoxic conditions. Therefore, their expected higher concentrations in the core 25020 are presented in Fig 2.7 while clear correlation of Cu vs Zn is visible from the Fig 2.8.

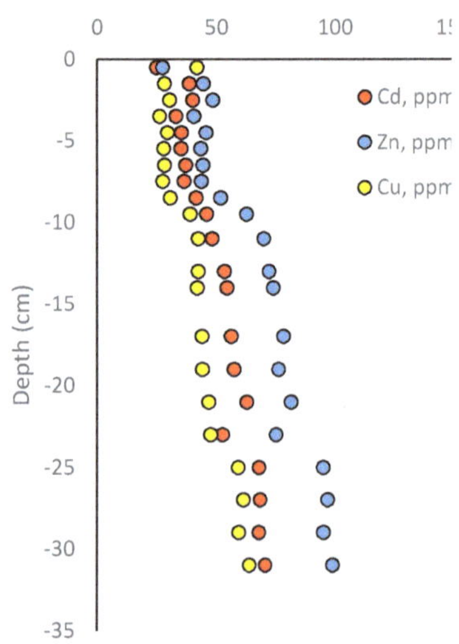

FIG. 3.7. Vertical profiles of Cd, Zn and Cu in sediment core 25020.

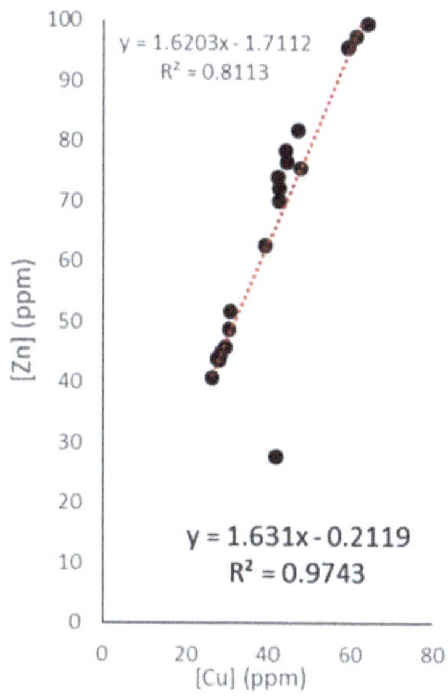

FIG. 3.8. Zn/Cu correlation in sediment core 25020.

It is necessary to emphasize the importance of a proper interpretation of heavy metals enrichment in specific sediment cores as it can't be simply associated with the anthropogenic activities in the investigated area during the last decades. Enrichment observed along the examined core 25020, even higher in the deepest layers reflects more than one century of deposition. It is a consequence of natural geochemical processes affecting the sampling area and supports the additional evidence regarding the possible artefacts and erroneous conclusions on anthropogenic contamination that can be reached if simple monitoring campaigns based only on the analysis of surface sediments are carried out.

3.4. CONCLUSION

The Namibian coastal zone, which is part of the dynamic nBUS is characterized by two types of surface sediment layers, transportation and accumulation bottoms, as confirmed by radiometric surveys studying surface layers of 21 collected sediment cores. The simple classification of the locations based on the data obtained from the study of surface sediments (confirmed by detailed study of sediment cores published elsewhere) will significantly simplify any further studies on the evaluation of the temporal evolution of potential contaminants in the last decades planned in the area. The proposed approach will concentrate efforts on locations with net and uniform sedimentation where the information on historical accumulation of potential contaminants is preserved in the different layers and consequently significantly reduce analytical efforts.

Specific areas along the Namibian coast are strongly influenced by anoxic conditions as a result of high productivity and intensive sulphur eruptions. Sediment cores collected from such areas are characterized by enrichment of ^{238}U that can be used as a fingerprint of anoxic conditions. Furthermore, such cores are also characterized by high deposition of organic matter that influences selective deposition of some heavy metals (e.g. Mn, Cd, Zn and Cu) with uncommon elemental profiles and ratios. Such enrichment is due to geochemical processes and not the consequence of anthropogenic activities in the region. Therefore, in general, it is necessary to emphasize the importance of careful interpretation of surface concentrations data that has to be supported by additional data for different core layers to avoid any misinterpretations in monitoring studies. Study of ^{210}Pb and ^{226}Ra profiles in cores affected by anoxic conditions indicated that they are not enhanced or depleted due to the redox conditions. As these two radionuclides are commonly used in the application of ^{210}Pb dating and as their profiles are not disturbed by anoxic conditions it can be concluded that ^{210}Pb dating method can be applied with confidence for sediment cores collected from the areas influenced by anoxic conditions.

3.5. ACKNOWLEDGEMENTS

Authors are thankful to the Namibian Ministry of Fisheries and Marine Resources (MFMR) for its support, as well as to all colleagues that were involved in the data collection-particularly the staff of the Environment Subdivision. Sampling was done on the RV Mirabilis and captain and crew members from the MFMR are also thanked for their support. The authors would also like to acknowledge the Regional Research Graduate Network in Oceanography (RGNO) a SCOR capacity and collaboration building project in Southern Africa, who bought the MC-400 Hedrick/Marrs Multi-Corer. This work has been partly financed through the project PGC2018-094546-B-I00 provided by the Spanish Government (Ministerio de Ciencia, Innovación y Universidades). The IAEA is grateful to the Government of the Principality of Monaco for the support provided to its Environment Laboratories.

3.6. REFERENCES

[3.1] GESAMP, Anthropogenic influences on sediment discharge to the coastal zone and environmental consequences, Rep. Stud. **52** (1994).

[3.2] GESAMP, Land/sea boundary flux of contaminants: contributions from rivers. Rep. Stud. **32** (1987).

[3.3] SANCHEZ-CABEZA, J.A., DRUFFEL, E.R., Environmental records of anthropogenic impacts on coastal ecosystems: an introduction, Mar. Pollut. Bull. **59** (2009) 87-90.

[3.4] SHILLINGTON, F.A., REASON, C.J.C., DUNCOMBE-RAE, C.M., FLORENCHIE, P., PENVEN, P., "Large scale physical variability of the Benguela Current large marine ecosystem (BCLME)", Benguela: Predicting a Large Marine Ecosystem. Large Marine Ecosystems 14. Elsevier, Amsterdam, (2006) 49–70.

[3.5] SHANNON, L.V., O'TOOLE, M.J., "Sustainability of the Benguela: ex Africa semper aliquid novi", Large Marine Ecosystems of the World – Trends in Exploitation, Protection and Research, Elsevier B.V. (2003).

[3.6] SHANNON, L.V., O'TOOLE, M.J., Integrated Overview of the Oceanography and Environmental Variability of the Benguela Current Region: Thematic Report 2, Synthesis and Assessment of Information on BCLME: October 1998, UNDP, Windhoek, Namibia. (1999)

[3.7] SHANNON, L.V., O'TOOLE, M.J., An Overview of the Benguela Ecosystem. Collected papers, First Regional Workshop, Benguela Current Large Marine Ecosystem (BCLME) Programme, UNDP. 22-24 July 1998, Cape Town, South Africa., (1998).

[3.8] LASS, H.U., MOHRHOLZ, V., On the interaction between the subtropical gyre and the Subtropical Cell on the shelf of the SE Atlantic, J. Mar. Syst. **74** (2008) 1–43.

[3.9] BUBNOV, V.A., Structure and characteristics of the oxygen minimum layer in the south eastern Atlantic, Oceanology,**12** (1972) 193–201.

[3.10] EMEIS, K.-C., BRÜCHERT, V., CURRIE, B., ENDLER, R., FERDELMAN, T., KIESSLING, A., LEIPE, T., NOLI-PEARD, K., STRUCK, U., VOGT, T., Shallow gas in shelf sediments of the Namibian coastal upwelling ecosystem, Cont. Shelf Res. **24** (2004) 627–647.

[3.11] BREMNER, J.M., Sediments on the Continental Margin off South West Africa between Sylvia Hill and the Kunene River. Unpublished PhD Thesis, University of Cape Town, Cape Town, South Africa (1977).

[3.12] BATURIN, G. N., Phosphorites on the sea floor, Developments in Sedimentation, **33** (1982). Elsevier, New York.

[3.13] BATURIN, G.N., Formation and Evolution of Phosphorite Grains and Nodules on the Namibian Shelf, from Recent to Pleistocene. Marine Anthogenesis: From Global to Microbial, SEPM Special Publication. **66** (2000) 185–199.

[3.14] COMPTON, J.S., BERGH, E.W., Phosphorite deposits on the Namibian shelf, Mar. Geol. **380** (2016) 290-314.

[3.15] ROZMARIC, M., CHAMIZO, E., LOUW, D., LOPEZ LORA, M., BLINOVA, O., LEVY, I., MUDUMBI, B., van DER PLAS, A., GARCIA TENORIO, R., MCGINNITY, P., OSVATH, I., Fate of anthropogenic radionuclides in seawater in the northern Benguela upwelling system off Namibia, Chemosphere, **286** 1 (2022) 131514.

[3.16] GUTIERREZ, J.M., LOUW, D.C., ROZMARIC, M., LÉRIDA-TOROA, V., Tracing the upwelling process in the northern Benguela upwelling system (nBUS) by ^{129}I, Chemosphere **265** (2021) 1-7.

[3.17] LÓPEZ-LORA, M., CHAMIZO, E., ROZMARIC, M., LOUW, D.C., Presence of ^{236}U and ^{237}Np in a marine ecosystem: the northern Benguela Upwelling System, a case study, Sci. Total Environ. **708** (2019) 1-10.

[3.18] INTERNATIONAL ORGANIZATION FOR STANDARDIZATION, Measurement of radioactivity in the environment - Soil - Part 3: Test method of gamma-emitting radionuclides using gamma-ray spectrometry, ISO 18589-3:2015, ISO, Geneva (2015).

[3.19] INTERNATIONAL ORGANIZATION FOR STANDARDIZATION, Water quality - Determination of the activity concentration of radionuclides - Method by high resolution gamma-ray spectrometry, ISO 10703:2007, ISO, Geneva (2007).

[3.20] INTERNATIONAL ORGANIZATION FOR STANDARDIZATION, Measurement of radioactivity - Gamma-ray emitting radionuclides - Generic test method using gamma-ray spectrometry, ISO 20042:2019, ISO, Geneva (2019).

[3.21] GENIE™ 2000 Basic Spectroscopy Software, Mirion (2021) https://www.mirion.com/products/genie-2000-basic-spectroscopy-software

[3.22] LABSOCS™ Calibration Software, Mirion (2021) https://www.mirion.com/products/labsocs-calibration-software

[3.23] SIMA, O., ARNOLD, D., DOVLETTE, C., Gespecor – a versatile tool in gamma-ray spectrometry, J. Radioanal. Nucl. Chem. **248 (2)** (2001) 359-364.

[3.24] VIDMAR, T., EFFTRAN - A Monte Carlo efficiency transfer code for gamma-ray spectrometry, Nucl Instrum Meth A **550** 3 (2005) 603-608.

[3.25] SILL, C. W., Preparation of Actinides for Alpha Spectroscopy without Electrodeposition, Anal. Chem. **53** (1981) 412-415.

[3.26] CREED, J.T., BROCKHOFF, C.A., MARTIN, T.D., Determination of Trace Elements in Waters and Wastes by Inductively Coupled Plasma-Mass Spectrometry, Revision 5.4., USEPA, Cincinnati, Ohio, Method 200.8 (1994).

[3.27] LEE. S-H., POVINEC, P., CHISHOLM, J.R.M., LEVY, I., MIQUEL, J-C., OH, J-S., Distribution of natural and anthropogenic radionuclides in northwest Mediterranean coastal sediments, J. Environ. Radioact. **172** (2017) 145-159.

[3.28] SWEERE, T., VAN DEN BOORN, S., DICKSON, A.J., REICHART, G-J., Definition of new trace-metal proxies for the controls on organic matter enrichment in marine sediments based on Mn, Co, Mo and Cd concentrations, Chem. Geol. **441** (2016) 235-245.

[3.29] TRIBOVILLARD, N., ALGEO, T.J., LYONS, T., RIBOULLEAU, A., Trace metals as paleoredox and paleoproductivity proxies: An update, Chem. Geol. **232** (2006) 12-32.

4. SPATIOTEMPORAL DISTRIBUTION OF ^{137}Cs IN MARINE SEDIMENT IN THE COASTAL WATERS AROUND JAPAN

M. Kusakabe
Marine Ecology Research Institute
Onjuku-Machi, Japan
Email: kusakabe@kaiseiken.or.jp

S. Kambayashi
Marine Ecology Research Institute
Onjuku-Machi, Japan

Abstract

Monitoring of radioactivity in seawater and sediment has been carried out regularly in Japanese coastal waters since 1984. After the Fukushima Daiichi Nuclear Power Plant accident in March 2011, an extended monitoring program started in the waters off Fukushima and nearby prefectures. ^{137}Cs concentrations in the surface sediment had declined over time before the accident: the effective half-life of ^{137}Cs in the surface sediment was 23.6 years on average, which was longer than that in seawater, 17.3 years. While the spatial variation of K_d values reflected the physical characteristics of the surface sediment such as specific surface area of sediment grains, the values increased with time on a long-term basis (> a few decades). In September 2011, the concentrations of ^{137}Cs in the surface sediment increased by two orders of magnitude to 580 Bq/kg-dry at most collection sites more than 30 km away from the accident site. Since September 2011, the ^{137}Cs concentrations in sediments in general have decreased with time with the effective half-life of 2.1 years. The vertical profiles of ^{137}Cs in the sediment column showed exponential decrease with depth, that was probably caused by vertical mixing of the sediment by bioturbation and bottom current induced by rough weather. The declines of ^{137}Cs concentration and inventory in the sediment were caused most likely by resuspension of sediment and subsequent lateral transportation. Total inventory of Fukushima-derived ^{137}Cs in the sediment was estimated to be 0.23 x 10^{15} Bq, which was approximately 6.4 % of the directly released ^{137}Cs activity to the ocean. Discharge of ^{137}Cs from the accident site altered the ^{137}Cs mass balance in the coastal environment so that K_d values have been converging to significantly higher values than the pre-accident values.

4.1. INTRODUCTION

Coastal areas are characterized by high biological productivity, fresh water input through riverine systems, vigorous water mixing in both horizontal and vertical directions, and relatively high loading of suspended materials caused by these processes. Once pollutants including radioactive materials are released to a coastal environment their behavior is expected to be complicated accordingly. In Japan all the nuclear power plants and a commercial reprocessing facility for spent nuclear fuel are located on the seashore. Monitoring programs of coastal waters near all the nuclear power plants have been in place since 1984, testing for radioisotopes such as ^{137}Cs (half-life of 30.2 years) and ^{90}Sr (half-life of 28.8 years) in their seawater, surface sediment and biota once a year (hereinafter referred to as the regular monitoring). The data collected for three decades at fixed sites are now truly indispensable data sets for the overall background level of radioactivity mostly derived from the atmospheric nuclear bomb tests in the coastal marine environment around Japan [e.g., 4.1–4.3].

The Fukushima Dai-ichi Nuclear Power Plant (FDNPP) accident in March 2011 released large amounts of radioactive materials to the ocean. Radioactive Cs, ^{137}Cs and ^{134}Cs (half-life of 2.06 years), are the main radionuclides with relatively long half-lives (> few years) discharged in the accident to the adjacent marine environment. Estimates have been made on the discharge of ^{137}Cs to the ocean; e.g., direct discharge of $3.6 \pm 0.7 \times 10^{15}$ Bq [4.4, 4.5] and atmospheric fallout on the ocean of $(11.7–14.8) \times 10^{15}$ Bq [4.6]. The activity ratio of ^{134}Cs to ^{137}Cs was estimated to be about 1 at the time of the accident. Discharge of such large amounts of artificial radionuclides to the limited area of the marine environment in a brief period was an unprecedented event. Right after the accident, a comprehensive marine monitoring scheme was added to the regular monitoring in the waters off Fukushima and nearby prefectures (hereinafter referred to as the extended monitoring).

In this document, results obtained from the regular and extended monitoring programs are summarized, placing emphases on distributions of ^{137}Cs and their temporal changes after the FDNPP accident in sediment in the coastal waters off Fukushima and nearby prefectures.

4.2. METHODS

As aforementioned, coastal waters near nuclear power plants have been monitored since 1984. Currently in 2020, four monitoring sites are located within each coastal area off 15 nuclear power plant sites. The monitoring targets at the sites are seawater and surface sediment. Seawater samples are collected once a year at the sea surface and about 10 m above the seafloor. Sediment samples are taken from the surface 0-3 cm depth.

After the FDNPP accident, extended monitoring began, including surface seawater sampling from March 2011 at select monitoring sites near the INAA location. The upper 3 cm of seabed sediment (designated as surface sediment samples) were sampled along with the surface seawater samples starting in May 2011. Since September 2011, 32 sites have been sampled four times per year as part of the extended monitoring campaign (Fig. 4.1). Sediment core samples with length of up to ~ 30 cm were taken at 9 sites for the extended monitoring during the period from May 2012 to October 2015.

A summary of radioactivity measurements can be found in references 4.1, 4.7–4.10.

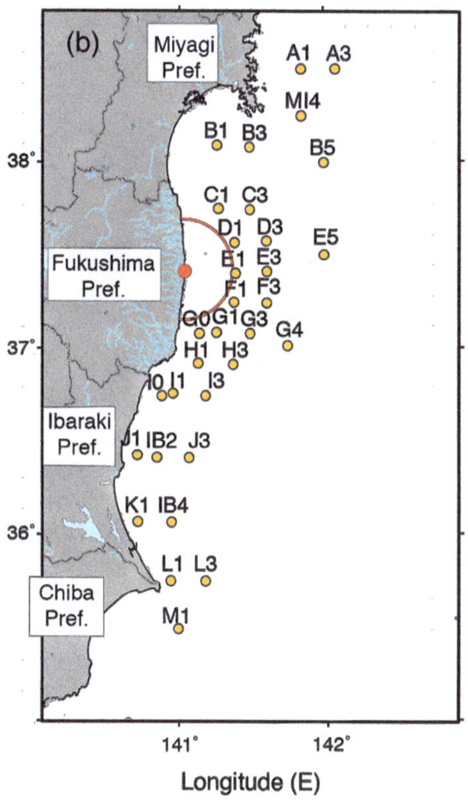

FIG. 4.1. Sampling sites for the extended monitoring program. Seawater and sediment samples are collected at each site. The red circle indicates the 30 km radius from the FDNPP site shown as the red dot.

4.3. RESULTS AND DISCUSSION

4.3.1. Distributions of ^{137}Cs prior to the FDNPP accident (1984-2010)

Cs-137 in the marine environment around Japan was basically of fallout origin before the FDNPP accident, except for some amount from the 1986 Chernobyl accident. The concentrations of ^{137}Cs in surface sediment and both surface and bottom seawaters had been decreasing with time from 1984 to 2010 (Fig. 4.2). While impact of the Chernobyl accident only appeared in the ^{137}Cs concentrations in the surface water all around Japan (Fig. 4.2(a)), those in bottom seawater and sediment did not show any increase at the time (Figs. 4.2(b) and (c)). The concentrations of ^{137}Cs in the surface water have been relatively constant horizontally with an average and standard deviation of 4.1±0.4 mBq/L in 1984-1985 and they decreased with time to 1.5±0.2 mBq/L in 2010. Those in bottom water have been lower and in a wider range than in surface water reflecting variable degrees of intrusion of intermediate open ocean water with significantly lower concentration: they were 4.2±0.4 mBq/L in 1984-1985 and 1.4±0.3 mBq/L in 2010. The ^{137}Cs concentrations in the surface sediment have been even more variable than those in seawater. They ranged from 0.7 to 16 Bq/kg-dry with a geometric mean of 2.6 Bq/kg-dry in 1984-1985 and decreased with time to 0.6 to 7.1 Bq/kg-dry with a geometric mean of 1.8 Bq/kg-dry in 2010. The spatial variation of the concentrations reflects the variable nature of physical characteristics of the surface sediment [4.11].

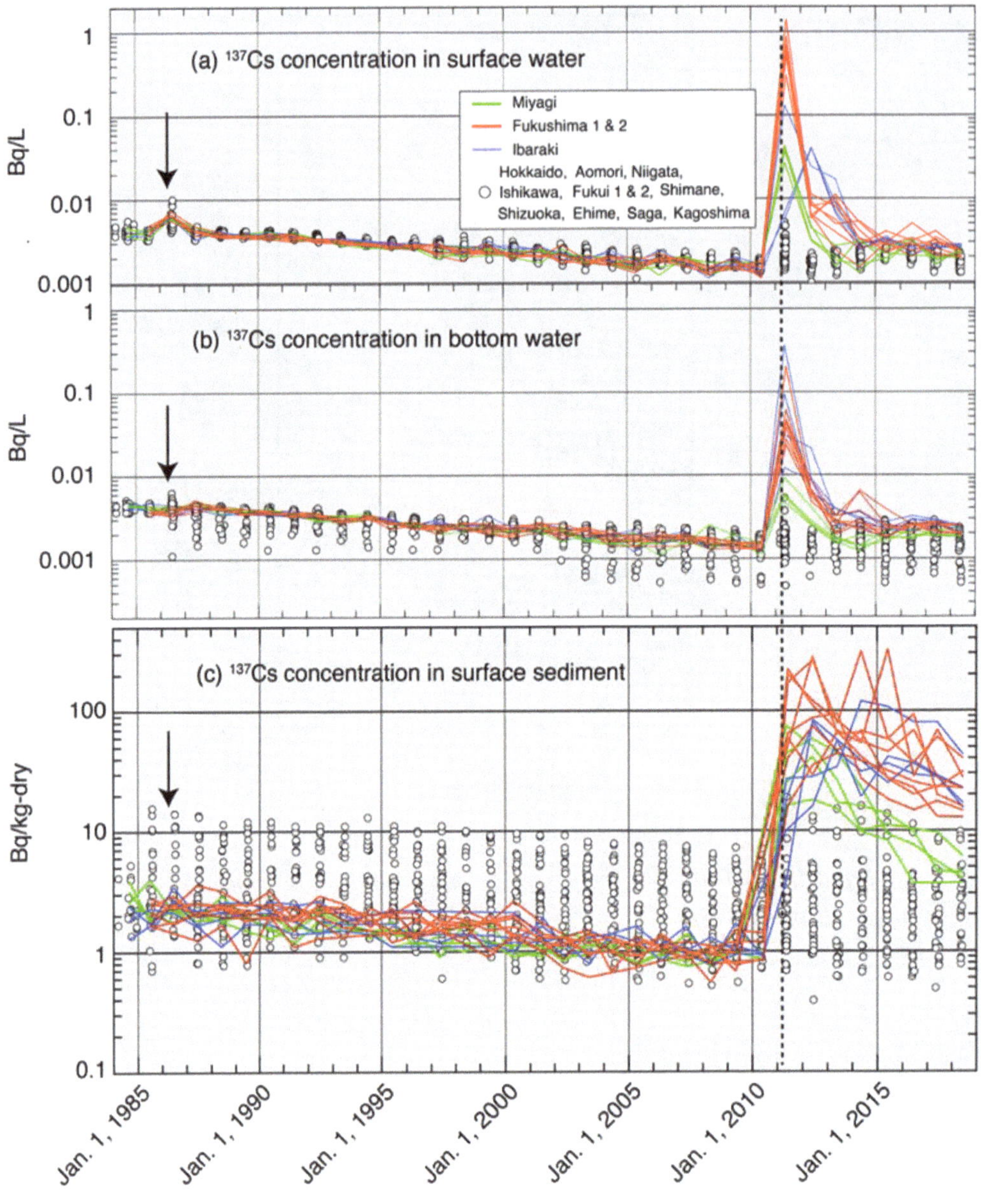

FIG. 4.2. Temporal changes of ^{137}Cs concentrations in surface seawater (a), bottom seawater (b) and surface sediment (c) in the monitored coastal water areas around Japan. The data obtained in the waters off Miyagi, Fukushima, and Ibaraki Prefectures are shown as lines. The data were adapted from Ref. [4.2]. Arrows and dotted lines indicate times of the Chernobyl accident and FDNPP accident, respectively.

The sediment distribution coefficient, K_d, of ^{137}Cs in a marine environment is defined as the ratio of ^{137}Cs concentration in seawater to that in surface sediment under the steady state condition [4.12] and it is used in radiological assessment models of the marine environment. The K_d values calculated for the coastal surface sediment samples collected around Japan between 1984-2010 ranged widely from 719 to 7540 L/kg with a geometric mean and median of 923 and 719 L/kg [4.2]. The K_d value for Cs reported by the IAEA [4.12] for an ocean margin was 4 x 10^3 L/kg. The significant difference between the suggested and measured values is ascribable mainly to the way they are estimated: the former was based on laboratory experiments using stable Cs, that is ^{133}Cs, and its field observations. Stable Cs in sediment tends to exist more as an immobile fraction, which is not involved in desorption-adsorption reactions with stable Cs in seawater, but all the stable Cs is used for the K_d calculation, resulting in a higher K_d value than the K_d for ^{137}Cs. Variability of the K_d values obtained in the coastal waters of Japan should be a function of the sediment grains' specific surface area available for the adsorption-desorption reactions between ^{137}Cs in seawater and surface sediment. A relationship between K_d (L/kg) and specific surface area, S (m^2/g), was derived as follows with a determination coefficient, r^2, of 0.78 (Fig. 4.3).

$$K_d = 11{,}200\ S^{1.66} \tag{4.1}$$

The plotted relationship suggested that ^{137}Cs of fallout origin in seawater had been equilibrated with that in sediment before the FDNPP accident. However, the ^{137}Cs concentrations in seawater and sediment also decrease over time as aforementioned so that the equilibrium here is not merely a static equilibrium in a closed system but rather a quasi-equilibrium or dynamic equilibrium with a part of the system, seawater, allowed to exchange with open ocean water.

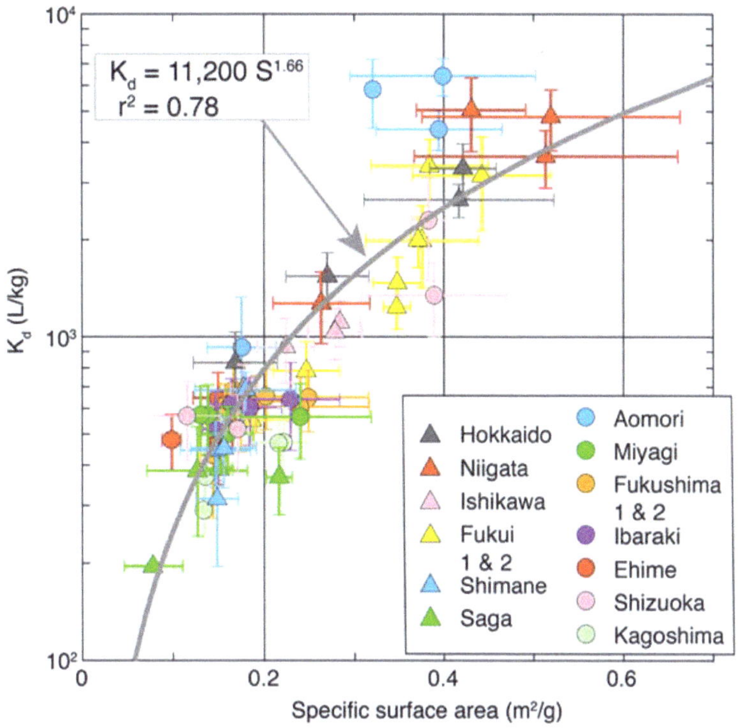

FIG. 4.3. Relationship between K_d value and specific surface area of sediment grains in the monitored coastal areas around Japan (Fig. 4.1(a)). The data were adapted from Ref. [4.2]. While the data for the specific surface values were derived from samples taken from 2016 to 2018, the K_d values were evaluated for the samples taken in 1984-2010.

The arithmetic means of effective half-lives of ^{137}Cs in surface and bottom seawaters calculated from the decreasing rates in the period of 1984-2010 were 16.0 and 17.3 years, respectively. These numbers agreed with the ^{137}Cs effective half-life of 16.5±0.9 years calculated for the north-western north Pacific Ocean seawater [4.13]. The effective half-lives for sediment were determined to range between 10.3 to 55.1 years with an arithmetic mean of 23.6 years. The difference in half-lives between seawater and sediment implies that while ^{137}Cs in seawater in Japanese coastal areas is mainly controlled by water mixing with the open ocean, that in sediment in general may not necessarily be in equilibrium with that in overlying seawater, and ^{137}Cs may stay longer in sediment than in seawater on a long-term basis. Based on a laboratory experiment using ^{134}Cs as a tracer, Børrentzen and Salbu [4.14] suggested that it takes about

1100 days for 50% of the ^{134}Cs ions to be fixed to the irreversible fraction of the sediment (silty sediment taken in the coastal water in the Cara Sea), while 99.7% of the Cs ions are fixed in about 12.5 years. Fixation of Cs to the sediment following adsorption onto the surface of the sediment grains may thus be one of the main causes for the longer residence time of ^{137}Cs in sediment. Kusakabe and Takata [4.2] presented a simple box model consisting of two boxes, coastal seawater and sediment, and they showed that average K_d values around Japan increased with time from 600 in March 1984 to a near-equilibrium value of 1030 by 2011. That is, concentrations of ^{137}Cs in sediment are predominantly controlled by the surface area of the sediment grains in a short time span (~ few years), but over the long-term, its fixation to the irreversible fraction of the sediment makes the K_d larger with time. Eventually decreasing rates for both seawater and sediment become identical to get a terminal dynamic-equilibrium k_d, $k_{d\infty}$ of 1030.

4.3.2. Distributions of ^{137}Cs after the FDNPP accident (2011-2019)

4.3.2.1. ^{137}Cs in the surface sediment

From May 10 to June 2, 2011, 60-83 days after the FDNPP accident started, surface sediment and seawater were collected at the monitoring sites labeled as Miyagi, Fukushima 1 & 2 and Ibaraki (Fig. 4.1(a)) close to the FDNPP site as a part of the regular monitoring program. Results after this time period are presented as the right-hand ends of the Fig. 4.2 curves. The four sites were heavily impacted by the accident with respect to the ^{137}Cs contents in seawater and surface sediment. Temporal variations of ^{137}Cs concentrations in sediment after the accident were quite different from those in the seawater. While the ^{137}Cs concentrations in the seawater increased by two orders of magnitude at most in May to June 2011 and readily decreased in the following years for Miyagi, Fukushima 1 & 2 and Ibaraki, those in the sediment did not follow the trends in the seawater: the concentration peaks appeared later with different timing and they were spatiotemporally fluctuating.

More comprehensive results on the effect of the accident to the coastal areas were obtained in the extended monitoring. Approximately one month after the start of the FDNPP accident, on April 15, 2011, the highest ^{137}Cs concentration ((86 Bq/L) was observed in surface water around 30 km to the east of the FDNPP (it is not shown in Fig. 4.1(b), Oikawa et al. [4.10]). The elevated concentrations decreased by about three orders of magnitudes until September 2011, and then declined more slowly. As of January 2019, the ^{137}Cs activity concentrations were similar to the pre-accident values, 1.6±0.3 mBq/L, averaged over 2006-2010 in the waters off Fukushima Prefecture. Although the bottom seawater layers from the sites having 3, 4, and 5 in the site code (e.g., G3, G4, and E5, Fig. 4.1 (b)) with depth of > 200 m were much less affected by the accident (Fig. 4.4(a)), the ^{137}Cs activity concentrations tracked the same declining trend over time as those of surface water. From May to September 2011, ^{137}Cs concentrations in surface sediment and seawater as well were measured 6 times at 12 sites for the extended monitoring. The highest ^{137}Cs concentration (10 Bq/L) in the surface water was found at site C1 on May 11, 2011.

The concentrations in surface sediment varied temporally and horizontally ranging from 2 Bq/kg-dry at site L1 to 580 Bq/kg-dry at site D1 for the period of May to September 2011 (Figs. 4.4 and 4.5). After that, two peaks of the concentration appeared in the northern and southern parts of the extended monitoring area, that is, around sites K1, D3, and B3 (Fig. 5). The horizontal variations were not necessarily related to proximity to the accident site nor physical characteristics of the sediment such as bulk density and water content. Kusakabe et al. [4.9] related the variation to the transport route of the radiocontaminated water from the

FDNPP; initially the radiocontaminated water migrated north after the accident, followed by some of the water moving to the south (Fig. 4.6). After September 2011 ^{137}Cs concentrations in the surface sediment were in general in a declining trend with time until January 2019. However, the trend involved a large fluctuation compared to that for seawater (Fig. 4.4). The fluctuation may reflect spatial heterogeneity of the ^{137}Cs concentrations in a small area. For example, six samples were taken at site D1 in succession on September 13, 2011. The concentrations varied in the range of 170-580 Bq/kg-dry with the average and standard deviation of 330±160 Bq/kg-dry, implying there was concentration heterogeneity [4.9]. The local heterogeneity of Fukushima-derived ^{137}Cs concentrations has been also reported by Thornton et al. [4.15]. They found local ^{137}Cs concentration anomalies which were an order of magnitude greater than concentrations of nearby sediment with the areal sizes of the anomalies ranging from a few to a few hundred meters. It was inferred that post-depositional migration of sediment due to bottom current lead to the formation of a pocket of ^{137}Cs-contaminated sediment at the base of vertical features of the terrain. Although positioning for the sampling has been done within at most ~1850 m from the pre-determined locations, the sporadic peaks in Fig. 4 may reflect the local heterogeneity of ^{137}Cs concentrations caused by the post-depositional migration of the sediment. Ikenoue et al. [4.16] used an autoradiography technique and found ^{137}Cs-enriched particles in the sediment samples taken at sites I1 and IB2 in May 2013 and May 2017, respectively. These particles included 9-64 % of the ^{137}Cs in the sediment samples. Because these ^{137}Cs-enriched particles are sparsely distributed in the sediment, they may be also one of the reasons for temporal and spatial fluctuations of the ^{137}Cs concentrations in sediment.

The temporal change of geometric mean of the ^{137}Cs concentrations in sediment (Fig. 4.4) indicates an exponential declining trend from September 2011 to January 2019 with an effective half-life of 3.1 years. However, the half-lives for the first half period and the latter half period were significantly different from each other, that is, 2.2 and 5.0 years, respectively. Further discussion of this is given in Sec. 4.3.2.3.

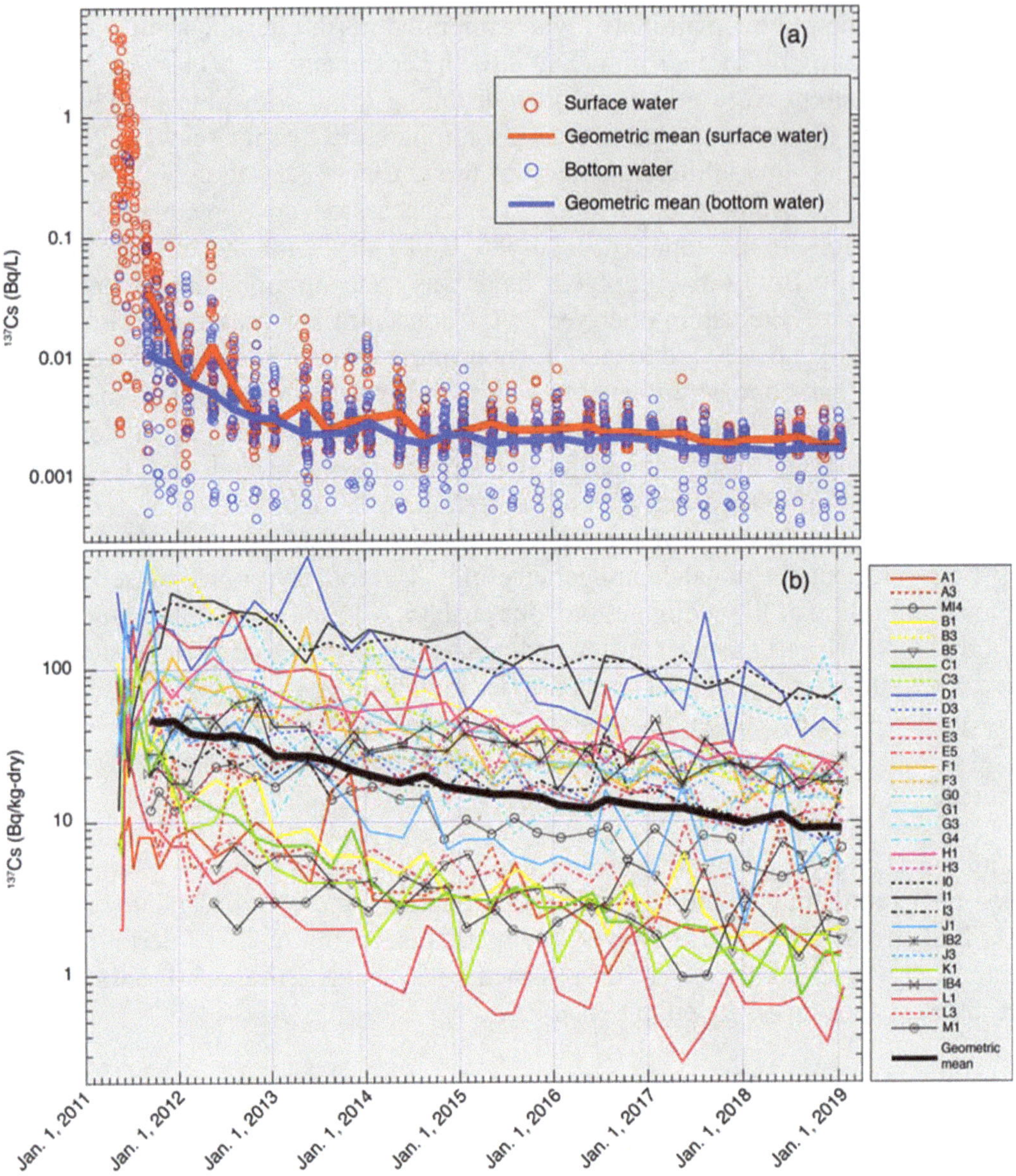

FIG. 4.4. Temporal changes of ^{137}Cs concentrations in surface and bottom water (a) and surface sediment (b) in the waters off Fukushima and nearby prefectures. The data were adapted from Ref. [4.2].

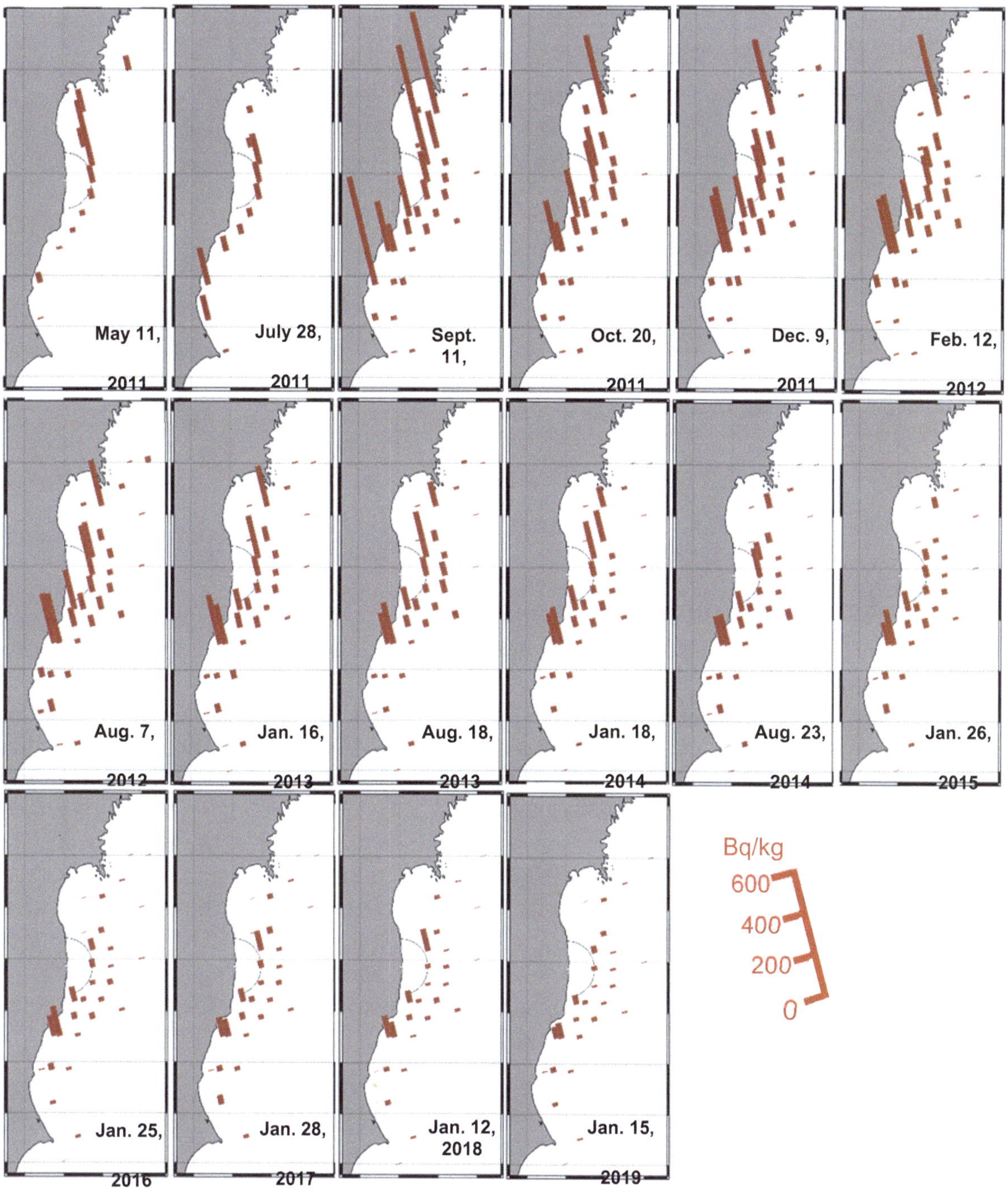

FIG. 4.5. Spatiotemporal change of ^{137}Cs concentration in the surface sediment. The data were adopted from Kusakabe et al. [4.9, 4.11] and Kusakabe and Takata [4.2]. Date shown in the panels are from the middle of each sampling period.

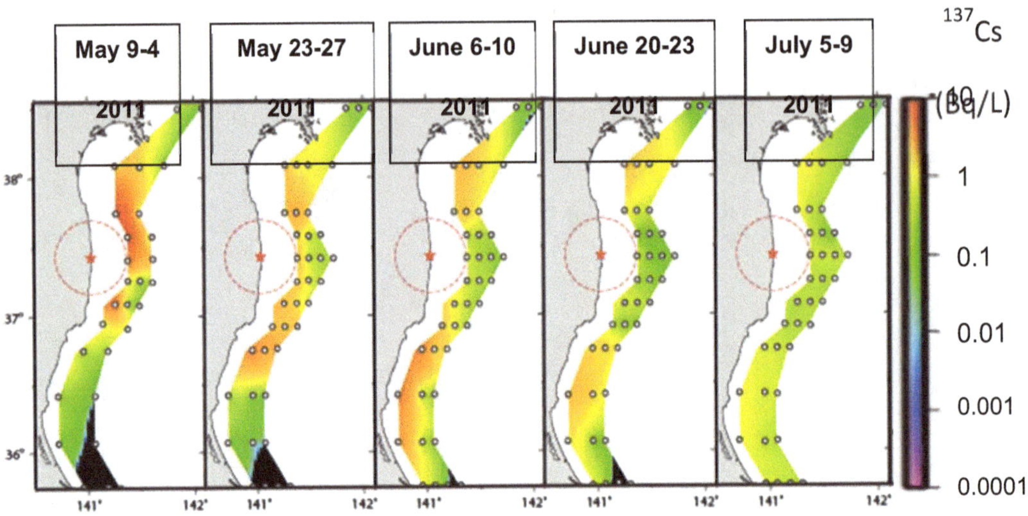

FIG. 4.6. Spatiotemporal change of ^{137}Cs concentration in the surface seawater. The data were adapted from Ref. [4.10].

Difference in responses to the additional input of ^{137}Cs to the seawater and surface sediment leads to the temporal change of the sediment distribution coefficient, K_d. After the accident, the steady-state condition should not be expected so that a new term, an apparent distribution coefficient, $K_{d(a)}$ is introduced: $K_{d(a)}$ is simply a ratio of ^{137}Cs concentration in seawater to that in surface sediment. Kusakabe and Takata [4.2] presented a schematic diagram for the temporal changes of ^{137}Cs concentration in seawater and surface sediment, and $K_{d(a)}$. Fig. 7 indicates three points. (1) The concentration of ^{137}Cs in seawater increased and reached the peak at time t_2 and then dropped to almost the pre-accident level. (2) The concentration of ^{137}Cs in surface sediment followed the trend with a time-lag fashion and slower pace, and the maximum value appeared at time t_3. (3) Consequently, their ratio, $K_{d(a)}$, dropped at t_1 and reached the maximum at t_4 approaching the terminal $K_{d(a)}$ value which was significantly higher than the pre-accident value. The drop of $K_{d(a)}$ value at time t_1 by a factor of ten was observed at the sites off Ibaraki Prefecture (data not shown here, Kusakabe and Takata [4.2]).

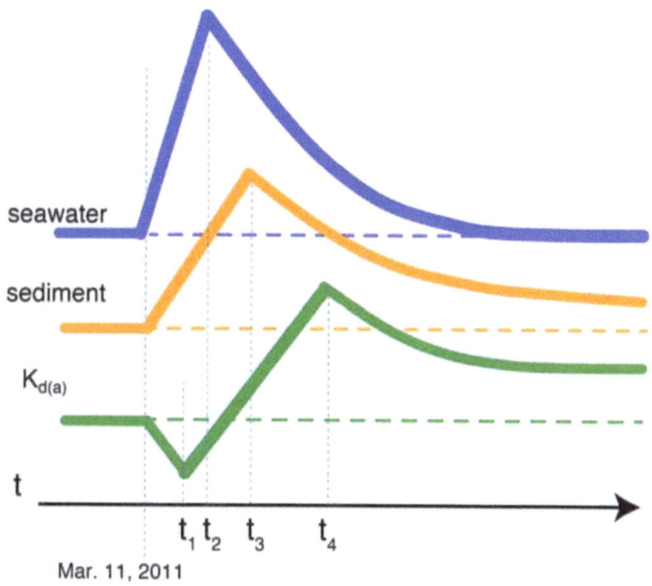

FIG. 4.7. Temporal evolution of ^{137}*Cs concentrations in seawater (blue/top line) and surface sediment (middle/orange line), and* K_d *(bottom/green line) in nearby waters following the Fukushima accident. The dashed lines indicated with "*t_1*", "*t_2*", "*t_3*"and "*t_4*" show temporal maximums for each parameter.*

A simple box model for migration of ^{137}Cs in the coastal area after the accident is schematically illustrated in Fig. 4.8. The model consists of two boxes, seawater and surface sediment. A transfer rate constant, k_1 is mainly controlled by the mixing of surface water with clean intermediate and deep seawater in the open ocean. Accident-derived ^{137}Cs in seawater is being removed by diffusion-advection to open water and by deposition-adsorption to the sediment with the rates, k_4, and k_2, respectively. Cs-137 in sediment migrates to overlying water via desorption-dissolution of Cs and/or resuspension and subsequent lateral transportation of sediment with a rate of k_3. By fitting the geometric means of ^{137}Cs concentrations in two boxes to the equations for mass balance between seawater and sediment, temporal changes of Cs concentrations in seawater and sediment, and $K_{d(a)}$ values were calculated for the period from September 2011 (t_3) to January 2019 [4.2].

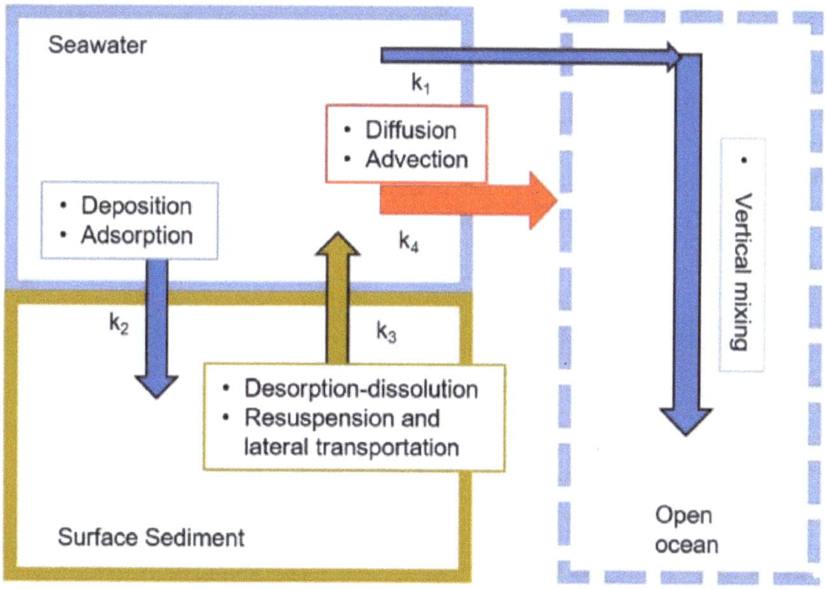

FIG. 4.8. Schematic diagram for migration of ^{137}Cs in coastal waters.

The results are plotted in Fig. 4.9. Data points in grey in the figure were omitted from the calculation because they were calculated based on smaller datasets than those for the rest of the points and thus had a larger error. The model presented here was a simplified one applied to a spatially variable environment in terms of the ^{137}Cs concentration. However, some general points can be drawn from the model calculation results. (1) ^{137}Cs concentrations in seawater decreased with time from 0.011 Bq/L in September 2011 to 0.004 Bq/L in September 2012, after that the decreasing rate slowed down. The average concentration was estimated to be 0.002 Bq/L in January 2019 and that was almost the same level as the pre-accident value. (2) On the other hand, the concentration in surface sediment reached the maximum of 45 Bq/kg in September 2011and decreased at a slower pace to 38 Bq/kg in September 2012. In January 2019 the average concentration was estimated to be 11 Bq/kg. (3) Consequently, the $K_{d(a)}$ value reached the maximum of 10,400 in January 2013 and decreased more slowly than the concentrations in sediment. The $K_{d(a)}$ value will eventually converge (or has already converged) to the permanent $K_{d(a)}$ value, $K_{d(a)\infty}$, of 6,240, which is significantly higher than the pre-accident value (~600).

Fates of radionuclides in the coastal area are generally site-specific and event-specific. Relevant factors include oceanographic settings, such as being a coastal sea facing the open ocean or being a semi-closed sea, and modes of radionuclide input, such as instantaneous or continuous discharges. In addition, characteristics of sediment such as grain size may also affect radionuclide fates in the coastal environment. In cases where these factors are different, the fates need to be changed accordingly.

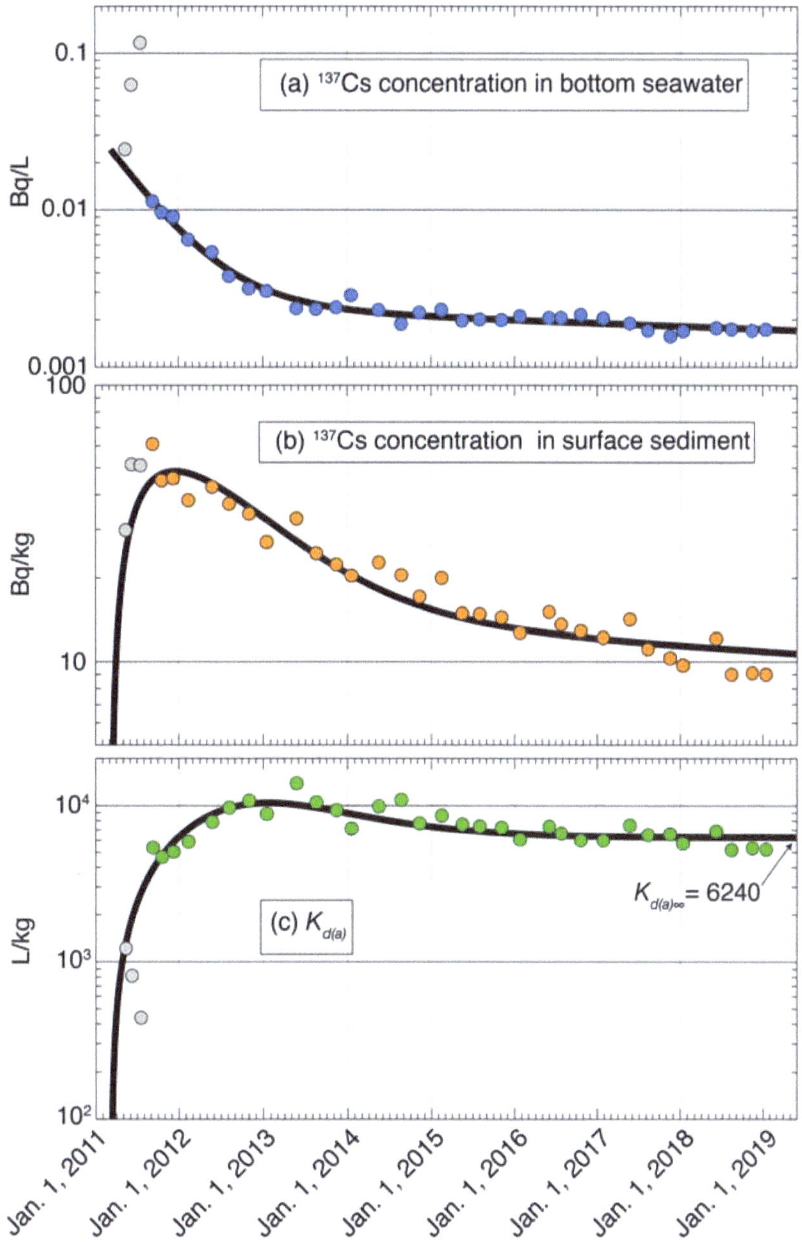

FIG. 4.9. Temporal changes of ^{137}Cs concentrations in geometric mean in bottom seawater (a) surface sediment (b), and the $K_{d(a)}$ values (c) measured in the extended monitoring zones. The best-fit curves determined by the model calculation are shown as black solid lines. The grey circles indicate data that were not used for the model calculation. See the text for details. The data were adapted from Ref. [4.2].

4.3.2.2. Vertical profiles of ^{137}Cs in the sediment

Vertical profiles of the ^{137}Cs concentrations in the sediment column collected at nine sites within the extended monitoring program are shown in Fig. 4.10. In general, the profiles were highest at the surface mixed layer or subsurface layer and decreased exponentially up to the depths of ~ 30 cm. It seemed that the sites D1 and E1 nearer to the FDNPP had higher concentrations detected not only at the surface mixed layer but also at layers of deeper than 3 cm (defined as intermediate and deeper layers). In addition, the higher concentrations were found at the sites located in a relatively flat bottom area such as G0 and I1, while low concentrations were found at the sites located on steep bottom slopes such as E5, G4, J1, and

J3. The ^{137}Cs concentrations in intermediate and deeper layers, except for site E1, decreased with time following the temporal trend in the surface layer.

Cs-134 was also detected together with ^{137}Cs in the sample cores (data not shown here, Kusakabe et al. [4.11]), and that suggested the FDNPP-derived radiocesium (^{134}Cs and ^{137}Cs) had penetrated to the 20 cm depth or below in the sediment by May 2013 at site I1, November 2013 at site B3, and May 2014 at sites E1 and G4. Penetration of the FDNPP-derived radiocesium into intermediate and deeper layers of the sediment column have been observed at other locations as well [4.17–4.20]. This penetration might mean that decrease of particulate ^{137}Cs input to the sediment such as from terrestrial material increased with time. However, since the sedimentation rate in the coastal area of northeastern Japan including near the FDNPP location was estimated to be 0.1-0.5 cm/year [4.18, 4.21, 4.22], meaning that it would take several decades to hundreds of years for the sediments to accumulate to a 20-30 cm depth, the possibility needs to be ruled out that the vertical profiles of ^{137}Cs in the sediment reflect the temporal trend of ^{137}Cs deposition to the sediment. Thus, it is most likely that the penetration of FDNPP-derived radiocesium into intermediate and deeper layers has been controlled by migration from an upper layer of sediment. The vertical mixing of sediment induced by rough weather and bioturbations might be possible mechanisms working in the sediment layer of a few tens of cm depth in a coastal region [4.19]. Teal et al. [4.23] estimated the bioturbation rates in the temperate zone ranged from 10 to 200 cm^2/year. Otosaka [4.19] demonstrated that the vertical profiles based on the numerical simulation with a bioturbation rate of 200 cm^2/year agreed with observed values except for the anomalous ^{137}Cs concentration in the shallow region. Therefore, here it was assumed that the bioturbation rate in the coastal waters off Fukushima was similar to the value reported by Teal et al. [4.23]. Given the above, the higher ^{137}Cs concentration might be observed in the intermediate and deeper layers of sediment column caused by such processes as vertical mixing of sediment due to rough weather and bioturbation.

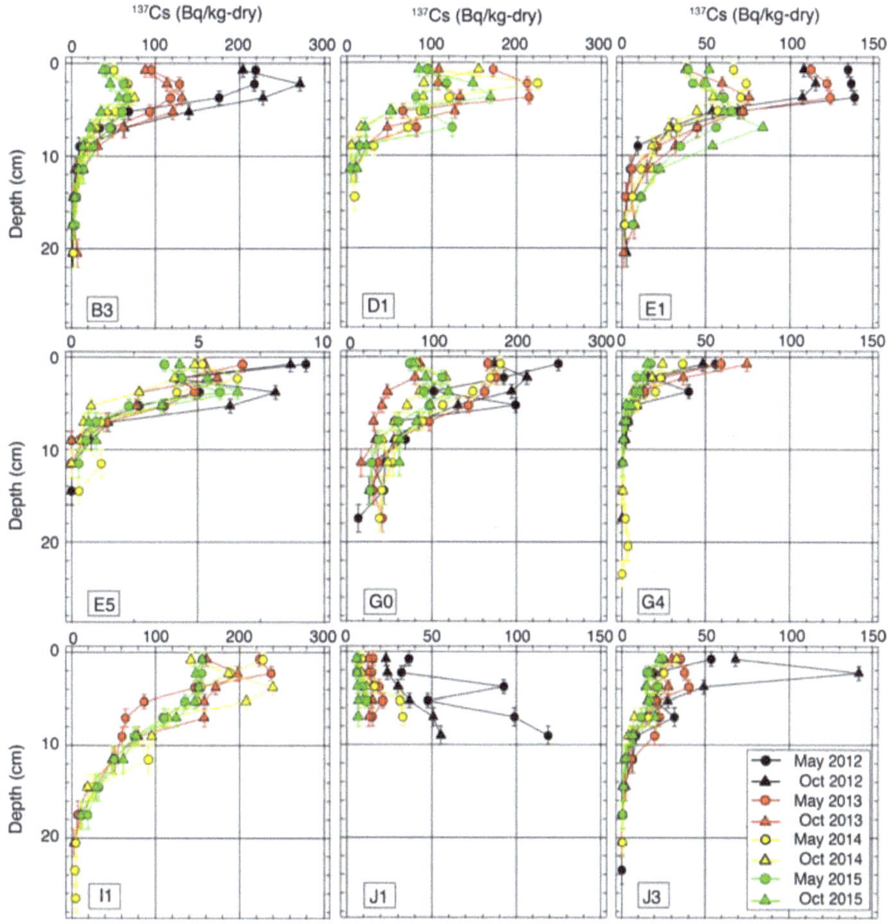

FIG. 4.10. Vertical profiles of ^{137}Cs concentrations in the bottom sediment. The data were adapted from Ref. [4.11].

As stated previously, the ^{137}Cs concentrations in the intermediate and deeper layers of sediment declined faster than the loss by physical decay from May 2012 to October 2015. Two kinds of processes may contribute to the declining trend: upward migration of ^{137}Cs desorbed from the sediment to porewater; and vertical mixing and lateral transport of sediment by physical agitation caused by rough weather. Radiocesium-enriched porewater has been observed in beach sand, lagoon and coastal sediment samples collected in Fukushima Prefecture [4.24–4.26] and in coastal sediment samples collected in the Irish Sea [4.27]. Otosaka [4.19] assumed that the sediment porosity was 0.41 and K_d was 3500 kg/L, and then calculated porewater ^{137}Cs porewater was less than 0.01% of the ^{137}Cs sediment inventory. And also, the loss of ^{137}Cs in sediment by the exchange of porewater could be negligible. These results suggest that the ^{137}Cs concentrations of the intermediate and deeper layers in sediment had changed by processes of vertical mixing of sediment. Some previous studies [4.28, 4.29] reported that re-suspension of mobile particles in the sediment and lateral transport of re-suspended particles occurred due to rough weather such as typhoons. Such weather conditions can change the vertical structures of the sediment not only in the shallow region but also in the offshore region. Therefore, re-suspension of the sediment and lateral transport of re-suspended particles need to be considered as two other main factors to shape the vertical distributions of radiocesium in the monitoring area (see Sec. 4.3.2.3). Thus, it can be considered that the vertical distributions of ^{137}Cs and decline of the ^{137}Cs concentrations in sediment were predominantly controlled by the vertical mixing and lateral transport due to physical agitation caused by rough weather such as typhoons in summer to fall and low air pressure in winter in the coastal area of Fukushima and nearby prefectures.

4.3.2.3. Inventory of ^{137}Cs in the sediment

Activity of ^{137}Cs in surface 0-3 cm sediment was calculated for the entire extended monitoring area, which is bounded by 38°40'N, 35°20'N, and the 800-m depth contour excluding the area within a 30 km radius from the FDNPP. The monitoring area was partitioned into Voronoi polygons (see the inset map of Fig. 4.11). Activity of ^{137}Cs in each polygon was calculated based on the concentration in the surface 0-3 cm sediment measured at the site inside the polygon, water content, and bulk density of the surface sediment, and area of the polygon. Total activity of ^{137}Cs in the surface sediment was calculated by summation of activities in each polygon. The calculation was made for the period from September 2011 to January 2019. The results are shown in Fig. 4.11. Following the general temporal trend of the ^{137}Cs concentration in the surface sediment (Fig. 4), the activity of ^{137}Cs in the surface 0-3 cm sediment in the monitoring area has been decreasing with time as well. In September 2011, it reached 4.6 x 10^{13} Bq, and decreased to 4.5 x 10^{12} Bq in January 2019. Thus, about 90 % of the ^{137}Cs present in the surface sediment was removed in 7.4 years, while 16 % decreased due to physical decay during the same period. Exponential decrease of the ^{137}Cs activity gives an effective half-life of 2.1 years (Fig. 4.11). It should be noted that the half-life based on the geometric mean of the concentrations (3.1 years) shown in Sec. 3.2.1 was longer than the half-life based on the activity because the former was calculated using data from the sites that were unevenly distributed in the monitoring area and the concentration was affected by the highly variable physical characteristics of sediment, so that the half-life based on the concentration should be somewhat biased. However, like the temporal trend of the concentration, the temporal change of activity seems also to have slowed down: the effective half-lives for the first half period and the latter half were 2.0 and 3.7 years, respectively. The effective half-lives calculated here should be an upper limit because ^{137}Cs that continues to be released directly from the FDNPP site or through rivers and ground water [4.24, 4.25, 4.30, 4.31] was not considered. Although accurate quantification of these impacts to the budget of ^{137}Cs especially in the coastal water sediment is necessary, the calculated half-lives might not be changed significantly after inputting these values to the calculation because the amounts of continuously released ^{137}Cs were estimated to be 2-3 orders of magnitude smaller than the amounts released right after the accident [4.6].

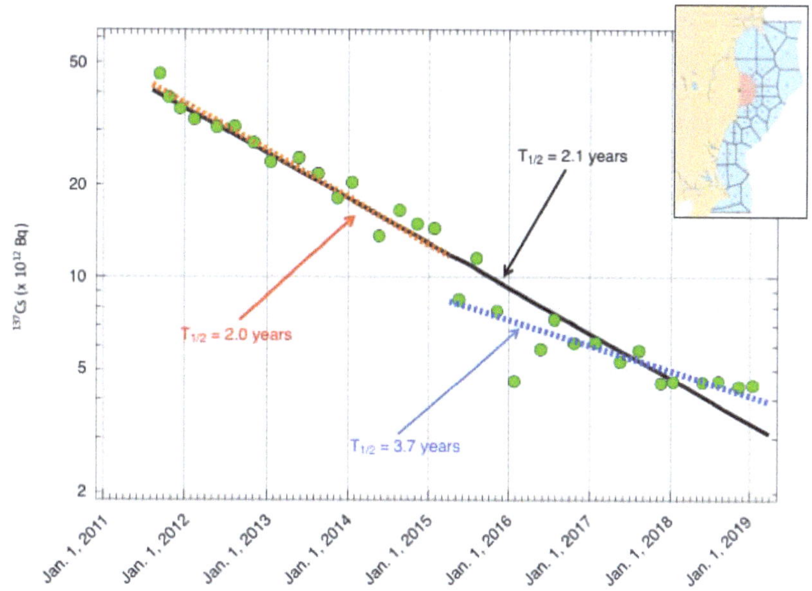

FIG. 4.11. Temporal change of ^{137}Cs activity in the surface sediment in the extended monitoring area. The inset map indicates the area used to calculate the activity which excluded the area close to the FDNPP site.

The main mechanisms to reduce the ^{137}Cs activity in the surface sediment would be: (1) desorption-dissolution of ^{137}Cs from the sediment; (2) downward migration by biological activity and/or vertical mixing of sediment induced by bottom current; and (3) resuspension and subsequent lateral transportation of the sediment. The first mechanism would work to a lesser extent because the reaction occurs only in a brief period (a few months) after the deposition and freshly deposited ^{137}Cs as mobile fraction in sediment tends to be transformed to immobile fraction [4.14]. In addition, based on a laboratory experiment, Takata et al. [4.32] suggested that the dissolvable ^{137}Cs in suspended sediment collected off the Fukushima coast was less than 10% of the total ^{137}Cs in particulates. The second mechanism should work to shape the vertical profiles as mentioned in Sec. 4.3.2.2, but it cannot explain fully the decrease of ^{137}Cs activity because an increase of ^{137}Cs concentration in the intermediate and deep layers of the sediment has been rarely observed. Kusakabe et al. [4.11] demonstrated that the rate of decrease was related to the physical characteristics of the sediment such as grain size. The higher decreasing rate was found in the sediment with larger grain size, where ^{137}Cs that is generally associated with more mobile, smaller grains has been more readily removed from the sediment. In other words, ^{137}Cs associated with larger grains is less mobile. This could also be a reason why the declining rate of ^{137}Cs activity in the surface sediment (Fig. 4.11) has been slowing down as with the concentrations shown in Figs. 4.4(b) and 4.9(b): small, mobile grains in sediment decrease with time leaving behind large, immobile grains leading to reduction of the decreasing rate of ^{137}Cs activity. Importance of grain migration induced by the bottom current in the post-depositional redistribution of ^{137}Cs in sediment has been examined using numerical simulation and in situ observation [4.15, 4.19, 4.33]. The lateral transportation of sediment crossing over the continental shelf should be the main mechanism to reduce the ^{137}Cs concentration and activity in the surface sediment in the monitoring area.

The activity calculated above is the lower limit of the total inventory of ^{137}Cs in the sediment in the coastal area because, in addition to the deep sediment layers (> 3 cm), two areas were omitted from the calculation: the area outside the extended monitoring area; the area within a

30 km radius from the FDNPP. There has been no systematic monitoring in any area outside of the extended monitoring area. However, the areas outside may not significantly contribute to the budget calculation due to the possible lower concentration further away from the FDNPP site. The area close to the FDNPP has been intensively surveyed since right after the accident by Tokyo Electronic Power Company Holdings (TEPCO), owner of the FDNPP. The data obtained by TEPCO have been presented through the company's web site. As shown in Sec. 4.3.2.2, ^{137}Cs was present at sediment layer depths of > 3 cm. The fraction of inventory in the surface 0-3 cm sediment to the total inventory was calculated to be 0.2-0.3 based on the vertical distributions of ^{137}Cs in the sediment obtained during the extended monitoring program (see Sec. 4.3.2.1). By using the data from the extended monitoring and TEPCO monitoring programs [4.34], and the surface fraction of 0.3 to the whole inventory, the inventory of ^{137}Cs in the area surrounded by the boundaries of 35°20'N, 38°40'N and the 800 m depth contour including the area inside the 30 km radius was estimated to be 0.23 x 10^{15} Bq as of September 2011 [4.11]. Directly released ^{137}Cs from the accident were estimated to be 3.6 x 10^{15} Bq [4.4, 4.5], of which 6.4 % deposited onto the sediment. Airborne transports of ^{137}Cs to the ocean have been estimated to be in a quite wide range of 5 x 10^{15} to 15 x 10^{15} Bq; this is due to the sparseness of datasets [4.6]. And airborne transports of ^{137}Cs to the monitoring area is even more difficult to estimate. Thus, sedimentary inventory of ^{137}Cs should be significantly lower than 6.4 % of the total ^{137}Cs released from the accident. Large amounts of radionuclides have been discharged from the Sellafield nuclear fuel reprocessing plant to the Irish sea. MacKenzie et al. [4.35] suggested that approximately 10% of the ^{137}Cs released from the Sellafield nuclear fuel reprocessing plant accumulated in the Irish Sea sediment. The difference in the fractions of ^{137}Cs preserved in sediment may be ascribed to differences in the sedimentary environment. The Irish sea sediment is rich in fine grains which tend to fix Cs from seawater. And ^{137}Cs has been released from the Sellafield plant continuously since 1960, which has kept transporting Cs in seawater to the sediment. In addition, the Irish sea is a semi-closed sea so that ^{137}Cs in seawater stays in the sea longer than in the coastal water off the FDNPP.

4.4. CONCLUSIONS

Since each coastal environment has its own characteristics such as oceanographic settings and bottom topography, it is quite difficult to predict fates of radionuclides in the environment in general. The results from the long-term monitoring of radioactivity at fixed sites in Japanese coastal waters provide essential information about the baselines of artificial radionuclides and their spatiotemporal changes in the coastal waters. Without the datasets derived from the regular monitoring, it would have been almost impossible to estimate the impacts of the FDNPP accident to the marine environment. Comprehensive monitoring after the accident has presented some interesting behaviors of artificial radionuclides. As of January 2019, while the ^{137}Cs concentration in seawater was almost at the pre-accident level, that in sediment was lower than the September 2011 level but significantly higher than the pre-accident level. As a result, the $K_{d(a)}$ value will converge to a terminal, elevated $K_{d(a)}$ value. However, quantification of factors controlling the declining trend of ^{137}Cs in the sediment as a whole have not been fully done. Further monitoring and data analysis are needed.

4.5. ACKNOWLEDGEMENTS

The authors are grateful to colleagues who have been involved in the monitoring programs for the last 35 years in the Marine Ecology Research Institute. Their help in the field and laboratory, and logistics support are greatly appreciated. N. Inatomi prepared graphs for Fig. 5. The data presented here were obtained from the programs under contract with the Ministry of Education, Culture, Sport, Science and Technology (1984-2013) and Secretariat of the Nuclear Regulation Authority (2013-present).

4.6. REFERENCES

[4.1] TAKATA, H., KUSAKABE, M., INATOMI, N., IKENOUE, T., Appearances of Fukushima Daiichi Nuclear Power Plant-Derived ^{137}Cs in Coastal Waters around Japan: Results from Marine Monitoring off Nuclear Power Plants and Facilities, 1983–2016, Environ. Sci. Technol. **52** 5 (2018) 2629–2637.

[4.2] KUSAKABE, M., TAKATA, H., Temporal trends of ^{137}Cs concentration in seawaters and bottom sediments in coastal waters around Japan: implications for the K_d concept in the dynamic marine environment, J. Radioanal. Nucl. Chem. **323** 1 (2020) 567–580.

[4.3] TAKATA, H., JOHANSEN, M. P., KUSAKABE, M., IKENOUE, T., YOKOTA, M., TAKAKU, H., A 30-year record reveals re-equilibration rates of ^{137}Cs in marine biota after the Fukushima Dai-ichi nuclear power plant accident: Concentration ratios in pre- and post-event conditions, Sci. Total Environ. **675** 20 (2019) 694–704.

[4.4] TSUMUNE, D., TSUBONO, T., AOYAMA, M., UEMATSU, M., MISUMI, K., MAEDA, Y., YOSHIDA, Y., HAYAMI, H., One-year, regional-scale simulation of ^{137}Cs radioactivity in the ocean following the Fukushima Dai-ichi Nuclear Power Plant accident, Biogeosciences **10** 8 (2013) 5601–5617.

[4.5] AOYAMA, M., HAMAJIMA, Y., HULT, M., UEMATSU, M., OKA, E., TSUMUNE, D., KUMAMOTO, Y., ^{134}Cs and ^{137}Cs in the North Pacific Ocean derived from the March 2011 TEPCO Fukushima Dai-Ichi Nuclear Power Plant accident, Japan. Part one: surface pathway and vertical distributions, J. Oceanogr. **72** 1 (2016) 53–65.

[4.6] BUESSELER, K., DAI, M., AOYAMA, M., BENITEZ-NELSON, C., CHARMASSON, S., HIGLEY, K., MADERICH, V., MASQUÉ, P., Morris, P. J., OUGHTON, D., SMITH, J. N., Fukushima Daiichi–Derived Radionuclides in the Ocean: Transport, Fate, and Impacts, Annu. Rev. Mar. Sci. **9** (2017)173–203.

[4.7] KASAMATSU, F, INATOMI, N., Effective environmental half-lives of ^{90}Sr and ^{137}Cs in the coastal seawater of Japan, J. Geophys. Res. Oceans **103** C1 (1998) 1209–1217.

[4.8] WATABE, T., OIKAWA, S., ISOYAMA, N., SUZUKI, C., MISONOO, J., MORIZONO, S., Spatiotemporal distribution of ^{137}Cs in the sea surrounding Japanese Islands in the decades before the disaster at the Fukushima Daiichi Nuclear Power Plant in 2011, Sci. Total Environ. **463–464** (2013) 913–921.

[4.9] KUSAKABE, M., OIKAWA, S., TAKATA, H., MISONOO, J., Spatiotemporal distributions of Fukushima-derived radionuclides in nearby marine surface sediments, Biogeosciences **10** 7 (2013) 5019–5030.

[4.10] OIKAWA, S., TAKATA, H., WATABE, T., MISONOO, J., KUSAKABE, M., Distribution of the Fukushima-derived radionuclides in seawater in the Pacific off the coast of Miyagi, Fukushima, and Ibaraki Prefectures, Japan, Biogeosciences **10** 7 (2013) 5031–5047.

[4.11] KUSAKABE, M., INATOMI, N., TAKATA, H., IKENOUE, T., Decline in radiocesium in seafloor sediments off Fukushima and nearby prefectures, J. Oceanogr. **73** 5 (2017) 529–545.

[4.12] INTERNATIONAL ATOMIC ENERGY AGENCY, Sediment distribution coefficients and concentration factors for biota in the marine environment, Technical Report Series 422, IAEA, Vienna (2004).

[4.13] POVINEC, P. P., AARKROG, A., BUESSELER, K. O., DELFANTI, R., HIROSE, K., HONG, G. H., ITO, T., LIVINGSTON, H. D., NIES, H., NOSHKIN, V. E., SHIMA, S., TOGAWA, O., ^{90}Sr, ^{137}Cs and ^{239,240}Pu concentration surface water time series in the Pacific and Indian Oceans – WOMARS results, J. Environ. Radioact. **81** 1 (2005) 63–87.

[4.14] BØRRENTZEN, P., SALBU, B., Fixation of Cs to marine sediments estimated by a stochastic modelling approach, J. Environ. Radioact. **61** 1 (2002) 1–20.

[4.15] THORNTON, B., OHNISHI, S., URA, T., ODANO, N., SASAKI, S., FIJITA, T., WTANABE, T., NAKATA, K., ONO, T., AMBE, D., Distribution of local ^{137}Cs anomalies on the seafloor near the Fukushima Dai-ichi Nuclear Power Plant, Mar. Pollut. Bull. **74** 1 (2013) 344–350.

[4.16] IKENOUE, T., ISHII, N., KUSAKABE, M., TAKATA, H., Contribution of ^{137}Cs-enriched particles to radiocesium concentrations in seafloor sediment: Reconnaissance experiment, PLOS ONE **13** 9 (2018) e0204289.

[4.17] AMBE, D., KAERIYAMA, H., SHIGENOBU, Y., FUJIMOTO, K., ONO, T., SAWADA, H., SAITO, H., MIKI, S., SETOU, T., MORITA, T., WATANABE, T., Five-minute resolved spatial distribution of radiocesium in sea sediment derived from the Fukushima Dai-ichi Nuclear Power Plant, J. Environ. Radioact. **138** (2014) 264–275.

[4.18] BLACK, E. E., BUESSELER, K. O., Spatial variability and the fate of cesium in coastal sediments near Fukushima, Japan, Biogeosciences **11** 18 (2014) 5123–5137.

[4.19] OTOSAKA, S., Processes affecting long-term changes in ^{137}Cs concentration in surface sediments off Fukushima, J. Oceanogr. **73** 5 (2017) 559–570.

[4.20] FUKUDA, M., AONO, T., YAMAZAKI, S., ISHIMARU, T., KANDA, J., NISHIOKA, J., OTOSAKA, S., Factors controlling ^{134}Cs activity concentrations in sediment collected off the coast of Fukushima Prefecture in 2013–2015, Geochem. J. **52** 2 (2018) 201–209.

[4.21] OKUMURA, Y., YAMASHITA, Y., KOHNO, Y., NAGASAKA, H., Historical trends of PCDD/Fs and CO-PCBs in a sediment core collected in Sendai Bay, Japan, Wat. Res. **38** 16 (2004) 3511–3522.

[4.22] OKUMURA, Y., NAGASAKA, H., KOHNO, Y., KAMIYAMA, T., SUZUKI, T., YAMASHITA, Y., Sedimentation rate of dioxins from the mid-1980s to 2002 in a sediment core collected off Ishinomaki in Sendai Bay, Japan, La mer **43** (2005) 33–42.

[4.23] TEAL, L. R., BULLING, M. T., PARKER, E. R., SOLAN, M., Global patterns of bioturbation intensity and mixed depth of marine soft sediments, Aquat. Biol. **2** 3 (2008) 207–218.

[4.24] SANIAL, V., BUESSELER, K. O., CHARETTE, M. A., NAGAO, S., Unexpected source of Fukushima-derived radiocesium to the coastal ocean of Japan, Proc. Natl. Acad. Sci. U.S.A. **114** 42 (2017) 11092–11096.

[4.25] KAMBAYASHI, S., ZHANG, J., NARITA, H., Significance of Fukushima-derived radiocaesium flux via river-estuary-ocean system. Sci. Total Environ. **793** (2021) 148456.

[4.26] OTOSAKA, S., KAMBAYASHI, S., FUKUDA, M., TSURUTA, T., MISONOU, T., SUZUKI, T., AONO, T., Behavior of radiocesium in sediments in Fukushima coastal waters: Verification of desorption potential through pore water, Environ. Sci. Technol. **54** 21 (2020) 13778–13785.

[4.27] SHOLKOVITZ, E. R., MANN, D. R., The pore water chemistry of ^{239,240}Pu and ^{137}Cs in sediments of Buzzards Bay, Massachusetts, Geochim. Cosmochim. Acta **48** 5 (1984) 1107–1114.

[4.28] OTOSAKA, S., NAKANISHI, T., SUZUKI, T., SATOH, Y., NARITA, H., Vertical and Lateral Transport of Particulate Radiocesium off Fukushima, Environ. Sci. Technol. **48** 21 (2014) 12595–12602.

[4.29] BUESSELER, K. O., GERMAN, C. R., HONDA, M. C., OTOSAKA, S., BLACK, E. E., KAWAKAMI, H., MANGANINI, S. J., PIKE, S. M., Tracking the fate of particle associated Fukushima Daiichi cesium in the ocean off Japan, Environ Sci. Technol. **49** 16 (2015) 9807–9816.

[4.30] KANDA, J., Continuing ^{137}Cs release to the sea from the Fukushima Dai-ichi Nuclear Power Plant through 2012, Biogeosciences **10** 9 (2013) 6107–6113.

[4.31] TAKATA, H., KUSAKABE, M., INATOMI, N., IKENOUE, T., HASEGAWA, K., The contribution of sources to the sustained elevated inventory of ^{137}Cs in offshore waters east of Japan after the Fukushima Dai-ichi Nuclear Power Station Accident, Environ. Sci. Technol. **50** 13 (2016) 6957−6963.

[4.32] TAKATA, H., HASEGAWA, K., OIKAWA, S., KUDO, N., IKENOUE, T., ISONO, R. S., KUSAKABE, M., Remobilization of radiocesium on riverine particles in seawater: The contribution of desorption to the export flux to the marine environment, Mar. Chem. **176** (2015) 51–63.

[4.33] HIGASHI, H., MORINO, Y., FURUICHI, N., OHARA, T., Ocean dynamic processes causing spatially heterogeneous distribution of sedimentary caesium-137 massively released from the Fukushima Daiichi Nuclear Power Plant, Biogeosciences **12** 23 (2015) 7107–7128.

[4.34] TOKYO ELECTRIC POWER COMPANY HOLDINGS, Monitoring of marine soil – Sea area close to TEPCO's Fukushima Daiichi NPS / Coastal area (2019), https://radioactivity.nsr.go.jp/en/list/205/list-1.html

[4.35] MACKENZIE, A. B., SCOTT, R. D., ALLAN, R. L., BEN SHABAN, Y. A., COOK, G. T., PULFORD, I. D., Sediment radionuclide profile: implications for mechanisms of Sellafield waste dispersal in the Irish Sea, J. Environ. Radioact. **23** 1 (1994) 39–69.

5. RECENT SEDIMENTATION RATES DETERMINED BY RADIOMETRIC TOOLS IN THE NORTHERN GULF OF AQABA, JORDAN

S.A. Al-Rousan
Department of Geology, School of Science, The University of Jordan
Amman, Jordan
Email: s.rousan@ju.edu.jo

M. Al Hseinat
Department of Geology, School of Science, The University of Jordan
Amman, Jordan

R. Garcia-Tenorio
Department of Applied Physics II, University of Sevilla
Centro Nacional de Aceleradores, Univ. Sevilla-J. Andalucía-CSIC
Sevilla, Spain

Abstract

The Gulf of Aqaba has significant strategic, economic, and recreational value to Jordan since it represents its only marine outlet. In the last few decades, the Gulf was subjected to a variety of serious environmental problems affecting its unique marine ecosystem. To evaluate the historical trends of pollution in such a stressed marine environment, two short sediment cores were collected from the Phosphate Loading Berth Port from depths between 200 to 330 m. The activity concentrations of some radionuclides (^{210}Pb, ^{226}Ra, ^{40}K, ^{137}Cs, ^{228}Ra, ^{234}Th) were determined along the cores. Sediment chronology of the cores was also constructed using ^{210}Pb age models (core bottom was estimated to date approximately 130 years before present) and temporal variation of sedimentation rates in the area was determined. The results showed high activity concentration of the measured uranium series radionuclides (^{210}Pb, ^{226}Ra, and ^{234}Th) in the topmost layer of the shallow sediment core (PLB1) close to the phosphate port as a result of the contribution of phosphate dust. Based on the different ^{210}Pb age models, recent sedimentation rates from the area varied between 0.15±0.07 and 0.22±0.03 cm yr^{-1}. Chronology obtained from ^{210}Pb was validated by the ^{137}Cs profile. The sedimentation rates in the area showed an increasing trend over time, especially since the year 1965. This date represents the start of a period that underwent increasing coastal activities, constructions, and fast urbanization around the port in Aqaba.

5.1. INTRODUCTION

The Jordanian coast of the Gulf of Aqaba represents the only seaport in Jordan. This short coastal area (27 km) serves all uses of the coastal area, including industrial, commercial, touristic, and urban developments and activities. In the past few decades, population density, industrial and harbors activities, and coastal development along the Jordanian coast have rapidly increased and resulted in detrimental effects to its unique, diverse and sensitive coral-reef ecosystems [5.1, 5.2, 5.3].

Globally, Jordan is the third largest exporter and fifth largest producer of phosphate rocks g [5.4]. The marketable phosphate powder is exported through the Aqaba port. During the loading, storage, and shipping phases, some phosphate powder escapes to the atmosphere, the land around the storage areas and loading docks, and to the sea. The fine sediment layers in the port berth show evidence substantial powder phosphate deposition (0.05-1%; [5.5, 5.6]). This phosphate powder is nearly insoluble in seawater [5.7] but contains numerous potentially toxic and radioactive elements of environmental concern [5.3]. Also, high sedimentation rates up to 8 $mg.m^{-2}\cdot day^{-1}$ were recorded in this port [5.3], which are responsible for the decreased health conditions of coral reefs in their vicinity [5.8] or even death to coral colonies by reducing light intensity, inhibiting coral calcification, and increasing sediment load [5.9]. Since heavy metals are released from the phosphate dust, relatively elevated concentrations of heavy metals were recorded in coral skeletons sampled from this port [5.2, 5.10].

Industrial and domestic activities are known to release a significant amount of hazardous heavy metals that are delivered to the coastal environment through rivers, waste effluents, and atmospheric emissions [5.11, 5.12, 5.13]. Several studies showed elevated concentrations of heavy metals in sediments near the Phosphate Loading Berth (PLB) compared to other areas along the Gulf [5.3, 5.5]. Because U concentrations are high in marine sediment that contain large amounts of organic matter, phosphate, or both [5.14], increased activity concentrations of ^{238}U, ^{235}U, ^{234}U and ^{226}Ra in sediments and seagrasses from the Phosphate Loading Berth were recorded [5.15, 5.16, 5.17]. These concentrations were 3-fold higher than in other coastal locations and were attributed to phosphate dust generated during transport [5.15].

The progression of the anthropogenic impact on the marine environment can be evaluated by analysing numerous parameters in dated sediment cores (if mixing is limited). Among these parameters, the sedimentation rate is very important since it allows reconstructing the environmental evolution and the human impact on the coastal environment [5.18]. The most common technique used to date recent sediments and calculate sediment deposition and accumulation rates during the last 100-150 years is based on ^{210}Pb profiles [5.19, 5.20].

The goal of the present study is to determine the activity concentrations of gamma-emitting radionuclides in deep sediment cores collected in front of the PLB area to study the effects of phosphate particles in the Gulf of Aqaba. Furthermore, the study aims to establish a sediment chronology (based on the ^{210}Pb dating method) showing the variations of sedimentation rate and the impact of human activities in the region over the last century.

5.2. MATERIALS AND METHODS

5.2.1. Study area

The Gulf of Aqaba (28°29'-30° N and 34°30'-35° E) is the north-eastern extension of the Red Sea. The Gulf is semi-enclosed basin, and the Straits of Tiran separate it from the Red Sea proper. The maximum depth of the Gulf is 1800 m, 180 km length, and variable-width ranges between 14-26 km. The climate in the Gulf is extremely arid with negligible precipitation (~2.2 cm yr^{-1}) and runoff and high evaporation (~400 cm yr^{-1}). However, flash floods after heavy rainfall events in winter might occur and transport terrestrial material into the Gulf. The frequency of these flash floods ranges from several times a year to once a decade [5.21]. Little is known regarding the contribution of terrestrial sediment that enters the marine environment during these floods.

The mean value of sea surface temperature is about 23.9°C with seasonal variations of about 6-7°C. Thermal stratification of the water column exists during summer, while the water is well mixed during winter [5.22]. Lack of regular freshwater and high evaporation causes particularly saline conditions (salinity 40-40.8‰) and the oligotrophic nature of the Gulf system [5.22]. The Gulf of Aqaba is influenced by both the airflows from the Mediterranean low-pressure system and the Indian monsoonal trough [5.23]. Winds year-round are similar in terms of magnitude and direction (90% northerly winds).

The study area is located within the Aqaba subbasin. According to Tibor [5.24], this subbasin has a short or non-existent tectonically based shelf with a limited sediment supply. However, the nearby Eilat subbasin in the western side of the Gulf is supplied with sediments from Araba Valley. Both subbasins connect in the south into the Eilat deep (900 m depth). Generally, the main sources of sediment to the area are restricted to dust deposition (negligible), material brought by flash floods, and earthquake-induced turbidity currents or resuspension [5.25].

5.2.2 Sediment core sampling and preparation

In February 2019, two short sediment cores were collected from the Jordanian coast of the Gulf of Aqaba in front of the PLB from different water depths (Table 5.1; Fig. 5.1). The samples were collected using UWITEC short sediment corer with PVC tubes of 60 cm long and 9 cm in diameter. One core (PLB1) was collected at a water depth of 202 m and another one (PLB3) at a water depth of 331 m.

TABLE 5.1. SUMMARY OF SAMPLING DATA IN FRONT OF THE PHOSPHATE LOADING BEARTH (PLB) SITE

Sediment core code	Longitude (E)	Latitude (N)	Water depth (m)	Core length (cm)
PLB1	29.505	34.985	202	17
PLB3	29.496	34.983	331	12

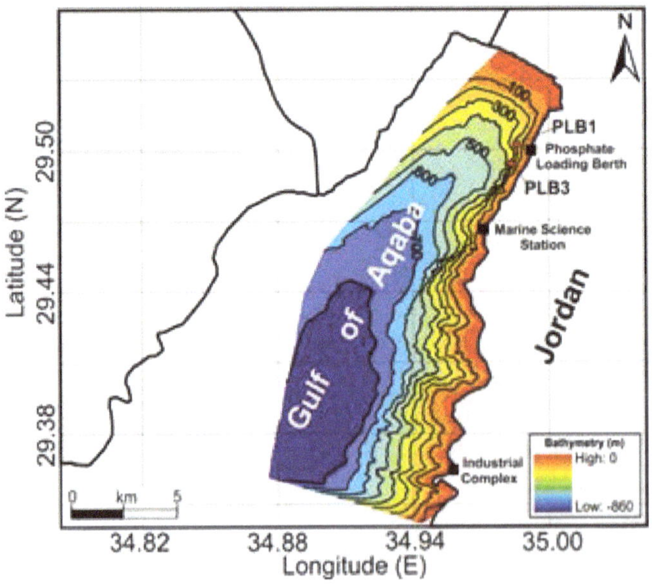

FIG. 5.1. Bathymetric map of the northern tip of the Gulf of Aqaba showing the locations of the sediment cores.

Immediately after extraction, the cores were carefully sliced every 0.5 cm for core PLB3, and 1 cm for core PLB1 down the length of the cores using slicer equipment. After that, the sediment samples were weighted (wet wt.) and oven-dried until a constant weight was reached and the dry bulk density was determined. The samples were then crushed, homogenized, and sieved through a 2-mm mesh. Contents of total organic matter were determined in each slice according to the method of loss in weight by ignition (LOI: 450°C, for 8 h) described by Byers [5.26].

5.2.3 Massic activity determination

For radiometric determination, subsamples were transported to Seville (Spain), and the activity analysis of ^{210}Pb, ^{226}Ra, ^{40}K, ^{137}Cs, ^{228}Ra, ^{234}Th was determined by high-resolution gamma-ray spectrometry using a low-background intrinsic High Purity Germanium (HPGe) detector with good configuration (nominal efficiency 60%, resolution 1.4 keV at 122 keV and 2.3 keV at 1332keV). Samples were kept in sealed plastic containers (presumably gas-tight) and stored for at least three weeks to allow equilibration of ^{226}Ra with its daughter nuclides [5.25].

5.3. RESULTS AND DISCUSSION

5.3.1. Massic activity

The activity concentrations of some radionuclides, including ^{210}Pb, ^{226}Ra, ^{40}K, ^{137}Cs, ^{228}Ra, and ^{234}Th (in Bq·kg^{-1} dry weight) in the uppermost layer from both sediment cores were first used to study the differences between the two locations (Table 5.2).

TABLE 5.2. ACTIVITY CONCENTRATION (BQ.KG^{-1} DRY WEIGHT) OF SURFACE SAMPLES FROM SEDIMENT CORES FROM THIS STUDY

		Activity concentrations of surface sample (Bq·kg^{-1} dry weight)					
Sediment core	Depth (m)	^{210}Pb	^{226}Ra	^{40}K	^{137}Cs	^{228}Ra	^{234}Th
PLB1	202	393±11	133±3	364±18	3±0.5	31±2	154±21
PLB3	331	74±4	26±1	297±13	ND	20±1	45±10

The surface samples from the sediment core PLB1 show much higher activity concentration for the radionuclides of the uranium series (^{210}Pb, ^{226}Ra, and ^{234}Th) than the samples from the core PLB3 (Table 2). This could be attributed to the contribution of phosphate dust lost to the sea during loading and shipment to core PLB1 since it is shallower and closer to the main phosphate port (Fig 1). However, the other radionuclides such as ^{40}K, ^{137}Cs, and ^{228}Ra showed moderately higher values in PLB1 (Table 2) that could be attributed to differences in sediment composition.

Similarly, Ababneh [5.15] found that the values of ^{238}U, ^{235}U, ^{226}Ra, and ^{40}K concentrations in the PLB location decrease in the direction from the beach towards the surf zone at greater depths, owing to e the diffusion of phosphate dust with increasing distance from the shore. Pittauerova [5.25] attributed the increased ^{226}Ra activities in core tops from Eilat to be caused by the contribution of phosphate dust from Eilat and Aqaba industrial ports. They found that ^{226}Ra is positively correlated with phosphorous concentrations, and the ^{226}Ra activity concentration in fertilizer samples is around 1200 Bq.kg^{-1}. The concentrations of ^{40}K in the present study are comparable to the results obtained by Pittauerova [5.25] from bottom sediments collected from Eilat deep (375-470 Bq.kg^{-1}), whereas the artificial radionuclide ^{137}Cs in this study was present in very low concentrations. The activity concentrations of ^{210}Pb, ^{226}Ra, ^{40}K, ^{137}Cs in shallow surface sediments (0-12 m) from the PLB were about 430 ± 73; 62 3± 65; 93.9 ± 13, BDL Bq/kg respectively [5.17], proving that the reduced effect of anthropogenic pollution is found in deeper cores [5.25].

5.3.2. Vertical profiles

The unsupported ^{210}Pb ($^{210}Pb_{ex}$) values were determined by subtracting the supported ^{210}Pb (estimated through ^{226}Ra) in each layer from the total ^{210}Pb (Fig. 5.2). In this way, the unsupported ^{210}Pb due to the contribution of phosphate dust was removed. The profiles of unsupported ^{210}Pb ($^{210}Pb_{ex}$) of PLB1 and PLB3 follow a relatively exponential decrease with depth (Fig. 5.2). The $^{210}Pb_{ex}$ activity profile in PLB1 has a maximum value in the first layer whereas it is beneath the surface layer in PLB3, indicating that the top layer in this core is subjected probably to disturbance and sediment mixing caused by biological (bioturbation) and physical processes (wave and current) [5.27, 5.28].

Total ^{210}Pb reaches equilibrium with the supporting ^{226}Ra activity at depths of 16 cm in PLB1 and 8.5 cm in PLB3, but above this depth, the activity declines more or less exponentially (Fig. 5.2). On the other hand, the supported ^{210}Pb of PLB3 has little variation with depth (Fig. 5.2).

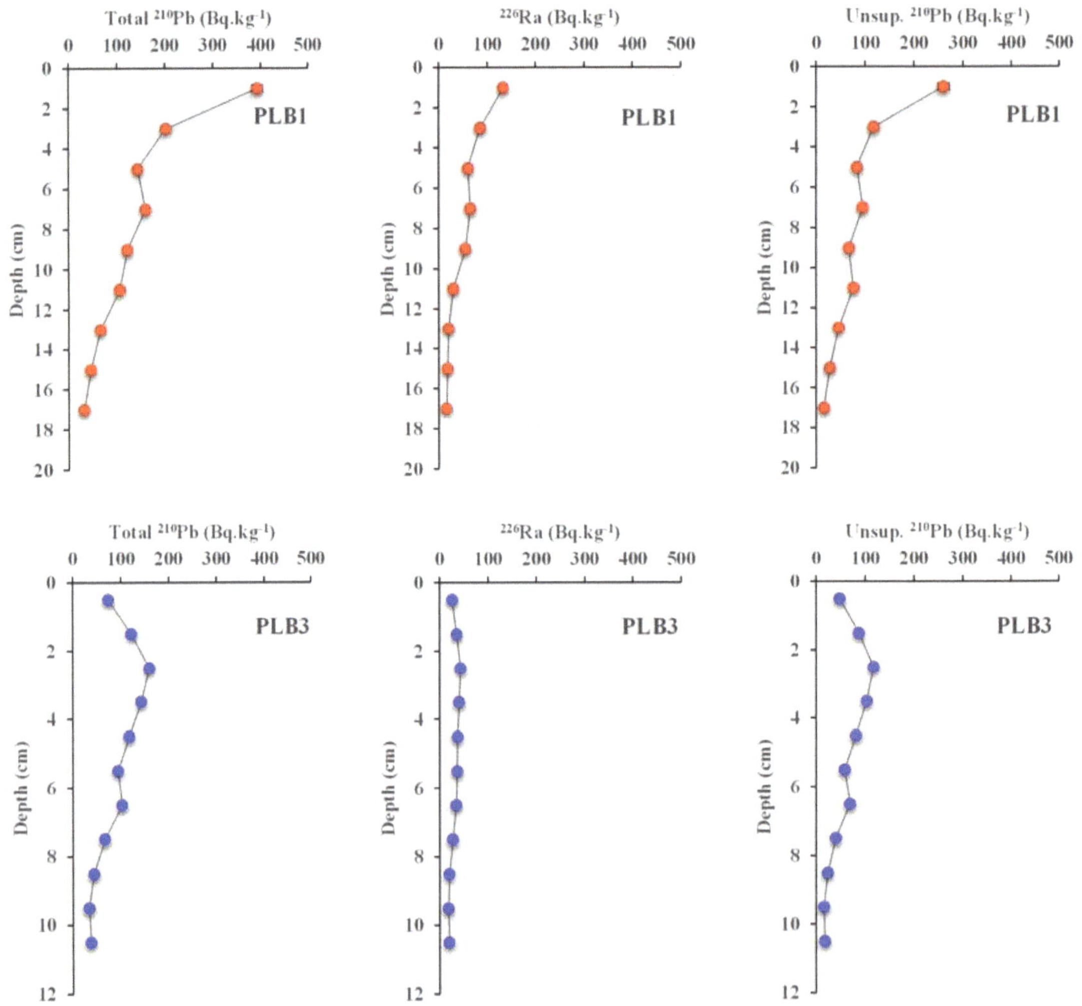

FIG. 5.2. Depth profiles of ^{210}Pb, ^{226}Ra, and excess ^{210}Pb in the sediment cores PLB1 and PLB3 from the northern Gulf of Aqaba. Error bars represent counting error.

The activity concentrations of ^{210}Pb and ^{226}Ra (Bq.kg^{-1}dry wt) varied among sites. They ranged from 32 to 393 Bq.kg^{-1}; 16 to 133 Bq.kg^{-1} dry wt. at station PLB1 and 34 to 160 Bq.kg^{-1}; 18 to 43 Bq.kg^{-1}dry wt. at station PLB3, respectively (Fig. 5.2). The ^{210}Pb activity concentrations in the sediment samples at station PLB1 were higher when compared to the activity at station PLB3. This is because PLB1 is the shallowest and closest to the mainland, which receives increased depositional flux of freshly high-activity sediments of ^{210}Pb supplied by phosphate powder. Conversely, PLB3 is from deeper water depth and far from the mainland (pollution sources). Furthermore, silt and organic carbon content may control ^{210}Pb activity concentrations; it was suggested that ^{210}Pb has a high tendency to be associated with fine particles and organic matter [5.29]. The organic carbon concentrations in core PLB1 ranged between 3.5-10.7%, and the profile showed a surface maximum and then an exponential decrease with depth (Fig. 5.3). An excellent correlation (r=0.93) was found between organic carbon content and unsupported ^{210}Pb activity concentrations. Organic matter in core PLB3 was not analysed.

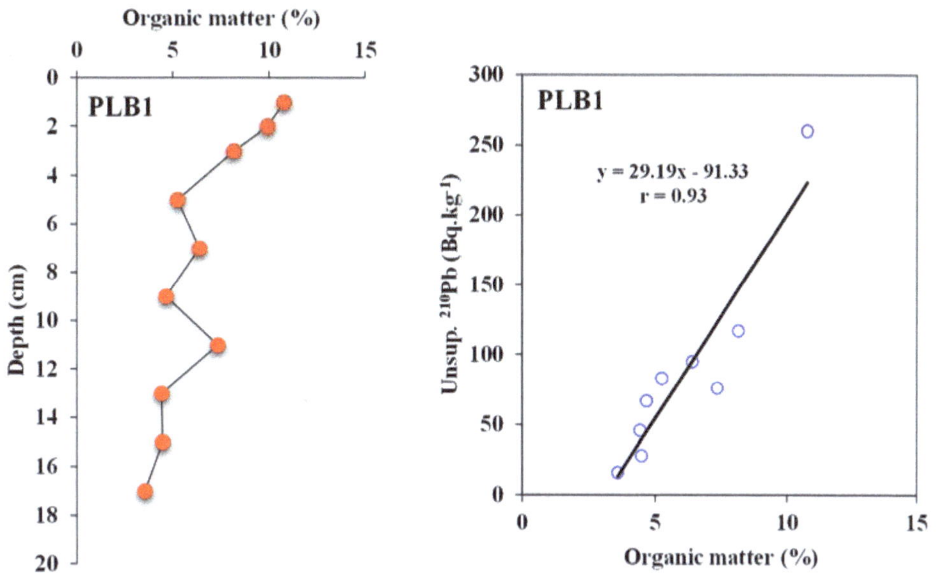

FIG. 5.3. Depth profile of organic matter content (%) and its correlation with ^{210}Pb activity concentrations in the sediment cores PLB1 from the northern Gulf of Aqaba.

Since ^{234}Th reflects the ^{238}U content, a strongly positive correlation between ^{226}Ra and ^{234}Th in both PLB1 and PLB3 sediment cores was obtained (Fig. 4), though the activity concentrations were different between the two sites. The correlation is stronger in PLB1 (r=0.99) compared to PLB3 (r=0.94) because the range covered by nuclides correlated is clearly smaller in PLB3 than in PLB1 (Fig. 5.4).

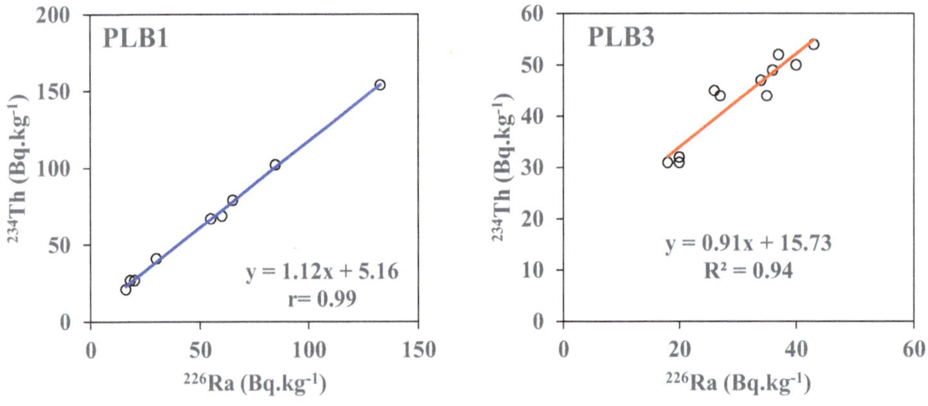

FIG. 5.4. Correlation between ^{234}Th and ^{226}Ra in sediment cores collected from the study area.

Profiles of ^{228}Ra, ^{40}K, and ^{234}Th showed minor variations of activity concentrations with depth in core PLB3. On the other hand, profiles of ^{228}Ra and ^{234}Th in core PLB1 showed an exponential decrease with depth in PLB1, and the ^{40}K profile showed minimal variation and increased values after 11 cm (Fig. 5.5). This could indicate the effect of the anthropogenic phosphate dust discharge to the upper part of the core (<11 cm) since phosphates and phosphogypsum are depleted in ^{40}K.

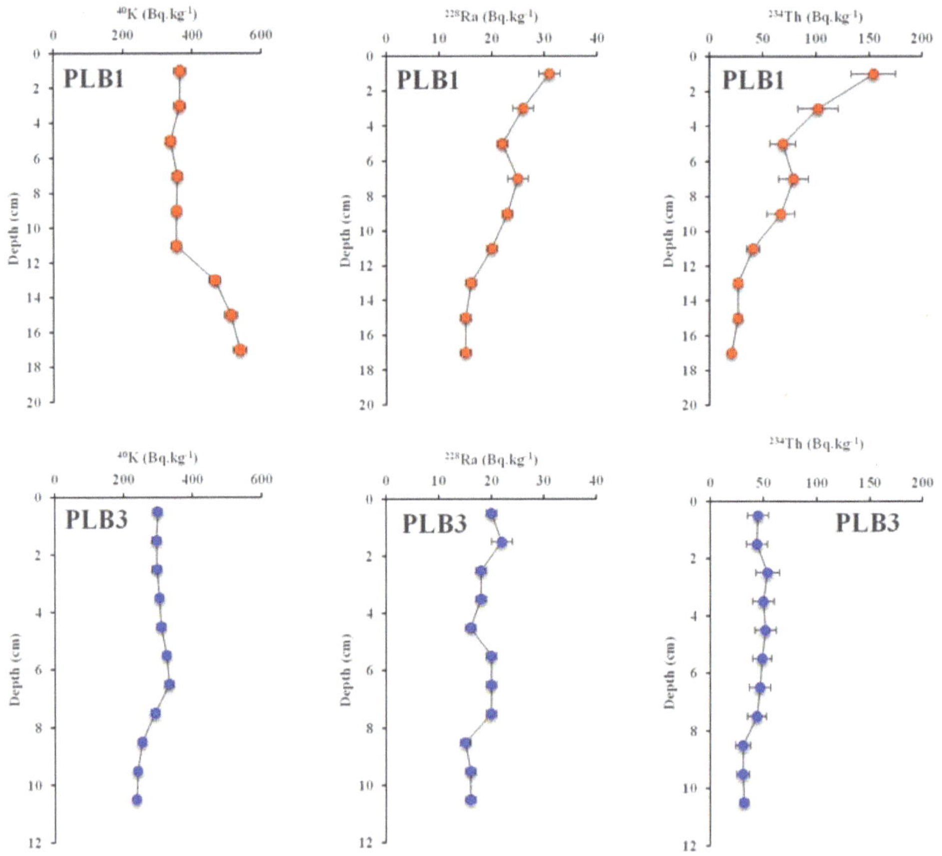

FIG. 5.5. Depth profiles of ^{228}Ra, ^{40}K, ^{234Th} ($Bq{\cdot}kg^{-1}$) in the sediment cores PLB1 and PLB3 from the northern Gulf of Aqaba. Error bars represent counting errors.

5.3.3. Age models and sedimentation rates

The vertical profiles of ^{210}Pb can be used to trace sediment accumulation that has occurred over the past 100 years [5.30, 5.31]. For the determination of the age model, and since the distribution of excess ^{210}Pb values follows a relatively exponential decrease with depth (Fig. 5.2), the CF:CS model, which assumes constant $^{210}Pb_{ex}$ flux and constant sedimentation rate, was applied to core PLB1.

According to the CF:CS model, the accumulation rates (MAR) and the sedimentation rate (SAR) in core PLB1 from the study area average 0.08±0.01 g.cm^{-2} yr^{-1} and 0.22±0.03 cm yr^{-1}, respectively. The sedimentation rate (SAR) value is close to those reported by Pittauerová [5.25] from the nearby Eilat subbasin. Their values using the CF:CS model vary between 0.10 ± 0.02 to 0.35 ± 0.23 cm yr^{-1}, with high values recorded in shallow depths (250 m depth) due to dilution of terrestrial sediments by authigenic carbonate. However, lower sedimentation rates ranging between 0.04 to 0.07 cm yr^{-1} were recorded from multicores during the last 1000 years (using ^{14}C dating method) from depths between 440–830 m [5.32]. This is consistent with ^{210}Pb results obtained and indicates that sedimentation rates from the study area increased in the last century and that the ^{14}C sedimentation rate could be affected by compaction effects.

Additionally, the sediment chronology from PLB1 was also constructed using Constant Flux (CF), widely known as the Constant Rate of Supply model (CRS). Results from both models are shown in Fig. 5.6. The CF:CS and CF ages agree well in the first 9 cm (last 40 yrs), and the studied core (PLB1) extended back to 1951 and 1893 according to CF:CS and CF models, respectively (Fig. 5.6).

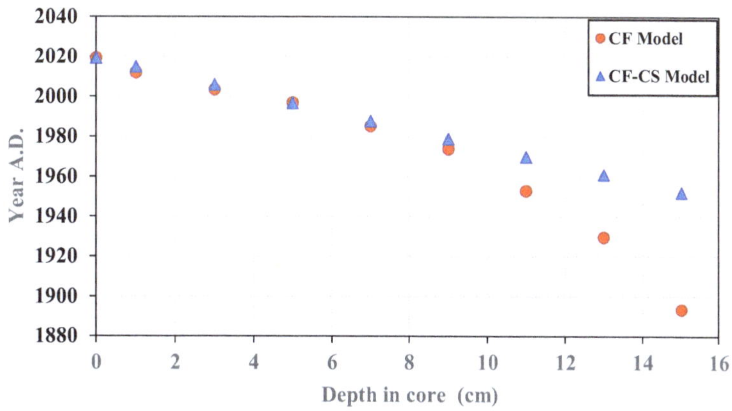

FIG. 5.6. Dates calculated from CF:CS and CF models versus depth.

The results for ages, sedimentation (SAR), and accumulation (MAR) rates obtained from the application of the CF model in core PLB1 are shown in Table 5.3 and represented in Fig. 5.7.

TABLE 5.3. MASS ACCUMULATION RATES (MAR, cm·yr^{-1}) AND SEDIMENTAION RATES (SAR, g·cm^{-2} yr^{-1}) DERIVED FROM $^{210}Pb_{ex}$ PROFILES USING CF AGE MODEL FROM CORE PLB1 FROM NORTHERN GULF OF AQABA

Year A.D.	Sedimentation rate (SAR) (cm yr^{-1})	Accumulation rate (MAR) (g.cm^{-2} yr^{-1})
2019	0.19±0.01	0.07±0.00
2011	0.17±0.01	0.07±0.00
2003	0.26±0.02	0.10±0.01
1996	0.24±0.02	0.09±0.00
1985	0.17±0.01	0.07±0.00
1973	0.14±0.01	0.05±0.00
1952	0.08±0.01	0.03±0.00
1929	0.07±0.01	0.02±0.00
1893	0.04±0.01	0.01±0.00

The sedimentation (SAR) and accumulation (MAR) rates have increased with time, from the minimum values in the bottom of the core to the maximum values at present (Fig. 5.7). The accumulation rate (MAR) increases from 0.01±0.00 in the year 1893 to 0.10±0.01 g.cm^{-2} yr^{-1} in the year 2003. Similarly, the sedimentation rates (SAR) increase from 0.04±0.01 cm yr^{-1} in the year 1893 to 0.26±0.02 cm yr^{-1} in the year 2003 by applying the CF model (Table 5.3, Fig. 5.7). These results are comparable to the previous result reported by Pittauerová et al. (2014) for cores collected from Eilat. Their values ranged between 0.11 ± 0.02 to 0.35 ± 0.23 cm yr^{-1}.

For the next step in this research, historical trends and evolution of anthropogenic pollutants profile will be constructed by analysing different heavy metals and contaminants in each layer of the core. The objective will be to analyse the temporal evolution of the possible increasing trends of the different metals toward the surface of the core due to the increase of anthropogenic activities in the area during the last decades.

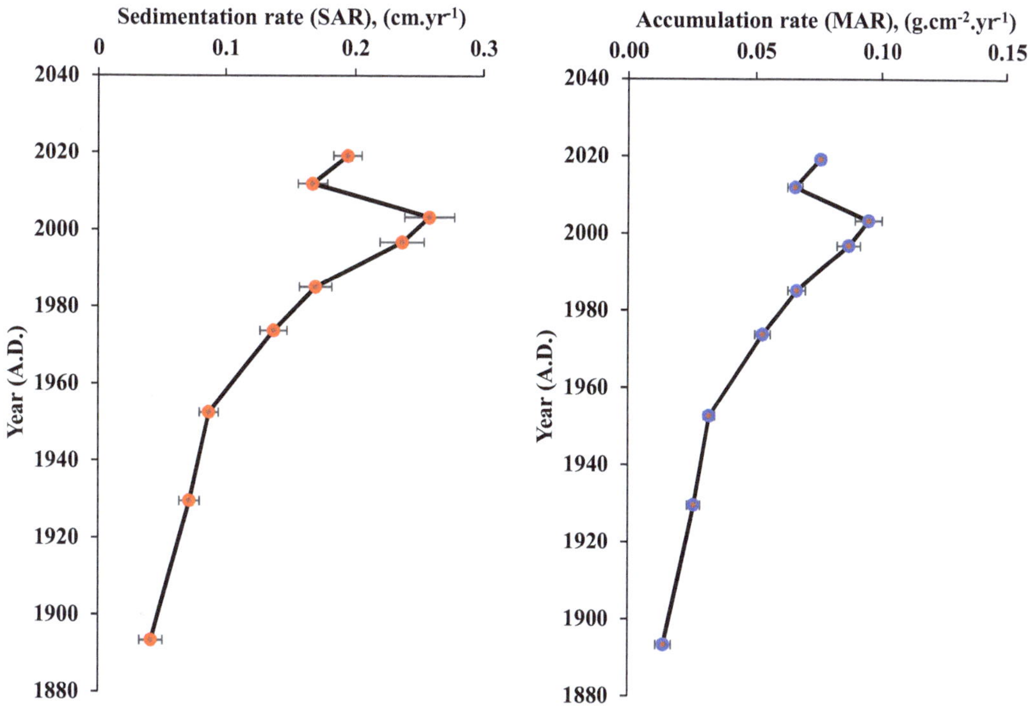

FIG. 5.7. Mass accumulation rates (MAR, g.cm^{-2} yr^{-1}) and sedimentation rates (SAR, cm yr^{-1}) derived from ^{210}Pb profiles by applying the CF model from the northern Gulf of Aqaba.

The sedimentation rate showed an important change of an increasing trend over time, particularly from the mid-1960s onwards (Table 5.3; Fig. 5.7). It is obvious from the CF model results (Table 5.3) that the sedimentation rates before 1965 were relatively stable, averaged 0.08 ± 0.04 cm yr^{-1}, and showed an accelerated increase to 0.21 ± 0.04 cm yr^{-1} after 1965. This change could be related to human activities and the start of increasing population, urban expansion in Aqaba, and the heavy industries as well in the mid-1960s. This date coincides also with the construction of the Phosphate Port in the mid-sixties (1960's) and the start of the exportation process.

Due to surface layer disturbances in core PLB3, no age model could be applied to this core. However, since the total equilibrium between ^{210}Pb and ^{226}Ra was not reached at the last layer (Fig. 5.2) and if it is assumed this will occur in depths >11 cm (correspond to 130 years), then a rough estimation of the sedimentation rates in this core should be higher than 0.09 cm yr^{-1}.

5.3.4. ^{137}Cs profile and sedimentation rate

In order to validate the chronology obtained from applying the ^{210}Pb models, the ^{137}Cs profile from the same cores were used. The radionuclide ^{137}Cs is widely used as an independent tracer to confirm the ^{210}Pb age dating method [5.31, 5.33]. The use of ^{137}Cs relies on detecting its atmospheric deposition peak from 1963 as a result of nuclear weapon test fallout.

According to Ruiz-Fernández [5.34], the sedimentation rate via ^{137}Cs data can be calculated as follows:

$$S = \frac{d}{(t_{(0)} - 1963)} \tag{5.1}$$

In which S is the average sedimentation rate (cm yr^{-1}), d is the depth (cm of the ^{137}Cs peak from the sediment profile), and $t_{(0)}$ is the collected sampling year

Referring to Fig. 5.6, the ^{210}Pb dating models indicates that the year 1963 value falls around 10 cm depth. Although the activity depth profile of ^{137}Cs shows a high value in the topmost part of core PLB1, it displays a small peak (maximum) in the 11 cm layer. However, core PLB3 displays this peak value around 4.5 cm layer (Fig. 5.8). In both cores, this peak could be attributed to the worldwide nuclear weapon test fallout. By assigning the year 1963 as the maximum ^{137}Cs concentration in the atmosphere to this layer (11 cm) in PLB1 core, the sedimentation rate of 0.20 cm yr^{-1} was determined. This value is consistent with that derived using the ^{210}Pb CF and CF:CS dating models (0.15-0.22 cm yr^{-1}), indicating that both radioactive isotopes yield comparable results. In a study by Pittauerová [5.25] from the nearby Eilat subbasin, the 1963 maxima was not present, rather ^{137}Cs activities decreased exponentially with depth. This depth distribution was attributed to either mixing at low sedimentation rates or a combined effect of mixing and sedimentation [5.25].

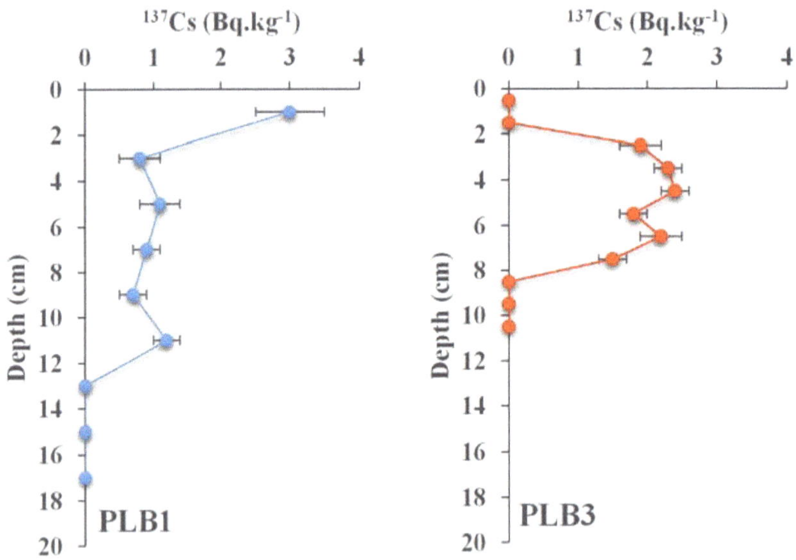

FIG. 5.8. Vertical distribution of ^{137}Cs activity (Bq.kg^{-1}) in the sediment cores PLB1 and PLB3 from the northern Gulf of Aqaba. Error bars represent counting errors.

5.4. CONCLUSIONS

The main conclusions from the present study can be summarized as follows:

— Higher activity concentrations of the uranium series radionuclides (^{210}Pb, ^{226}Ra, and ^{234}Th) were recorded in the topmost layer of the shallow sediment cores compared to deep cores that could be attributed to the contribution of phosphate dust lost to the sea during loading and shipment.

- The profiles of excess ^{210}Pb ($^{210}Pb_{ex}$) from both cores follow a relatively exponential decrease with depth. However, the top layer in core PLB3 showed signs of sediment disturbance due to mixing caused by bioturbation and physical processes.
- Profiles of ^{228}Ra, ^{40}K, and ^{234}Th showed minor variations of activity concentrations in the deep core. However, they showed an exponential decrease with depth in the shallower core.
- Based on the CF:CS age model, recent sedimentation rate averages 0.22±0.03 cm yr^{-1}, whereas the accumulation rate averages 0.08±0.01 g.cm^{-2} yr^{-1}.
- By applying the CF model, the sedimentation rate showed an important change of an increasing trend over time, particularly from mid-1960s onwards (from 0.08 ± 0.04 to 0.21 ± 0.04 cm yr^{-1}). This change could be related to human activities, the starting of the increasing population, urban expansion in Aqaba, and the heavy industries in the mid-1960s.
- The sedimentation rate value derived from the ^{137}Cs profile from the study area (0.20 cm yr^{-1}) were consistent with values obtained from ^{210}Pb profiles (0.15-0.22 cm yr^{-1}).

5.5. ACKNOWLEDGEMENTS

The authors would like to thank the administrative and technical staff of the Marine Science Station in Aqaba/Jordan. Special thanks are due to M. Al-Omor, H. Abu-Taleb, I. Zghol, and M. Almousa for assistant in samples collection. This work was supported by the IAEA, Coordinated Research Project (CRP) K41016/Research Contract No: 20866.

5.6. REFERENCES

[5.1] AL-HORANI, F., AL-ROUSAN, S., AL-ZIBDEH, M., KHALAF, M., The status of coral reefs on the Jordanian coast of the Gulf of Aqaba, Red Sea. Zool, Middle East **38** (2006) 99–110.

[5.2] AL-ROUSAN, S., Skeletal extension rate of the reef building coral Porites species from Aqaba and their environmental variables, Nat. Sci. **4** (2012) 731–739.

[5.3] AL-ROUSAN, S., AL-TAANI, A., RASHDAN, M., Effects of pollution on the geochemical properties of marine sediments across the fringing reef of Aqaba, Red Sea, Mar Pollut Bull. **110** (2016) 546–554.

[5.4] JORDAN PHOSPHATE MINES COMPANY, Personal Communication. Aqaba Office, Aqaba, Jordan (2015).

[5.5] ABU-HILAL, A., Phosphate pollution in the Jordan Gulf of Aqaba. Mar, Pollut. Bull. **16** (1985) 281–285.

[5.6] FREEMANTLE, M., HULINGS, N., MULQI, M., WATTON, E., Calcium and phosphate in the Jordan Gulf of Aqaba. Mar. Pollut. Bull., **9** (1978) 79–80.

[5.7] RASHEED, M., AL-ROUSAN, S., BADRAN, M., Phosphate enrichment in the northern Gulf of Aqaba: Regulation by carbonate sediments and impact on nitrogen elevation, Chem. Ecol., **21** (2005) 199-208.

[5.8] KTEIFAN, M., WAHSHA, M., AL-HORANI, F. Assessing stress response of *Stylophora pistillata* towards oil and phosphate pollution in the Gulf of Aqaba, using molecular and biochemical markers, Chem. And Ecol. **33** 4 (2017) 281–294.

[5.9] WALKER, D.I., ORMOND R.F.G., Coral death from sewage and phosphate pollution at Aqaba, Red Sea, Mar. Pollut. Bull., **13** 1 (1982) 21-25.

[5.10] AL-ROUSAN, S., AL-SHLOUL, R., AL-HORANI, F., ABU-HILAL, A., Heavy metal contents in growth bands of Porites corals: record of anthropogenic and human developments from the Jordanian Gulf of Aqaba, Mar. Pollut. Bull. **54** (2007) 1912–1922.

[5.11] STEVENSON, A., Metal concentrations in marine sediments around Scotland: a Baseline for environmental studies, Cont. Shelf Res. **21** (2001) 879-897.

[5.12] MAANAN, M., Heavy metal concentrations in marine molluscs from the Moroccan coastal region, Environ. Pollut. **153** 1 (2008) 176–183.

[5.13] AL-ABSI, E., MANASRAH, R., ABUKASHABEH, A., WAHSHA A. Assessment of heavy metal pollutants at various sites along the Jordanian coastline of the Gulf of Aqaba, Red Sea, Int. J. Env. Anal. Chem. **99** 8 (2019) 726-740.

[5.14] ORDONEZ-REGIL, E., ALMAZÁN-TORRES, M.G., SANCHEZ-CABEZA, J.A., RUIZ-FERNÁNDE, A.C., Presence of uranium and plutonium in marine sediments from gulf of tehuantepec, Mexico, J Radioanal Nucl Chem., **298** (2013) 981-986.

[5.15] ABABNEH, Z.Q., AL-OMARI, H., RASHEED, M., AL-NAJJAR, T., ABABNEH, A.M., Assessment of gamma-emitting radionuclides in sediment cores from the Gulf of Aqaba, Red Sea, Radiat. Prot. Dosim. 141 (2010) 289–298.

[5.16] WAHSHA, M., AL-ABSI, E., MANASRAH, R., AL-ZYOUD, W., BINI, C., The impact of uranium neat a phosphate mining port on the environment in the northern Gulf of Aqaba, Red Sea, Environ. Qual., **22** (2016) 25-32.

[5.17] AL-ABSI, E., MANASRAH, M., WAHSHA, M., AL-MAKAHLEH, M., Radionuclides levels in marine sediment and seagrass in the northern Gulf of Aqaba. Red Sea, Fresenius Environmental Bulletin **25** (2016) 3461-3474.

[5.18] MARY, Y., EYNAUD, F., ZARAGOSI, S., MALAIZÉ, B., CREMER, M., SCHMIDT, S., High frequency environmental changes and deposition processes in a 2 kyr-long sedimentological record from the Cap-Breton canyon (Bay of Biscay), The Holocene **25** (2015) 348–365.

[5.19] APPLEBY, P.G., OLDFIELD, F., "Application of ^{210}Pb to sedimentation studies", Uranium-series Disequilibrium: Application to Earth, Marine, and Environmental Sciences. Clarendon Press, Oxford, UK (1992).

[5.20] ZHANG,Y., LU, X., SHAO, X., CHEN, C., LI, X., ZHAO, F., LI, G., MATSUMOTO, E., Temporal variation of sedimentation rates and potential factors influencing those rates over the last 100 years in Bohai Bay, China, Sci. Total Environ. **572** (2016) 68–76.

[5.21] KATZ, T., GINAT, H., EYAL, G., STEINER, Z., BRAUN, Y., SHALEV, S., GOODMAN-TCHERNOV, B.N., Desert flash floods form hyperpycnal flows in the coral-rich Gulf of Aqaba, Red Sea, Earth Planet. Sc. Lett. **417** (2015) 87-98.

[5.22] MANASRAH, R., BADRAN, M., LASS, U., FENNEL, W., Circulation and winter deep-water formation in the northern Red Sea, Oceanologia **46** (2004) 5-23.

[5.23] HULINGS, N., A review of marine science research in the gulf of Aqaba, Marine Science Station, University of Jordan Press (1989)

[5.24] TIBOR, G., NIEMI, T., BEN-AVRAHAM, Z., AL-ZOUBI, A., SADE, R., HALL, J., HARTMAN, G., AKAWI,E., ABUELADAS, A., & AL-RUZOUQ, R. (2010). Active tectonic morphology and submarine deformation of the northern Gulf of Eilat/Aqaba from analyses of multibeam data. Geo-Marine Letters, 30 , 561–573.

[5.25] PITTAUEROVÁ, D., KIRCHNER, G., GARBE-SCHÖNBERG, D., HERUT, B., NISHRI, A., FISCHER, H. W., Radionuclides and recent sedimentation and mixing rates in Northern Gulf of Eilat/Aqaba, Red Sea. J. Mar. Sys. **139** (2014) 1–8.

[5.26] BYERS, S. C., MILLS, L. E., STEWART, P. L., A comparison of methods of determination of organic matter in marine sediments, with suggestions for a standard method, Hydrobiologia **58** 1 (1987) 43–47.

[5.27] HANCOCK, G.J. AND HUNTER, J.R., Use of excess ^{210}Pb and ^{228}Th to estimate rates of sediment accumulation and bioturbation in Port Philip Bay, Australia, Mar Fresh. Wat. Res. **50** (1999) 533-545.

[5.28] MIL-HOMENS, M., STEVENS, R. L., BOER, W., ABRANTES, F., CATO, I., Pollution history of heavy metals on the Portuguese shelf using ^{210}Pb-geochronology, Sci. Total Environ. **367** (2006) 466-480.

[5.29] MAHMOOD, Z.U.Y.W., MOHAMED, C.A.R., ISHAK, A.K., BAKAR, N.S.A., ISHAK, K., Vertical distribution of ^{210} Pb and ^{226}Ra and their activity ratio in marine sediment core of the East Malaysia coastal waters, J. Radioanal. Nucl. Chem. **289** (2011) 953–959.

[5.30] APPLEBY, P.G., OLDFIELD, F., "Application of ^{210}Pb to sedimentation studies Uranium-series Disequilibrium: Application to Earth, Marine, and Environmental Sciences, Clarendon Press, Oxford, UK (1992).

[5.31] SANCHEZ-CABEZA, J.A., MASQUE, P., ANI-RAGOLTA, I., MERINO, J., FRIGNANI, M., ALVISI, F., PALANQUES, A., PUIG, P., Sediment accumulation rates in the southern Barcelona continental margin (NW Mediterranean Sea) derived from ^{210}Pb and ^{137}Cs chronology, Prog. Oceanog. **44** (1999) 313–332.

[5.32] AL-ROUSAN, S., PÄTZOLD, J., AL-MOGHRABI, S., WEFER, G., Invasion of anthropogenic CO_2 recorded in planktonic foraminifera from the northern Gulf of Aqaba, Int. J. Earth Sci. **93** (2004) 1066–1076.

[5.33] KUZYK, Z. Z. A., GOBEIL, C., MACDONALD, R.W., ^{210}Pb and ^{137}Cs in margin sediments of the Arctic Ocean: controls on boundary scavenging, Global Biogeochem Cycles **27** 2 (2013) 422–439.

[5.34] RUIZ-FERNÁNDEZ, A.C., FRIGNANI, M., HILLAIRE-MARCEL, C., GHALEB, B., ARVIZU, M.D., RAYGOZA-VIERA, J.R., PÁEZ-OSUNA, F., Trace metals (Cd, Cu, Hg, and Pb) accumulation recorded in the intertidal mudflat sediments of three coastal lagoons in the Gulf of California, Mexico, Estuar. Coasts. **32** 3 (2009) 551-564.

6. DEPOSITION HISTORY OF ARSENIC, BARIUM CHROMIUM AND ZINC AT PERAI INDUSTRIAL AREA OF PENANG, MALAYSIA – A CASE STUDY USING NUCLEAR TECHNIQUES

YII MEI-WO
Malaysian Nuclear Agency
43000 KAJANG, MALAYSIA
Email: yii@nuclearmalaysia.gov.my

ANA CAROLINA RUIZ-FERNANDEZ
Universidad Nacional Autónoma de México
82000 MAZATLAN, MEXICO

ZAL U'YUN WAN MAHMOOD
Malaysian Nuclear Agency
43000 KAJANG, MALAYSIA

WILFRED PAULUS
Malaysian Nuclear Agency
43000 KAJANG, MALAYSIA

Abstract

The Perai Industrial Area is one of the densest industrial and heavily populated zones in Malaysia. Concentration levels of arsenic, barium, chromium and zinc were investigated in the two main rivers, i.e, Juru River and Perai River in this area. From each river, one sediment core was collected at the river estuary and another one within the river, about 2-3 km away from the estuary, using a KC™ Kajak corer. Concentrations of arsenic, barium, chromium and zinc were performed using Neutron Activation Analysis (NAA), complemented with Energy Dispersive X-ray Fluorescence Spectrometer (EDXRF), whereas the sediment deposition rates were determined through Lead-210 dating, which was measured by using a Gamma-ray Spectrometry System. Using continental crust values and iron concentration as the normalization metal, enrichment factors for elements were calculated. The results showed arsenic having severe enrichment at the study area. Zinc enrichment was mostly at a moderate level, although the enrichment was more serious at Juru River. Meanwhile, no enrichment for barium and chromium was found in most of the cases, except chromium was found severely enriched in one location. Sedimentation rates indicated the contaminants were deposited quickly. Analyses of metal concentration in surface sediments collected along the river indicate contaminants were released from a diffuse source upstream.

6.1. INTRODUCTION

In nature, environmental issues are complex processes and difficult to understand. Interactions occur both between the abiotic environment and the biosphere and within themselves [6.1]. Around one-third of the world's population lives near the coastal area (less than 100 km), where they utilise the coastal and marine resources to create job and business opportunities [6.2-6.3]. Development of the coastal zone, exploitation of its natural resources, and its use as a receptacle for societal wastes promote sediment contamination and endanger the living marine resources [6.4]. Remediation of a marine environment is not a standalone process but has to be taken into consideration the ecological interdependence of the oceans, coastal areas and the freshwater systems associated with them [6.5].

Pollutants released from land will be transported to rivers that finally end up in marine ecosystem. Inorganic pollutants such as heavy metals, that are released in the river water can be accumulated in the sediments and aquatic food chain [6-7], which will bring health effects to the aquatic life [8-9] and also to human being through fish consumption. . Few important sources of inorganic anthropogenic pollution are agriculture activities, animal waste, domestic, industrial, mining and petroleum activities, and also from the others industrial emissions [6.10-6.11].

Sediment is a mixture of complex materials resulting from industrial, biological and chemical waste brought by rivers and the coastal erosion. Weathering is one of the major sources of sediments; however, the fast growing economy and population growth also contribute large portion of the waste materials into the coastal area that stock up into the sediment over time [16.2]. Natural occurring radionuclides in the sediment such as ^{210}Pb could then be utilized as a powerful tool for the determination of mass fluxes in the marine environment, particularly for investigation of recent sedimentation. The deposition rate of contaminants in sediment can be estimated using the natural radionuclide profiles of ^{210}Pb in a sediment core, where the sedimentation rates can be estimated by employing different mathematical models. Pb-210 ($T_{1/2}$ = 22.20±0.22 years), a member of the ^{238}U decay series, is present in the sediments through two main routes. First, there is a ^{210}Pb fraction continuously produced *in situ* from ^{226}Ra ($T_{1/2}$ = 1600±7 years); this fraction, known as background or supported ^{210}Pb, is assumed to be in radioactive secular equilibrium with ^{226}Ra. On the other hand, ^{222}Rn ($T_{1/2}$ = 3.8235±0.0003 days) which is a noble gas, emanates from the Earth's surface and goes into the atmosphere where it decays into ^{210}Pb, which is removed from the atmosphere back to the Earth's surface (by wet precipitation and/or dry fallout) contributing to the unsupported or excess fraction of ^{210}Pb [6.13]. This atmospheric addition of ^{210}Pb is in excess of the amount permanently supplied by the *in situ* decay of ^{226}Ra. As time goes by, the activity of excess ^{210}Pb will decay until reaching equilibrium with the supported ^{210}Pb fraction (^{226}Ra). The activity differences between the initial excess ^{210}Pb (at the surface of the sediment core) and the subjacent core sections can be used to estimate the time when the sediment at this section was deposited [6.14-6.18].

With regards to the elemental analysis, in general, Neutron Activation Analysis (NAA) is a favourite nuclear analytical technique that have multi-application and non-labour-intensive. NAA technique is a non-destructive method where it does not have to perform most chemical preparation steps such as heating, digestion, usage of reagent and etc. This can help to ensure the integrity of the sample and minimise the potential contamination on sample due to reagent/chemical add in [6.19]. Besides these, the NAA method is also having other advantages such as fast and simple preparation, ability to determine multi-element simultaneously and can determine element concentration down to sub-picogram levels [6.20]. NAA operational is working based on the properties of the nucleus, radioactivity and the interaction of radiation with matter. When performing NAA, neutron source emitted from the reactor will be absorbed by the sample's (soil, sediment, water, etc) element nucleus after a non-elastic collision and turn into a new nucleus at highly excited state. This target nucleus will become radioactive which later start decaying and emitting gamma rays [6.20]. Emitted gamma rays can be measured by using a Gamma Spectrometry system, where the peak energies positions are unique for different radionuclides (element) and the intensities are proportional to their concentrations.

Energy dispersive X-ray fluorescence (EDXRF) spectrometry is another widely used non-destructive nuclear analytical technique that can also measure simultaneous elemental information from different types of materials. It is employed in a wide range of industries and applications including cement production, cosmetics, food and environmental, forensics, glass production, healthcare products, iron, mining and mineral beneficiation, petroleum and petrochemicals, pharmaceuticals, polymers and related industries, steel and non-ferrous metals. The spectrometer contains a radiation source, a sample and the detector. In an EDXRF, the radiation source is the X-ray tube that irradiates a sample, and an energy dispersive detector measures the fluorescence light produced by the sample. This detector can determine the different energies of a sample's characteristic radiation and assign them to different elements that are present in the sample. This separation phenomenon is known as dispersion.

The Perai Industrial Estate is one of the major industrial parks in Penang state and in Malaysia. Penang is located on the northwest coast of Peninsular Malaysia, next to the Malacca Strait. It consists of two parts: Seberang Perai (early well known as Wellesley Province) on the Malay Peninsula mainland, and the Penang Island. The Perai Industrial Estate is about 15 km^2 and is bordered by the Perai River to the northwest, Juru River to the east, the Penang Channel to the southwest and Baru Perai road to the north (Fig. 6.1). The Perai Industrial Estate was established in the 1970s in response to the need of industrial expansion of earlier rubber estates, such as Mak Mandin. With limited land area on the Penang Island, the choice seems right to turn the coastal area of Perai into a major industrial area. This was further boosted when the 13.5 km long Penang Bridge was completed in 1985. With improved access to the Penang international airport on the Penang Island, many more companies set up factories in this area. Perai Bulk Cargo Terminal that located at the estuary of Perai river encourage larger container ship to deliver huge cargo. As rapid population growth and heavy industrial development occur continuously surroundings Perai industrial area, the household and industrial discharges were suspected to be released into the nearby rivers; Juru River, for instance, is reported to be one of the most polluted rivers in the country [6.21].

This study is aimed to assess the changes in sediment deposition rates, as well as, the concentration and enrichment of As, Cr, and Zn, in sediment cores sampled from Perai River and Juru River, Penang, that might have been promoted by the population growth and economic development in the Perai Industrial area.

6.2. EXPERIMENTAL

6.2.1. Collection and preparation of samples

Sediment cores were collected during the dry season between 2017 and 2018 within the estuary zone and a few km upriver, in Juru River and Perai River (Fig. 6.1) to avoid massive mud transportation into the river sediment by the heavy downpour during rainy season and the water turbulence that disturbs the sediment surface. The cores were collected using a KC™ Kajak Core Sampler (Fig. 6.2) with an 8 cm diameter PVC tube. The sediment core was extruded from the bottom of the tube and sliced at 2 cm thick sections at the sampling site; samples were put into plastic bags and cooled in icebox prior elemental analysis. Back to the laboratory, the sediments were weighed, then a small scope of sample (1-2 g) were taken for particle analysis, balance samples were re-weighed and dried in an oven at 80°C until constant weight [22]. The dried samples were grounded to powder size of less than 74µm (200 mesh) and homogenised prior to further analysis All data is reported on a dry weight basis.

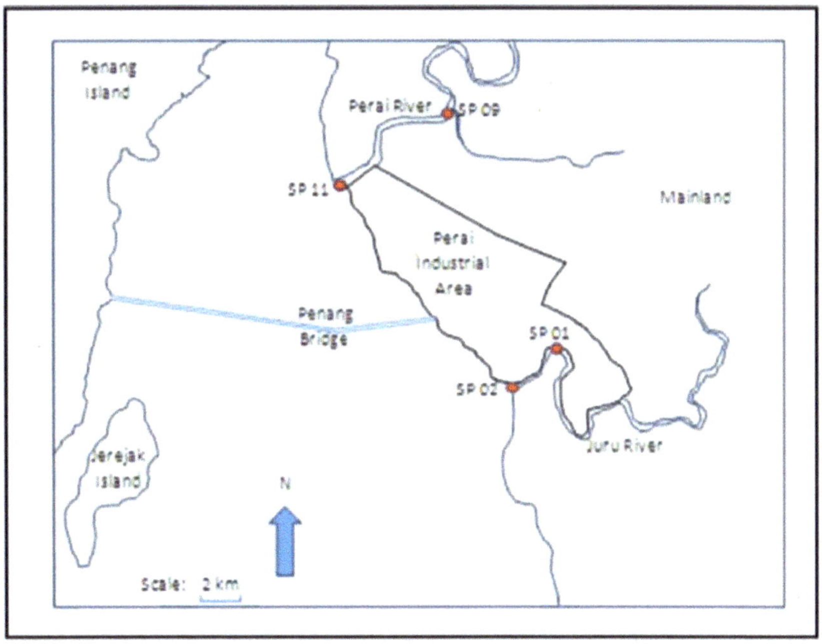

FIG 6.1. Sampling location.

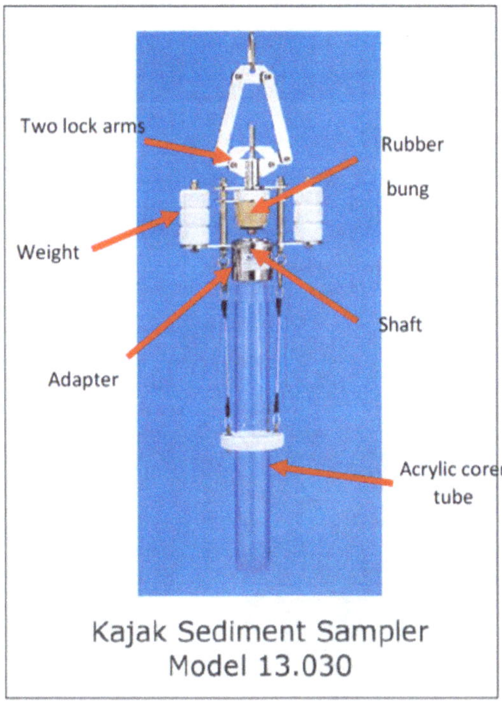

FIG. 6.2 Tools for sampling.

TABLE 6.1 SAMPLING LOCATION DETAILS

Date	Station	Coordinates	Core Length (cm)
Feb 2017	SP 01 Juru River	5.3464 ° N 100.4189 ° E	74
Feb 2017	SP 02 Juru River	5.3395° N 100.4048° E	78
July 2018	SP 09 Perai River	5.3992 °N 100.3927° E	62
July 2018	SP 11 Perai River	5.3832 °N 100.3697° E	30

6.2.2. Particle size analysis

Particle size distribution was analysed with a Microtrac™ X100 laser diffraction (Fig. 6.3). Using the Udden–Wentworth scale, particles were characterized into percentage of clay (< 4 um), silt (4 – 63 um) and sand (> 63 um) [6.23].

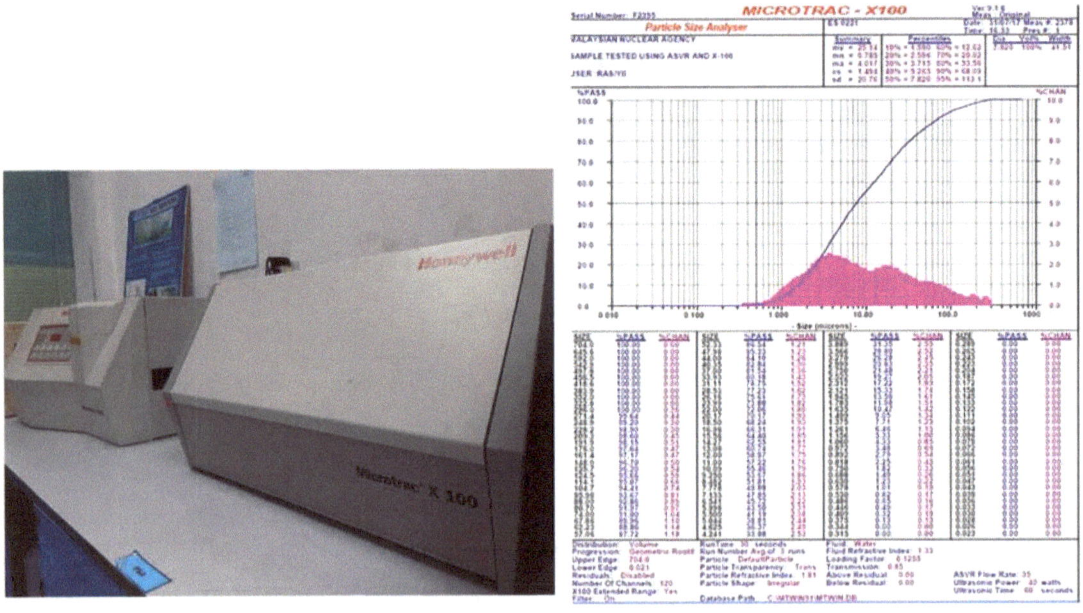

FIG. 6.3. Particle size analyser and sample output.

6.2.3. Loss on Ignition

Loss on Ignition (LOI) is an approach to estimate the percentage of organic matter content (% OM) in soil/sediment samples. This approach only requires the use of muffle furnace and does not need to use any chemicals (Fig. 6.4). The % OM is calculated by comparing the sample weight before and after ignition. Organic matter present in the sample will be oxidised to carbon dioxide before further ashed at temperatures between ~200 to 500 °C. The weight losses are due to water evaporation and carbon dioxide evolutions can then be calculated using the following two equations by weighing the sample before and after controlled heating (ignition at 550 and 1000 °C), that can be correlated to water content, the organic matter and carbonate content [24]. So, by using equation (6.1) and equation (6.2), the percentages of organic carbon and carbonate were calculated for every layer of sediment core.

$$\text{\% Organic Matter} = \frac{(Weight\ Post\ 105^{o}C\ Dry\ sample)-(Weigh\ tPost\ 550^{o}C\ Ash)}{(Weight\ Post\ 105^{o}C\ Dry\ sample)} \times 100 \quad \text{eq (6.1)}$$

$$\text{\% } CaCO_3 = \frac{(Weight\ Post\ 550\ ^{o}C\ Ash)-(Weight\ Post\ 1000\ ^{o}C\ Ash)}{(Weight\ Post\ 105^{o}C\ Dry\ sample)} \times 2.274 \times 100 \quad \text{eq (6.2)}$$

where 2.274 = 100.088 / 44.009 = molecular weight $CaCO_3$ / molecular weight of CO_2.

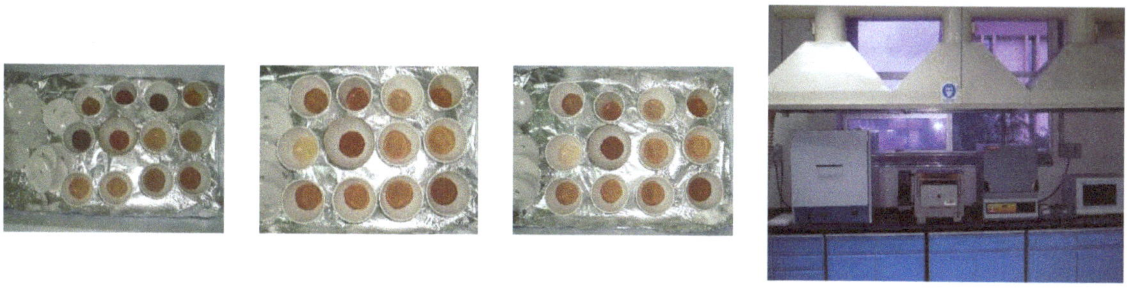

FIG. 6.4. a) LOI (from left to right) at 105°C, 550°C, and 1000°C. b) Furnace for LOI.

6.2.4. Radionuclides analysis

Radioactivities for Pb-210,^{226}Ra and ^{137}Cs were determined by gamma-ray spectrometry (Fig. 6.5) consisting of a high-purity germanium (HPGe) setup and a multichannel analyser (16,384 channels). The detector used is a closed end coaxial well-detector from CANBERRA™ that operated at 2,000 HV bias supply (detector: 62 mm diameter, 49 mm, and 5 mm distance from the window; well: 35.5 mm depth, 23.5 mm diameter, active volume of 8.8 cc). This 25% relative efficiency p-type well detector having 820 eV FWHM resolution at 122 keV (^{57}Co) gamma-ray line and 1.85 keV at 1332 keV (^{60}Co) gamma-ray line. The shielding chamber is made of lead, copper and cadmium with the total thickness up to 11 cm to reduce cosmic radiation. This system was calibrated following procedure reported by Yii et al. [6.25]] using a customized gamma multi-nuclide standard source (source no. 1290-84 and 1755-30 from Isotope Products Laboratories, USA) comprised of ^{210}Pb, ^{241}Am, ^{109}Cd, ^{57}Co, ^{123m}Te, ^{51}Cr, ^{113}Sn, ^{85}Sr, ^{137}Cs, ^{88}Y and ^{60}Co in the same counting geometry. Fine-ground dry sediments from each section of the core were sealed in a 6mL cylindrical HDPE container using thick PVC tape to prevent radon from escaping [6.26]. Net weight of each sample was divided by its sample volume to obtain sample density. To reach secular equilibrium between ^{226}Ra and their progenies, all samples were stored aside for at least a month before gamma counting [6.27-6.28].

All samples were counted for 15 hours and corrected for density and sampling date. An inert material in a container of the same geometry as the samples was counted over a weekend to determine the background counts. The counting times were long enough to achieve below

10%2σ counting error [6.29-6.31]. The ^{226}Ra were measured through the gamma transitions of their progenies; ^{214}Pb (295.21 and 351.92 keV) and ^{214}Bi (609.31 keV, 1120.29 keV and 1764.49 keV). Meanwhile, ^{210}Pb was measured directly *via* energy 46.5 keV peak and ^{137}Cs was measured directly *via* energy 661.62 keV peak [6.28, 6.29, 6.32]. The sample activities were calculated using the equation reported by Yang et al. [6.28] and Chen et al. [33]. The minimum detectable activity (MDA) for the radionuclides of interest was: ^{226}Ra (1 Bq/kg), ^{210}Pb (5 Bq/kg) and ^{137}Cs (1 Bq/kg). The reference materials IAEA-Soil-6 and IAEA-412 were used for quality control evaluation. If measurement values for the reference materials were within the 95% confidence level (as stated in the certificate), the results of samples will be accepted.

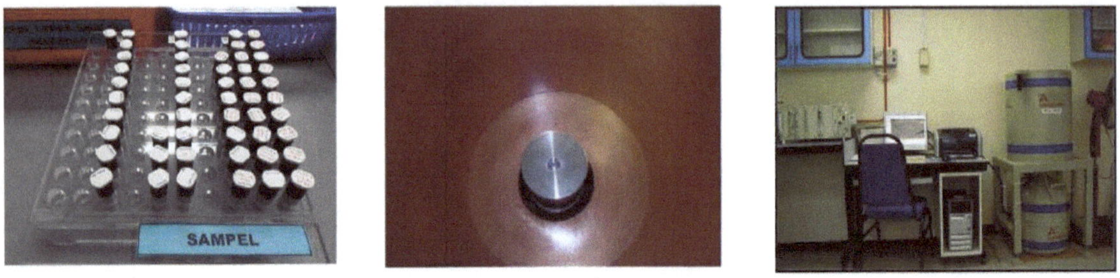

FIG. 6.5. Radioactivity measurement (from left to right): samples, well detector, gamma spectrometry system.

6.2.5. Trace elements analysis

Trace elements present in sediment samples were analysed using two different methods, i.e. by NAA or EDXRF. To perform NAA analysis, duplicate samples of ~0.1 g powdered sediment aliquots were weighted in polyethylene vials and heat sealed (Fig. 6.6). IAEA-Soil-7 reference material was used as quality control material. All samples were placed into a rotating rack inside a pneumatic transport facility and irradiated up to 6 hours with thermal neutron flux (4.0×10^{12} n cm^{-2} s^{-1}) produced by Malaysian Nuclear Agency's TRIGA Mark II research reactor. Before counting the irradiated samples using a Gamma Spectrometry System, the samples were allowed to cool for duration between 2 to 4 days (first counting), or 3 to 4 weeks (second counting). Table 6.2 shows the radioisotope counted and their properties for different elements [6.34-6.35].

TABLE 6.2. NEUTRON ACTIVATION ANALYSIS (NAA) RESULTS (*Adapted from [6.66]*)

Element	Radioisotope	Half-life	Energy (keV) γ-ray
Cr	^{51}Cr	27.8 days	320
Fe	^{59}Fe	45.1 days	1099,1292
Ba	^{131}Ba	11.5 days	496
Zn	^{65}Zn	244 days	1115
As	^{76}As	26.4 hours	559

The gamma-ray detector used for the NAA analysis was a closed end coaxial 3" x 3" HPGe detector connected to a multichannel analyser that was previously calibrated from low to high energies by using a mix standard solution including ^{241}Am (59.5 keV), ^{109}Cd (88.1 keV), ^{57}Co (122.1, 136.5 keV), ^{133}Ba (81.0, 303.0, 356.0, 384.0 keV), ^{137}Cs (661.7 keV), ^{60}Co (1173.2, 1332.5 keV) and ^{88}Y (898.0, 1836.1 keV) [6.36-6.37]. The gamma-ray spectrometry counting process of the irradiated samples, mix standard solution and the reference material were performed for one hour each. If measurement values for the reference materials were within the 95% confidence level (as stated in the certificate), the results of samples will be accepted.

FIG. 6.6. NAA metal analysis (from left to right): Samples, neutron irradiation, gamma counting.

Meanwhile, for the elemental analysis using EDXRF, in-house loose powder method was applied. Approximately 1.0 – 2.0 g ground sample aliquots were placed into a small Teflon ring cap cup, covered with a 6 μm polypropylene Mylar® to be analysed using a Rhodium x-ray tube. The reference material SRM 2709a was used as quality control material and calibration of the method. The results of samples were accepted when the reference material achieved the 95% confidence level as mentioned in the certificate.

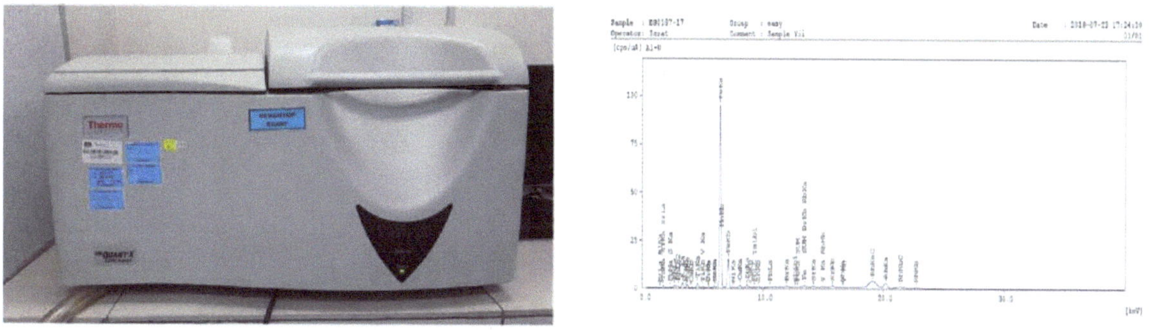

FIG. 6.7. Energy Dispersive X-Ray Fluorescence Spectrometer.

6.2.6. Enrichment factor (EF)

In order to evaluate potential contamination caused by trace elements, the enrichment factor (EF) was calculated based on the Eq. (3) [6.35]:

$$\text{EF} = \frac{(E/R)_{measure}}{(E/R)_{CC}} \qquad \text{eq. (6.3)}$$

in which $(E/R)_{measure}$ is the concentration ratio between the target element (E) and the reference element (R) measured in the sample, and $(E/R)_{CC}$ is the concentration ratio in the continental crust [6.38]. In order to reduce the minerology and particle size variations, trace element concentrations was normalised using iron (Fe) serving as the reference element [39]. Iron was selected as normalisation agent because it was found didn't correlate significantly with other trace elements. Iron distribution also being reported not affected by other elements [40]. Since iron usually present at a relative high concentration naturally, therefore it will not be enriched very much by anthropogenic sources in estuarine sediment [41-42]. If the element found in the sediment have EF value less than 2, it is consider originally coming from the lithogenous materials, otherwise it is coming from the anthropogenic sources [43]. EF values classification were used according to Ayari et al. [44].

6.3. RESULTS AND DISCUSSION

6.3.1. Particle size analysis

The sediments in all cores are mostly composed of fine-grained particles (Fig. 6.8). In most core sections, the percentage of silt is higher than the clay, and the clay is higher than that of sand. Overall, for all the cores, the silt content is around 40 – 70 %, the clay content between 10 – 60 % whilst the sand content around 0 – 40 %. Clay content shows a decreasing trend from the top towards the bottom of cores SP 02, SP 09 and SP 11 and also from 20 cm onwards for core SP 01. These trends could be due to the re-suspension of fine particles when the river water flows through the area. Meanwhile, the silt content shows a slight increase from the upper section of the core towards the bottom for cores SP 01 and SP 09 (collected upriver areas) whilst the cores of SP 02 and SP 11 (collected in the estuarine areas) have higher contents of silt at the upper part of the core. On the other hand, the sand contents are low at the top sections of the cores SP 02, SP 09 and SP 11 whilst increasing values are observed at the top portions of core SP 01 (Fig. 6.8). Movement of sand might be related to erosion due to nearby river activities and also the volume and speed of water flow.

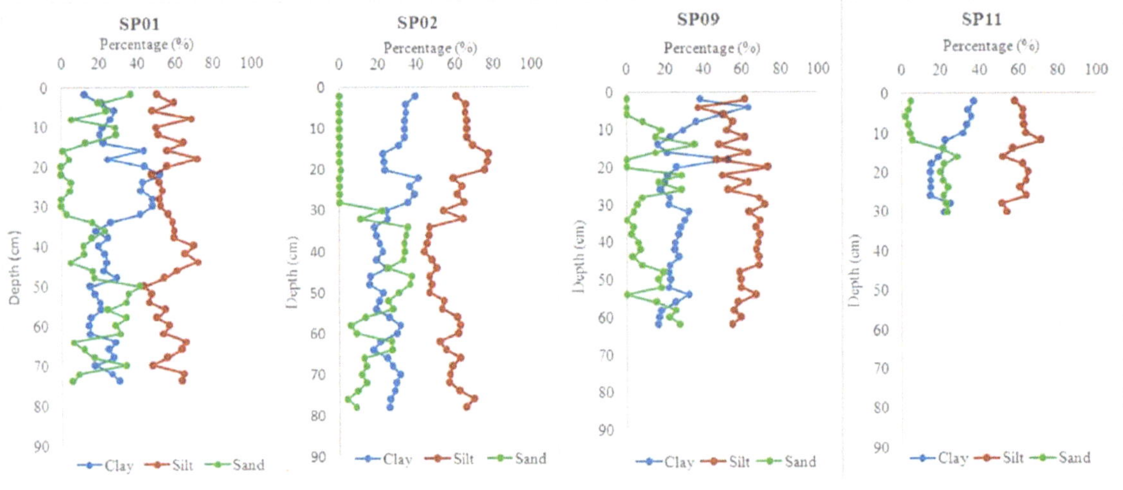

FIG. 6.8. Particle size distribution in sediment cores at study area.

6.3.2. Loss on Ignition

Generally, percentages of organic carbon decrease gradually from the surface sediment towards the bottom of the core whilst carbonate contents are almost constant (Fig. 6.9). The organic carbon content in sediment cores SP 01, SP 02, SP 09 and SP 11 at the study site are between 7.7 – 17.8 %, 7.8 – 14.3%, 6.1 – 19.9%, 9.5 – 13.0 %, respectively. The organic content for the estuary cores (SP 02 and SP 11) are comparable with the finding reported by Wan Mahmood et al. [6.18] of between 4.1 – 13.6 %. Significant correlations ($p<0.05$) are noted between the clay percentage and the organic matter in all sediment cores, although the correlation coefficients in the cores from Perai River (SP 09, r = 0.71 and SP 11, r = 0.90) are higher than in those from the Juru River (SP 01, r = 0.56 and SP 02, r = 0.52).

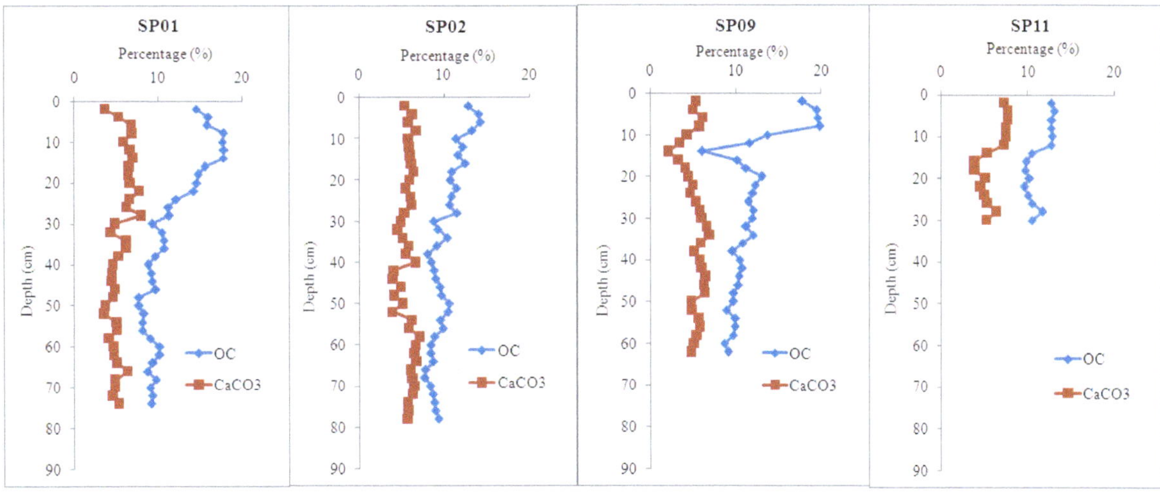

FIG. 6.9 Loss on ignition (LOI) in sediment cores at study area.

6.3.3. Radionuclide activities

The activity concentrations for ^{137}Cs, ^{210}Pb and ^{226}Ra in each core are illustrated in Fig. 6.10 and summarized in Table 6.3. The ^{137}Cs activities are below the detection limit of 1 Bq/kg for all sediment cores. Meanwhile, ^{226}Ra activity ranges (Bq/kg) are highly variable among cores, with highest values observed in the cores collected upriver (Table 6.3). These values are higher than those found in sediment cores from previous studies in Asia (e.g. the Economic Exclusive Zone of the east coast of Peninsular Malaysia, 16 – 46 Bq/kg [6.45], east coastal area of Peninsular Malaysia, 46.2 – 121.5 Bq/kg [6.46] and Vietnam coast, 23.1 – 40.2 Bq/kg [6.47]), or elsewhere (e.g. Gulf of Mexico, 11.8 – 97.3 Bq/kg [6.48]; Estuary of Coatzacoalcos River, 15.57±1.4 Bq/kg [6.49]; Santos - Cubatão drainage basin, 28.2 – 80.0 Bq/kg [6.50]), but much lower than values reported for Krka River estuary (45 – 662 Bq/kg [6.51]). A single factor ANOVA analysis revealed that ^{226}Ra activities are significantly different among cores: SP 01 (97.0±32.8 Bq/kg) = SP 11 (79.3±6.8 Bq/kg) > SP 02 (53.9±9.9 Bq/kg) =SP 09 (57.5±26.1 Bq/kg).

Cores SP 01 and SP 09 (both collected upriver) show high variability of ^{226}Ra activities (Fig. 6.10) which could be due to the input of radioactivity from the mainland, likely due to human activities, for instance, the use of high phosphate content fertiliser in agriculture. Transported radionuclides eluted from soil and contained in dry and wet atmospheric precipitation (^{210}Pb) are deposited together onto sediment [6.52]. Ra-226 was found to have poor correlation with the particle size fraction or organic matter content. Only in the two upriver cores are ^{226}Ra activities correlated with clay percentage (SP 01, r = 0.44 and SP 09, r = 0.65, $p<0.05$) and in SP 09 there is a significant correlation with organic matter (r = 0.92, $p<0.05$) (Table 6.4). This could be because the inner river sediment received more input of ^{226}Ra from the mainland due to soil erosion, and as radium is soluble in water, as fluvial sediments reach the estuarine zone, radium isotopes desorb owing to ion exchange competition with the major cations present in seawater [6.53].Activities of ^{210}Pb are also higher than those previously reported in the region such as at Linggi River Estuary (south west Peninsular Malaysia), 37.9 – 176.2 Bq/kg [6.18], at Kuala Muda area (north west Peninsular Malaysia), 11.9 – 78.8 Bq/kg [54] and at Brunei Bay, 2.6 – 32.3 Bq/kg [6.12]. Again, the single factor ANOVA analysis revealed that there are significant differences between cores for the ^{210}Pb activities among all cores in this study: SP01 (154.7±51.5 Bq/kg) = SP 11 (159.2±23.1 Bq/kg) > SP02 (100.2±20.0 Bq/kg) = SP 09 (100.9±54.1 Bq/kg).

The activities of ^{210}Pb are mostly higher than ^{226}Ra along all cores, and the equilibrium between both radionuclides was only achieved at cores SP01 and SP09. The $^{210}Pb_{ex}$ (excess ^{210}Pb) activities were obtained by subtracting the ^{226}Ra values from ^{210}Pb activities (Table 6.3). Pb-210_{ex} activities are significantly ($p<0.05$) correlated with organic carbon content (r= 0.76 in SP 01; r = 0.74 in SP02; r= 0.87 in SP 09: and r = 0.79 in SP 11. Also, significant ($p<0.05$) correlations are observed between $^{210}Pb_{ex}$ and clay content in most cores (with r = SP 02: 0.67, SP 09: 0.73 and SP 11: 0.78). It has been shown that ^{210}Pb may be associated with biogenic particles and high ^{210}Pb activities may be derived from the lithogenic (detrital) inclusion [6.55]. Lead-210 is easily attach to particle and is readily captured by organic matter and clay size particles [6.56], although under anoxic conditions, ^{210}Pb moved back into the water bodies [6.57].

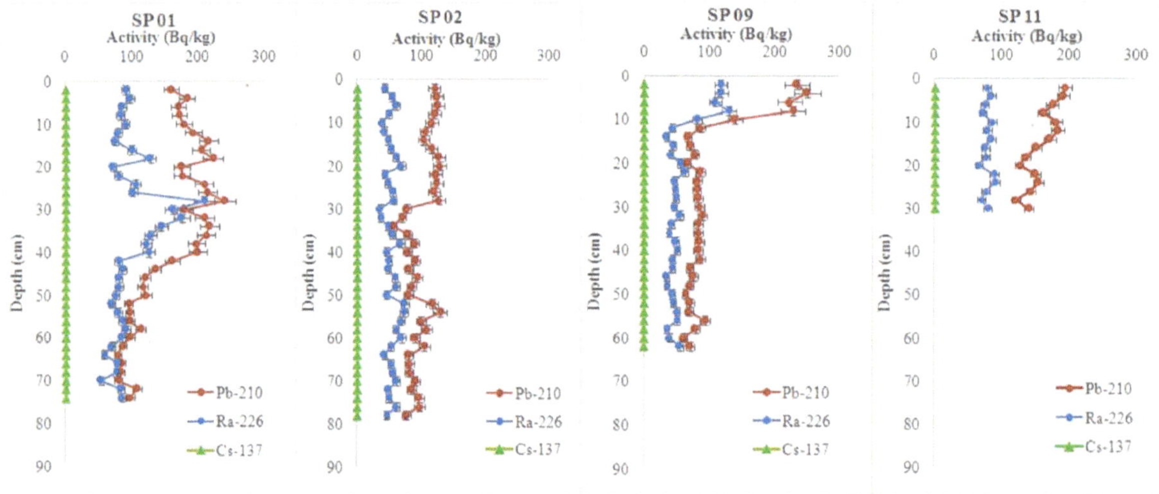

FIG. 6.10. Activity concentrations of ^{137}Cs, ^{210}Pb and ^{226}Ra in sediment cores at study site.

The logarithmic values of $^{210}Pb_{ex}$ activity depth profiles in all cores show significant departures from linearity (Fig. 6.11), characteristic of a profile resulting only from radioactive decay, in aquatic environments where steady state sedimentation process occur. In most cases, such divergences could be the result of strong variations in the sediment accumulation process, sediment sources and mixing [6.58]. As previously described, $^{210}Pb_{ex}$ activities in the cores of this study are strongly influenced by the organic matter content and differences in grain size. In addition, the uppermost 20 cm segment in core SP 01 showed $^{210}Pb_{ex}$ activities decreasing upward near the surface, which could have resulted from the accumulation of $^{210}Pb_{ex}$-depleted (older) sediments, from the dilution of $^{210}Pb_{ex}$ activities by high sediment loads, or from the accumulation of sediments from a different source (as suggested by the changes in particle size distribution) (Fig. 6.8); whereas between 20 and 40 cm depth, where $^{210}Pb_{ex}$ activities are almost homogeneous, the sediments might be mixed, although, looking at the high ^{226}Ra activities in this segment, likely this core segment accumulated sediments from a different source. Core SP02 showed almost homogenous $^{210}Pb_{ex}$ activities in the upper 30 cm (corresponding to increasing values of silt (Fig. 8)) and a slump deposit between 30 and 50 cm depth, characterized by coarser sediments (Fig. 6.8); the sediments in the uppermost ~12 cm segment of core SP 09 seems to have been transported from a difference source (as indicated by the higher ^{226}Ra activities and high content of clay) and $^{210}Pb_{ex}$ activities near the surface are nearly homogenous.

The $^{210}Pb_{ex}$ activities were normalized by the clay content and organic matter in the cores, with the purpose to evaluate if the $^{210}Pb_{ex}$ profile could be improved; however, results are not satisfactory. Nonetheless, despite all the complex features observed in the $^{210}Pb_{ex}$ activity depth profiles, all of them showed significant ($p<0.05$) dropping trends with depth, with high slope values, which would account for high accumulation rates [6.59]; although, owing to the non-monotonic $^{210}Pb_{ex}$ depth profiles, it is very difficult to derive an age model from them. However, attempts are made to approximate a preliminary mean sedimentation rate for each core, by using the constant rate-constant sedimentation (CF:CS) model [6.60], which assumes a constant $^{210}Pb_{ex}$ flux to the sediment surface and a sediment cumulative rate [6.17]. Thus, the logarithm of excess ^{210}Pb (ln $^{210}Pb_{ex}$) data obtained in this study was plotted against the depth of sediment core; and the slope of the regression between both variables was used to estimate the mean sediment accumulation rate (SAR, in cm/y). The mean SAR values obtained (Table 6.3) ranged from 0.94±0.11 cm/y in SP01 to 2.70±0.68 cm/y in SP02. The highest value recorded at station SP 02 corresponds to the estuarine area of Juru River, reportedly affected by heavy loads of waste water discharges [6.21].

The mean sedimentation rate interval from in this study of 0.94 – 2.70 cm/y are comparable to those reported by Joseph et al. [6.12] for Brunei Bay where the sedimentation rates ranged between 0.47 –2.13 cm/y and higher sedimentation was attributed to rapid urban and industrial development. The study results are also comparable to the sedimentation rates found at Linggi River estuary between 0.70 – 1.97 cm/y [6.61]. However, the sedimentation rate found in this study is higher than some other areas as reported by Cheevaporn et al. [62] at Bang Pakong River, Thailand (0.47 – 0.72 cm/y), and by Xu et al. [6.63] at Nile River Delta, Egypt (0.42 – 0.56 cm/y).

TABLE 6.3. RADIOACTIVITY CONCENTRATIONS OF ^{226}Ra, ^{210}Pb, THEIR RATIOS AND SEDIMENTATION RATE AT STUDY SITE

Sampling Station	Activity concentration (Bq/kg dw.)			Pb-210 excess (Bq/kg dw.)	Sedimentation accumulation rate (cm/y)
	^{137}Cs	^{226}Ra	^{210}Pb		
SP 01	< MDA	54.0 – 209.9	78.8 – 241.0	5.9 – 142.0	0.94 ± 0.11
SP 02	< MDA	36.0 – 74.3	56.5 – 129.9	8.4 – 79.3	2.70 ± 0.68
SP 09	< MDA	35.8 – 131.0	62.5 – 250.3	10.4 – 131.4	1.79 ± 0.49
SP 11	< MDA	66.8 – 90.6	121.4 – 193.9	50.6 – 115.1	1.15 ± 0.14

TABLE 6.4. SUMMARY OF CORRELATION ANALYSIS OF RADIOACTIVITIES CONCENTRATIONS WITH THE PERCENTAGE OF ORGANIC MATTER AND PARTICLE SIZE FRACTIONS IN SEDIMENTS FROM JURU AND PERAI RIVERS, MALAYSIA

Station	Element	Correlation coefficient (r)			
		Organic (%)	Clay (%)	Silt (%)	Sand (%)
SP 01	^{210}Pb	**0.56**	**0.44**	0.11	**-0.45**
	^{226}Ra	-0.02	**0.44**	0.05	-0.41
	$^{210}Pb_{ex}$	**0.76**	0.21	0.11	-0.25
SP 02	^{210}Pb	**0.65**	**0.53**	**0.56**	**-0.63**
	^{226}Ra	-0.11	-0.23	0.02	0.10
	$^{210}Pb_{ex}$	**0.74**	**0.67**	**0.57**	**-0.71**
SP 09	^{210}Pb	**0.93**	**0.72**	**-0.45**	**-0.36**
	^{226}Ra	**0.92**	**0.65**	**-0.41**	-0.33
	$^{210}Pb_{ex}$	**0.87**	**0.73**	**-0.46**	**-0.37**
SP 11	^{210}Pb	**0.70**	**0.69**	0.34	**-0.77**
	^{226}Ra	-0.09	-0.07	0.12	0.00
	$^{210}Pb_{ex}$	**0.79**	**0.78**	0.34	**-0.85**

Note: r values in Bold are significant at 95%, "-" value indicate negative correlation.

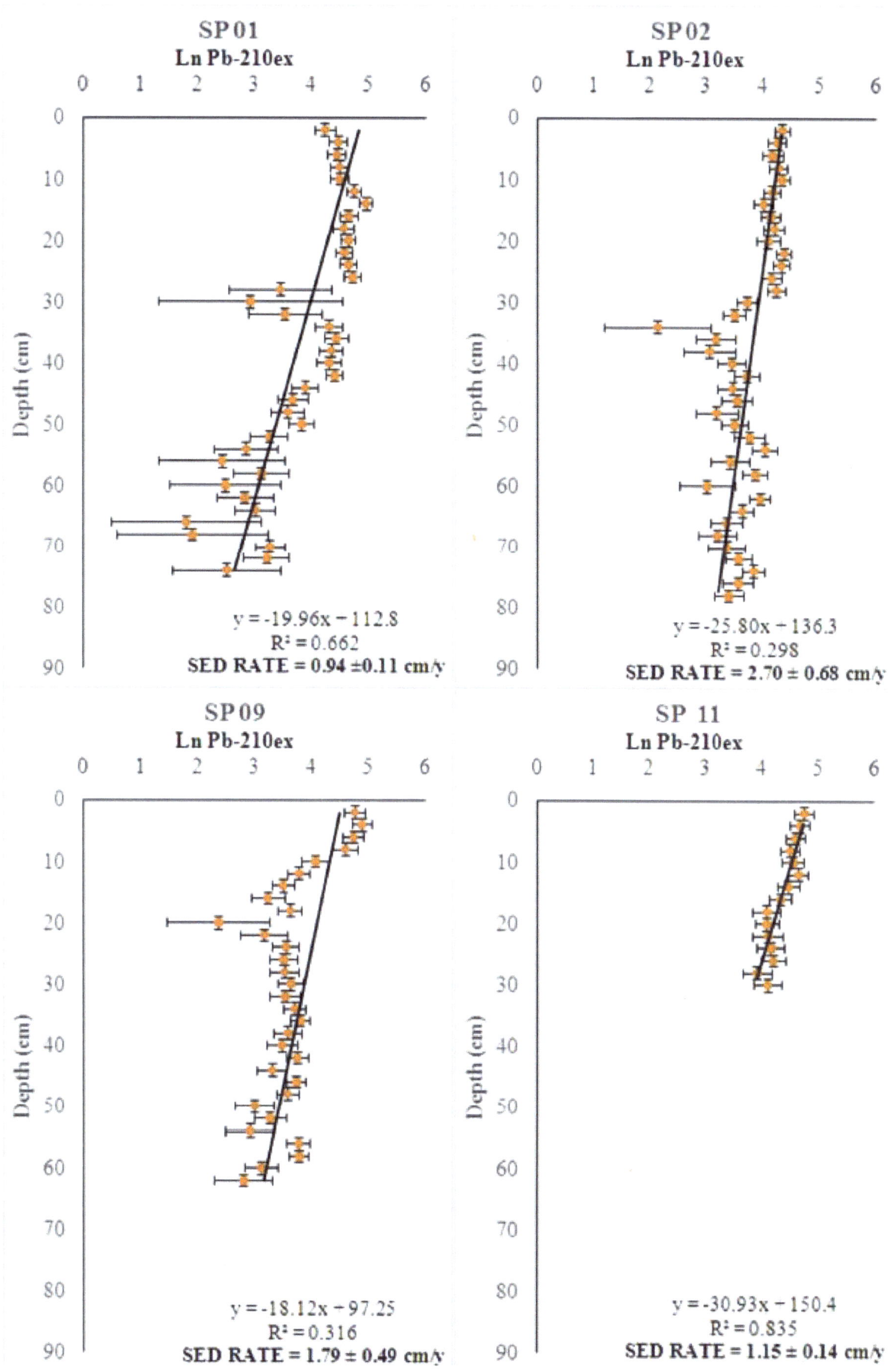

FIG. 6.11. Sedimentation rates in core sediment at study site.

6.3.4. Trace elemental concentrations

Table 6.5 summarizes the concentration of the trace metals As, Ba, Cr, Fe and Zn in the study area. Arsenic (As) concentrations in all the cores are less than 50 µg/g, whilst Ba, Cr, Fe and Zn have maximum concentrations higher than 100 µg/g. Overall, the concentration ranges for As, Cr and Zn in this study are higher than those reported by other researchers in the same area/region. For instance, a study at Juru River conducted by Wood et al. [6.64] reported a range from 0.90 – 12.3 µg/g (As), 7.0 – 77.7 µg/g (Cr) and 36.7 – 83.7 µg/g (Zn); whereas Zn concentrations are also higher than those (2 – 483 ppm) reported for surface sediment by Lim and Kiu [6.65] at the same river. Analysis of metal concentration in surface sediment collected along the Juru River indicated that contaminants are released from a diffuse source upstream from the sample sites, as reported by Yii et al. [6.66]. A study conducted by Cheevaporn et al. [6.62] reported a maximum of 61µg/g for Zn and 14 µg/g for Cr found at Bang Pakong River, Thailand.

TABLE 6.5. CONCENTRATIONS AND ENRICHMENT FACTOR (EF) RANGES OF POTENTIALLY TOXIC ELEMENTS (µg g^{-1}) IN SEDIMENT CORES FROM JURU AND PERAI RIVERS, MALAYSIA

Element	CC[1]	Location	Concentration Range[2]	Concentration Range[3]	EF range
As	1.7	SP 01	9.7 – 25.9 (15.7)	-	5.7 – 14.7 (11.3)
		SP 02	10.2 – 24.0 (14.5)	-	6.8 – 17.5 (10.0)
		SP 09	-	-	-
		SP 11	9.2 – 23.9 (18.0)	-	6.6 – 19.6 (13.2)
Ba	584	SP 01	211 – 390 (280)	-	0.3 – 1.2 (0.6)
		SP 02	179 – 383 (277)	-	0.4 – 0.8 (0.6)
		SP 09	-	193 – 583 (420)	0.4 – 1.5 (1.1)
		SP 11	319 – 466 (397)	-	0.7 – 1.1 (0.8)
Cr	126	SP 01	52.8 – 100 (74.0)	-	0.4 – 1.1 (0.7)
		SP 02	91.3 – 134 (110)	-	0.8 – 1.2 (1.0)
		SP 09	-	65.3 – 2088 (555)	0.8 – 23.5 (6.7)
		SP 11	84.0 – 129 (107)	-	0.7 – 1.4 (1.1)
Fe	43200	SP 01	25250 – 52560 (36366)	-	-
		SP 02	33900 – 42100 (37062)	33047 – 40644 (37353)[4]	-

Element	CC[1]	Location	Concentration Range[2]	Concentration Range[3]	EF range
		SP 09	-	18300 – 33708 (29296)	-
		SP 11	30970 – 39235 (35125)	-	-
Zn	65	SP 01	53.6 –808 (229)	-	1.3 – 13.5 (3.9)
		SP 02	84.7 – 648 (252)	-	1.6 – 10.8 (4.4)
		SP 09	-	58.2 – 576 (148)	1.4 – 12.0 (3.3)
		SP 11	86.4 – 199 (141)	-	1.6 – 3.8 (2.7)

[1]CC = Continental crust values as published by Wedepohl in 1995 [38].
[2]Determined by Neutron Activation Analysis (NAA)
[3]Determined by Energy dispersive X-ray fluorescence (EDXRF) spectrometry
[4]Concentrations of Fe for core SP 02 were determined using both NAA and EDXRF techniques for comparison.
Note: Values given in the parenthesis are the average values

Correlations of elemental concentration with organic matter and particle size are summarized in Table 6.7. The zinc concentrations in all cores are significantly correlated ($p<0.05$) with organic matter content (r = values of 0.87 in SP 01, 0.71 in SP 02, 0.93 in SP 09, and 0.74 in SP 11). The correlation between zinc with clay content is also significant ($p<0.05$), with r values of 0.59 in SP 02, 0.69 in SP 09, and 0.74 in SP 11. This observation implies that organic matter, attached to finest particles of clay, is the predominant carrier of zinc to the sediments. Iron (Fe) also shows significant correlation ($p<0.05$) with the clay content in three of the cores (r = values of 0.59 in SP 01, 0.54 in SP 02, and 0.65 in SP 11). Meanwhile, no significant correlations are observed among arsenic, barium and chromium with the organic matter or particle size fractions.

6.3.5. Enrichment factors

The enrichment factors obtained in this study (Table 6.5 and Fig. 6.12) indicated that sediments are enriched by arsenic in the entire cores collected from SP01, SP02 and SP11 (moderate severe to severe levels); by zinc at all four cores (from moderate to severe level at the top portion of the cores, but at minor level and/or not enriched down core) and by chromium only in core SP 09 from 10 cm onwards until the bottom. Meanwhile, no enrichment is observed by barium in all four cores.

At Juru River, for core SP 01, the EFs of arsenic showed severe enrichment throughout most of the sediment core especially at the bottom part of the core. Arsenic concentrations can be categorized from moderately severe to severely enriched. Barium and Cr showed no enrichment throughout the whole sediment core. On other hand, Zn shows an attractive profile which varied from severe enrichment at the top portion of the core down to minor and/or no enrichment at the bottom after 20 cm core depth. The enrichment factor values for Zn ranges from 13.5 down to 1.1 (average, 3.8). Meanwhile, for core SP 02, arsenic shows severe enrichment at the centre part of the core while the top and bottom portions of the core are moderately severely enriched with an average EF value of 10.0. The Zn profile up to the depth of 30 cm shows moderately severe enrichment while the lower part of the core had minor and/or no enrichment. Once again, Ba and Cr showed no enrichment in this core.

On the other hand, cores at the Perai River, at station SP 09, zinc varies from severe enrichment at the top part of the core down to minor and/or no enrichment at the bottom after 10 cm core depth. Chromium shows a very fluctuating profile with enrichment found throughout the core except at the top portion of the core and in some layers, this element is severely enriched. Barium remains not enriched throughout the core. Lastly, at station SP 11, arsenic shows an increasing trend of enrichment from moderate severe to severe moving down the core. Zinc is moderately enriched at the top half of the core but has no enrichment at the bottom half. Meanwhile, Ba and Cr remain not enriched throughout the core.

In summary, the EFs obtained for the element concentrations in the sediments from the two rivers are comparable (Table 6.5). The most relevant difference is the high EF by Cr observed in the upriver area of Perai River. The Perai Industrial area is densely populated and is comprised of many heavy industries, and the most possible sources of pollution might be coming from agriculture, sewage and manufacturing industries emissions [6.67]. Another possible source of arsenic pollution could be due to the use of arsenical herbicides in plantations along the river [6.68]. Electroplating, leather tanning, and textile industries release, as well as leaching from topsoil and rocks could be the potential source of chromium at the study side [6.69]. High zinc content found in the river could be contributed by the widely used zinc foil roofing for houses and restaurants along the river.

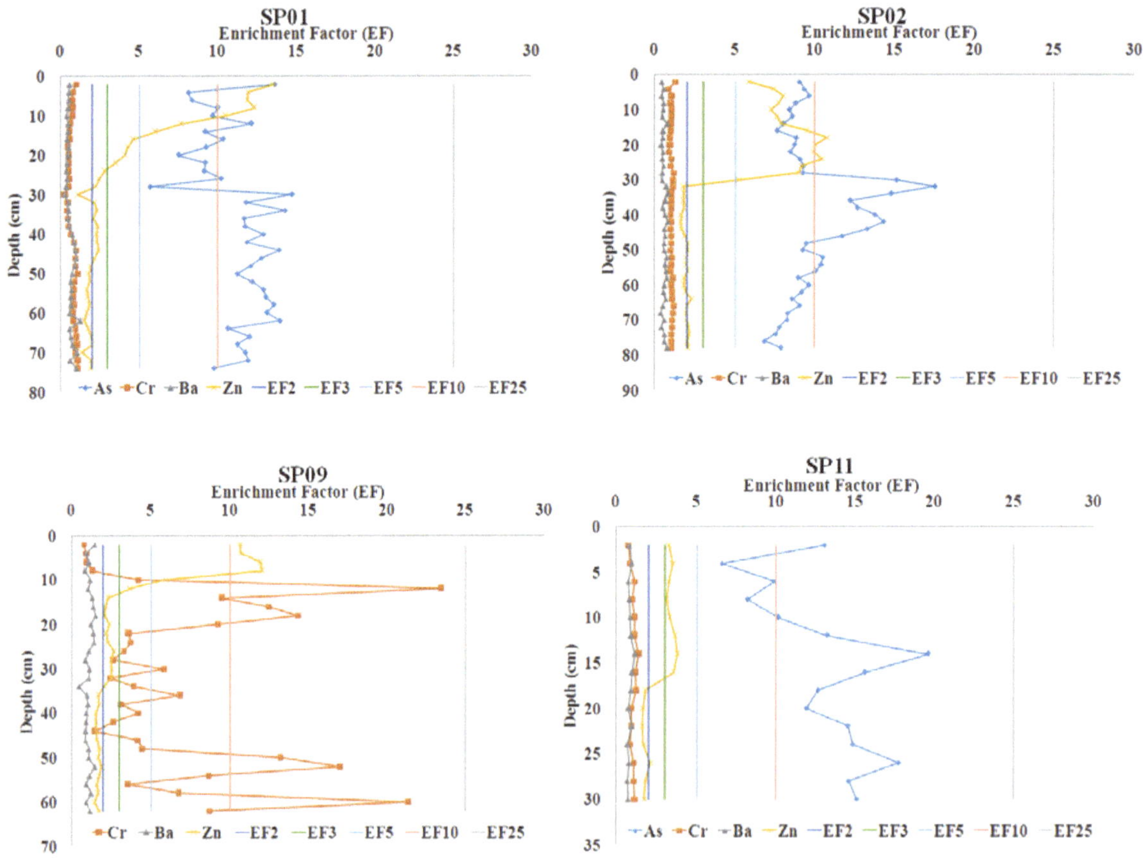

FIG. 6.12. Elemental Enrichment Factor versus depth for core SP01, SP02, SP 09 and SP11, respectively.

TABLE 6.7. SUMMARY OF CORRELATION ANALYSIS OF TRACE ELEMENT CONCENTRATIONS WITH THE PERCENTAGE OF ORGANIC MATTER AND PARTICLE SIZE FRACTIONS IN SEDIMENTS FROM JURU AND PERAI RIVERS, MALAYSIA

Station	Element	Correlation coefficient (r)			
		Organic (%)	Clay (%)	Silt (%)	Sand (%)
SP 01	As	0.32	0.22	-0.09	-0.13
	Ba	-0.05	0.00	0.24	-0.14
	Cr	**0.46**	-0.10	-0.17	0.19
	Fe	**0.69**	**0.59**	-0.12	**-0.43**
	Zn	**0.87**	-0.01	-0.05	0.04
SP 02	As	-0.20	-0.31	**-0.45**	**0.45**
	Ba	-0.21	**-0.37**	-0.20	0.31
	Cr	-0.09	**0.34**	0.26	**-0.34**
	Fe	0.31	**0.54**	**0.59**	**-0.66**
	Zn	**0.71**	**0.59**	**0.66**	**-0.73**
SP 09	Ba	-0.01	0.03	-0.15	0.09
	Cr	**-0.47**	-0.30	0.03	0.28
	Fe	0.10	0.13	0.24	-0.33
	Zn	**0.93**	**0.69**	**-0.43**	-0.35
SP 11	As	**-0.59**	**-0.57**	-0.30	**0.65**
	Ba	0.26	0.31	0.27	-0.42
	Cr	0.07	0.04	0.06	-0.07
	Fe	**0.67**	**0.65**	0.37	**-0.76**
	Zn	**0.74**	**0.74**	0.19	**-0.74**

Note: r values in Bold are significant at 95%, "-" value indicate negative correlation.

6.4. CONCLUSION

During dry season, at the Perai Industrial Area, two sediments cores were collected from both two main rivers (Perai and Juru river) adjacent to the area. These cores were analysed using three nuclear techniques, in order to determine the sediment accumulation rates and to assess the enrichments by arsenic, barium, chromium and zinc at the study area. Concentrations of arsenic, barium, chromium and zinc were quantified using Neutron Activation Analysis (NAA) technique and complimented with Energy Dispersive X-ray Fluorescence Spectrometer (EDXRF). By using the gamma spectrometry system, ^{210}Pb and ^{226}Ra activity profiles were established for each individual core; unfortunately, due to the complexity of the excess ^{210}Pb activity profiles found in this study and lack of supporting ^{137}Cs activity profiling, an accurate age model was not able to be produced from them. However, preliminary apparent sediment can still be calculated and found to be between the range of 0.94 – 2.70 cm/y with the core collected from the Juru River's estuary giving the highest sedimentation rate. The level of trace element enrichment by As, Ba and Zn was comparable among the cores from Perai and Juru rivers; but significantly higher Cr was found upriver at Perai River. Considering that high enrichment levels of As, Cr and Zn observed in the area, further monitoring and mitigation strategies for trace element contamination are needed.

6.5. ACKNOWLEDGEMENTS

The authors like to thank the International Atomic Energy Agency for funding this study under research grant (K41016 – RC 20884). The authors would also like to thank the Technical Officer, Ms. Martina Rozmaric Macefat and Member States participating the K41016 project for their guidance and are also grateful for the support and cooperation given by staff from Malaysian Nuclear Agency throughout this study.

6.6. REFERENCES

[6.1] HUMBIRD, R., The Biosphere, (2019), access at https://cnx.org/contents/OkLEsdYf@1.1:OyQ8wf02@2/THE-BIOSPHERE.

[6.2] BURKE et al., AAAS Atlas of Population and Environment, AAAS, University of California Press, Berkeley (2001).

[6.3] UNITED NATION, "Factsheet: People and Oceans", The Ocean Conference, United Nations, New York (2017).

[6.4] COCHRAN, J.K., MASQUE, P., "Natural radionuclides applied to coastal zone process", Marine Radioactivity Elsevier, Amsterdam (2004)

[6.5] JELEFF, S., OCEANS, ISBN 92-871-3876-1, (1999), access at https://books.google.com.my/books?id=edZm0Lb6xCAC&pg=PA169&lpg=PA169&dq=Jeleff+ocean&source=bl&ots=lCgPszVP0l&sig=ACfU3U2iTNztAfvY52YlHPgLFhdecmmZ2A&hl=en&sa=X&ved=2ahUKEwij9sOJ56bjAhUF6nMBHV_SDnQQ6AEwCHoECAkQAQ#v=onepage&q=Jeleff%20ocean&f=false.

[6.6] ISLAM, M.A., AL-MAMUN, A., HOSSAIN, F., QURAISHI, S.B., NAHER, K., KHAN, R., DAS, R., TAMIM, U., HOSSAIN, S.M., NAHID, F., Contamination and ecological risk assessment of trace elements in sediments of the rivers of Sundarban mangrove forest, Bangladesh, Mar. Pollut. Bull. **124** 1 (2017) 356–366.

[6.7] LI, H., YE, S., YE, J., FAN, J., GAO, M., GUO, H., Baseline survey of sediments and marine organisms in Liaohe Estuary: heavy metals, Polychlorinated biphenyls and organochlorine pesticides, Mar. Pollut. Bull. **114** 1 (2017) 555–563.

[6.8] ELIAS, M.S., HAMZAH, M.S., RAHMAN, S.A., WEE, B.S., SALIM, N.A.A., Assessment of sediment quality collected from Tunku Abdul Rahman Park, Sabah, Nucl. Sci. J. Malays. **24** 1 (2012) 59–70.

[6.9] YI, Y.J., YANG, Z.F., ZHANG, S.H., Ecological risk assessment of heavy metals in sediment and human health risk assessment of heavy metals in fishes in the middle and lower reaches of the Yangtze River basin, Environ. Pollut. **159** 10 (2011) 2575–2585.

[6.10] PANDEY, J., SINGH, R., Heavy metals in sediments of Ganga River: up and down stream urban influences, Appl. Water Sci. **7** (2017) 1669–1678.

[6.11] ZHU, L.M., XU, J., WANG, F., LEE, B., An assessment of selected heavy metal contamination in the surface sediments from the South China Sea before 1998, J. Geochem. Explor., **108** 1 (2011) 1–14.

[6.12] JOSEPH, B., ADIANA, G., SHAZILI, N.A.M., ONG, M.C., JUAHIR, H., SHAARI, H., YII, M.W., KAMARUDIN, M.K.A., GASIM, M.B., MAULUD, K.N.A., ISLAM, M.S., The Evaluation of Brunei Bay Sediment Cores Sedimentation Rate Using ^{210}Pb Radiometric Dating Technique, Int. J. Res. Eng. Technol. **7** 3.14 (2018) 107–114.

[6.13] KRISHNASWAMI, S., LAL, D., “Radionuclide alimnochronology”, LAKES, CHEMISTRY, GEOLOGY, PHYSICS, Springer Verlag, New York (1978).

[6.14] CHAKRABARTY, A., JHA, S.K., GOTHANKAR, S.S., TRIPATHI, R.M., KHAN, A.H., PURANIK, V.D., “Sedimentation rate of Nagarjuna Sagar Dam using a natural radiotracer”, Application of Radiotracers in Chemical, Environmental, and Biological Sciences, Vol. 2, Saha Institute of Nuclear Physics, Kolkata (2006).

[6.15] ALIEV, R.A., BOBROV, V.A., KALMYKOV, St.N., MELGUNOV, M.S., VLASOVA, I.E., SHEVCHENKO, V.P., NOVIGATSKY, A.N., LISITZIN, A.P., Natural and artificial radionuclides as a tool for sedimentation studies in the Arctic region, J. Radioanal. Nucl. Chem. **274** (2007) 315–321.

[6.16] LIKUKU, A.S., Factors influencing concentrations of ^{210}Pb and ^{7}Be over the city of Edinburgh (55.9°N, 03.2°W), J. Environ. Radioact. **87** 3 (2006) 289–304.

[6.17] SANCHEZ-CABEZA, J.A., RUIZ-FERNANDEZ, A.C., ^{210}Pb sediment radiochronology: an integrated formulation and classification of dating models, Geochim. Cosmochim. Acta **82** (2012) 183–200.

[6.18] WAN MAHMOOD, Z., YII, M.W., ISHAK, A.K., ^{210}Po, ^{210}Pb and ^{210}Po/^{210}Pb in sediment core from surrounding Sungai Linggi estuary, Nucl. Sci. J. Malays. **28** 1 (2016) 1–8.

[6.19] ELIAS, M.S. et al., Development of Fingerprint Characteristic for forensic Investigation of Elemental Pollution in Sediment from Linggi Area, NUKLEARMALAYSIA/L/2014/102, internal report, Malaysian Nuclear Agency, Kajang, 2014.

[6.20] IAEA, PRACTICAL ASPECTS OF OPERATING A NEUTRON ACTIVATION ANALYSIS LABORATORY, TECDOC-564, The International Atomic Energy, Vienna (1990).

[6.21] ZAINUDIN, Z., Benchmarking River Water Quality in Malaysia, Jurutera (2010) 12–15.

[6.22] IAEA, COLLECTION AND PREPARATION OF BOTTOM SEDIMENT SAMPLES FOR ANALYSIS OF RADIONUCLIDES AND TRACE ELEMENTS, TECDOC-1360, The International Atomic Energy, Vienna (2003).

[6.23] KRUMBEIN, W.C., ABERDEEN, E., The Sediments of Barataria Bay, J. Sediment Petrol. **7** 1 (1937) 3–17.

[6.24] NELSON, D.W., SOMMERS, L.E., “Total carbon, organic carbon, and organic matter”, Methods of Soil Analysis, Part 2, Chemical and Microbiological Properties, 2nd Edition, Series NO. 9, ASA SSSA, Madison, Wisconsin (1982).

[6.25] YII, M.W., AHMAD, Z., ISHAK, A.K., Distribution of naturally occurring radionuclides activity concentration in East Malaysian marine sediment, Appl. Radiat. Isot. **67** 4 (2009) 630–635.

[6.26] YII, M.W. et al., Ambient radioactivity and radiological studies in the vicinity of Lynas Rare-Earth Processing Plant, Gebeng Industrial Estate, Kuantan, Pahang, NUCLEARMALAYSIA/L/2016/96, internal report, Malaysian Nuclear Agency, Kajang, 2016.

[6.27] DOWDALL, M., O'DEA, J., Ra-226/^{238}U disequilibrium in an upland organic soil exhibiting elevated natural radioactivity, J. Environ. Radioact. **59** 1 (2002) 91–104.

[6.28] YANG, Y.X., WU, X.M., JIANG, Z.Y., WANG, W.X., LU, J.G., LIN, J., WANG, L.M., HSIA, Y.F., Radioactivity concentrations in soils of the Xiazhuang granite area, China, Appl. Radiat. Isot. **63** 2 (2005) 255–259.

[6.29] EL-REEFY, H.I., SHARSHAR, T., ZAGHLOUL, R., BADRAN, H.M., Distribution of gamma-ray emitting radionuclides in the environment of Burullus Lake: I. Soils and vegetation, J. Environ. Radioact. **87** 2 (2006)148–169.

[6.30] AHMED, N.K., EL-ARABI, A.G.M., Natural radioactivity in farm soil and phosphate fertilizer and its environmental implications in Qena governorate, upper Egypt, J. Environ. Radioact. **84** 1 (2005) 51–64.

[6.31] AROGUNJO, A.M., OFUGA, E.E., AFOLABI, M.A., Levels of natural radionuclides in some Nigerian cereals and tubers, J. Environ. Radioact. **82** 1 (2005)1–6.

[6.32] MISHRA, U.C., SADASIVAN, S., Gamma spectroscopic measurements of soil radioactivity, Int. J. Appl. Radiat. Isot. **22** (1971) 256–257.

[6.33] CHEN, S.B., ZHU, Y.G., HU, Q.H., Soil to plant transfer of ^{238}U, ^{226}Ra and ^{232}Th on a uranium mining-impacted soil from south eastern China, J. Environ. Radioact. **82** 2 (2005) 223–236.

[6.34] ASHRAF, A.R., SAION, E., GHARIBSHAHI, E., YAP, C.K., KAMARI, H.M., ELIAS, M.S., RAHMAN S.A., Distribution of heavy metals in core marine sediments of coastal East Malaysia by instrumental neutron activation analysis and inductively coupled plasma spectroscopy, Appl. Radiat. Isot. **132** (2018) 222–231.

[6.35] ELIAS, M.S., IBRAHIM, S., SAMUDING, K., RAHMAN, S.A., YII, M.W., Assessment of toxic elements in sediments of Linggi river using NAA and ICP-MS techniques, Methods X **5** (2018) 454–465.

[6.36] ALNOUR, I.A., WAGIRAN, H., IBRAHIM, N., HAMZAH, S., WEE, B.S., ELIAS, M.S., New approach for calibration the efficiency of HPGe detectors, AIP Conf. Proc. **1584** (2014) 38–44.

[6.37] SAION, E., WOOD, A.K.H., SULAIMAN, Z.A., ALZAHRANY, A.A., ELIAS, M.S., WEE, B.S., Determination of heavy metal pollution in depth profile of marine sediment samples from the Strait of Malacca, Fresenius Environ. Bull. **16** 10 (2007) 1279–1287.

[6.38] WEDEPOHL, K.H., The composition of the continental crust, Geochim. Cosmochim. Acta **59** 7 (1995) 1217–1232.

[6.39] LORING, D.H., RANTALA, R.T.T., Manual for the geochemical analyses of marine sediments and suspended particulate matter, Earth-Sci. Rev. **32** 4 (1992) 235–283.

[6.40] DEELY, J.M., FERGUSSON, J.E., Heavy metal and organic matter concentration and distributions in dated sediments of a small estuary adjacent to a small urban area, Sci. Total Environ. **153** 1–2 (1994) 97–111.

[6.41] ABRAHIM, G.M.S., PARKER, R.J., Assessment of heavy metal enrichment factors and the degree of contamination in marine sediments from Tamaki Estuary, Auckland, New Zealand, Environ. Monit. Assess. **136** (2008) 227–238.

[6.42] NIENCHESKI, L.F., WINDOM, H.L., SMITH, R., Distribution of particulate trace metal in Patos Lagoon estuary (Brazil), Mar. Pollut. Bull. **28** 2 (1994) 96–102.

[6.43] SZEFER, P., SZEFER, K., GLASBY, G.P., PEMPKOWIAK, J., KALISZAN, R., Heavy – metal pollution in surficial sediments from the Southern Baltic Sea off Poland, J. Environ. Sci. Health **31** 10 (1996) 2723–2754.

[6.44] AYARI, J., AGNAN, Y., CHAREF, A., Spatial assessment and source identification of trace metal pollution in stream sediments of Oued El Maadene basin, northern Tunisia, Environ. Monit. Assess. **188** 7 (2016) 397–408.

[6.45] WAN MAHMOOD, Z., YII, M.W., Marine radioactivity concentration in the Exclusive Economic Zone of Peninsular Malaysia: ^{226}Ra, ^{228}Ra and ^{228}Ra/^{226}Ra, J. Radioanal. Nucl. Chem. **292** (2012) 183–192.

[6.46] WAN MAHMOOD, Z., AHMAD, Z., ISHAK, A.K., YII, M.W., MOHAMED, N., SHARIB, J., ISHAK, K., RAZALI, K.N., MAHMUD, M., Kajian Awal Ke Atas Taburan Radionuklid Tabii Di Perairan Pantai Timur Semenanjung Malaysia, Malaysian J. Anal. Sci. **9** 2 (2005) 325–337. ISSN: 1394–2506. (in Malays)

[6.47] NGUYEN, T.N., NGUYEN, T.B., NGUYEN, V.P., LE, N.S., TRUONG, Y., MAI, T.H., NGUYEN, T.L., NGUYEN, M.S., PHAN, S.H., LE, N.C., DANG, D.N., NGUYEN, Q.L., NGUYEN, H.Q., TRAN, T.M., Radionuclides concentration in marine environmental samples along the coast of Vietnam, Nucl. Sci. J. Malays. **21** 2 (2009) 12–20.

[6.48] JAMES, W.D., BOOTHE, P.N., PRESLEY, B.J., Compton suppression gamma-spectroscopy in the analysis of radium and lead isotopes in ocean sediments, J. Radioanal. Nucl. Chem. **236** (1998) 261–266.

[6.49] RUIZ-FERNANDEZ, A.C., SANCHEZ-CABEZA, J.A., ALONSO-HERNANDEZ, C., MARTINEZ-HERRERA, V., PEREZ-BERNAL, L.H., PREDA, M., HILLAIRE-MARCEL, C., GASTAUD, J., QUEJIDO-CABEZAS, A.J., Effect of land use change and sediment mobilization on coastal contamination (Coatzacoalcos River, Mexico), Cont. Shelf Res. **37** (2012) 57–65.

[6.50] SANDERS, C.J., SANDERS, L.M., SMOAK, J.M., PATCHINEELAM, S., MACHADO, W., LUIZ-SILVA, W., "Radium-226 and lead-210 ratios along a sediment core profile implying fertilizer industry source", AGU Meeting of the Americas, Foz do Iguacu (2010).

[6.51] CUCULIC, V., CUKROV, N., BARISIC, D., MLAKAR, M., Uranium in sediments, mussels (Mytilus sp.) and seawater of the Krka River estuary, J. Environ. Radioact. **85** 1 (2006) 59–70.

[6.52] SKWARZEC, B., JAHNZ, A., The inflow of polonium ^{210}Po from Vistula River catchments area, J. Environ. Sci. Health Part A **42** 14 (2007) 2117–2122.

[6.53] LI, Y.H., MATHIEU, G.G., BISCAYE, P., SIMPSON, H.J., The flux of ^{226}Ra from estuarine and continental shelf sediments, Earth Planet. Sci. Let. **37** 2 (1977) 237–241.

[6.54] YII, M.W., WAN MAHMOOD, Z., SHARIB, J., AHMAD, Z., Comparison of ^{210}Pb level in Kuala Muda tsunami affected marine sediment core measured using two different techniques, Nucl. Sci. J. Malays. **22** 1 (2010) 29–41.

[6.55] BRALOWER, T.J., THIERSTEIN, H.R., "Organic carbon and metal accumulation in Holocene and mid-Cretaceous marine sediments: Paleoceanographic significance", MARINE PETROLEUM SOURCE ROCKS, Special volume 26, Blackwell Scientific Publications, Oxford London Edinburgh (1987).

[6.56] ONTIVEROS-CUADRAS, J.F., RUIZ-FERNÁNDEZ, A.C., SANCHEZ-CABEZA, J.A., WEE-KWONG, L.L., PÉREZ-BERNAL, L.H., Geochemical fractionation of ^{210}Pb in oxic estuarine sediments of Coatzacoalcos River, Gulf of Mexico, J. Radioanal. Nucl. Chem. **292** (2012) 947–956.

[6.57] BENOIT, G., "The biogeochemistry of ^{210}Po and ^{210}Pb in fresh waters and sediments", Ph.D. Dissertation, Massachusetts Institute of Technology, Cambridge, Massachusetts (1988).

[6.58] UNESCO, UNESCO/SCOR WORKSHOP ON THE BIOGEOCHEMISTRY OF ESTUARINE SEDIMENTS, UNESCO, New York (1978).

[6.59] ALONGI, D.M., PFITZNER, J., TROTT, L.A., TIRENDI, F., DIXON, P., KLUMPP, D.W., Rapid sediment accumulation and microbial mineralization in forests of the mangrove Kandelia candel in the Jiulongjiang Estuary, China, Estuar. Coast. Shelf Sci. **63(4)** (2005) 605–618.

[6.60] KRISHNASWAMY, S., LAL, D., MARTIN, J., MEYBECK, M., Geochronology of lake sediments, Earth Planet. Sci. Lett. **11(1)** (1971) 407–414.

[6.61] WAN MAHMOOD, Z., YII, M.W., Sedimentation rate in the Sungai Linggi estuary using excess ^{210}Pb and ^{137}Cs. J. Radioanal. Nucl. Chem. **298** (2013) 1727–1732.

[6.62] CHEEVAPORN, V., JACINTO, G.S., SANDIEGO-McGLONNE, M.L., History of heavy metal contamination in Bang Pakong River estuary, Thailand, J. Sci. Soc. Thailand **20** (1994) 9–22.

[6.63] XU, Z., SALEM, A., CHEN, Z.Y., ZHANG, W.G., CHEN, J., WANG, Z.H., SUN, Q.L., YIN, D.W., Pb-210 and Cs-137 distribution in Burullus lagoon sediments of Nile River delta, Egypt: sedimentation rate after Aswan High Dam, Front Earth Sci-PRC. **2** (2008) 434–438.

[6.64] WOOD, A.K., MUHAMMAD, N., MAHMOOD, C.S., AHMAD, Z., SHAZILI, N.A., LAW, A.T., YAAKOB, R., Corer sampling and the use of neutron activation analysis in evaluating pollution at the Juru waterway, Penang, Nucl. Sci. J. Malays. **11(2)** (1993) 105–128.

[6.65] LIM, P.E., KIU, M.Y., Determination and speciation of heavy metals in sediments of the Juru River, Penang, Malaysia, Environ. Monit. Assess. **35** (1995) 85–95.

[6.66] YII, M.W., WAN MAHMOOD, Z., ELIAS, M.S., ABDULLAH, Y., The distribution of heavy metals and natural radionuclides within the surface sediments of the Juru river, Penang, AIP Conf. Proc. **2295** (2020) 020006.

[6.67] SHAZILI, N.A.M., YUNUS, K., AHMAD, A.S., ABDULLAH, N., Heavy metal pollution status in the Malaysian aquatic environment, Aquat. Ecosys. Health **9(2)** (2006) 137–145.

[6.68] SARMANI, S.B., "The determination of heavy metals in water, suspended materials and sediments from Langat River, Malaysia", SEDIMENT/WATER INTERACTIONS: DEVELOPMENTS IN HYDROBIOLOGY, VOL 50, Springer, Dordrecht (1989).

[6.69] AGENCY FOR TOXIC SUBSTANCES & DISEASE REGISTRY (ATSDR), ENVIRONMENTAL HEALTH AND MEDICINE EDUCATION, (2008), access at https://www.atsdr.cdc.gov/csem/csem.asp?csem=10&po=5.

7. STUDY ON THE SEDIMENTATION RATE AND CONTAMINATION HISTORY IN THE HA LONG BAY, VIETNAM

B.D. DUNG, D.D.THANG
Institute for Nuclear Science and Technology
Hanoi, Vietnam
Email: dacdung@gmail.com

J.M. GODOY
Chemistry Department
Pontifícia Universidade Católica do Rio de Janeiro, Brazil

P.G. APPLEBY
Department of Mathematical Sciences
University of Liverpool, United Kingdom

Abstract

The study was conducted on coastal areas of Ha Long Bay, the north-west part of the Gulf of Tonkin (Vietnam). The aim of the study was to find suitable dating method in using ^{210}Pb and ^{137}Cs techniques for sediment chronology, and to evaluate the temporal trends of heavy metals and nutrients inputs into the Bay during the last century. Two sediment cores were taken and analysed for excess ^{210}Pb and ^{137}Cs. Several ^{210}Pb dating models were applied for sediment chronology, and it was found that the CIC model would not be suitable for dating the cores. CF:CS model was also found not applicable to either of the cores as a whole because the model would give results with large uncertainties. CF:CS model, however, could be applied to estimate initial $^{210}Pb_{ex}$ concentration and the missing $^{210}Pb_{ex}$ inventory below the bottom layer. CRS model was used in combination with the use of ^{137}Cs fallout peak as an independent reference date. With the assumption that fallout ^{137}Cs in sediments was not redistributed by post-depositional processes, the use of the 1964 date for the ^{137}Cs fallout peak for correcting CRS model data could be the most suitable for dating of the two cores. Sedimentation rates in the survey area were found in the range from 0.17 to 1.08 g cm^{-2} y^{-1} (0.17 to 1.75 cm yr^{-1}). The sedimentation rate in the Bay has been increasing in recent years, possibly could be due to agricultural and coal mining activities as well as urbanization along the coast. Coastal sediments were found to be below regulatory standards for heavy metal concentrations. However, TOC, N, and P profiles show increasing concentrations trend since 1990 onward, which could be due to increased agricultural activities in the continent along the Bay's coast.

7.1. INTRODUCTION

The marine environment in many areas of the world, including Vietnam, is facing many challenges and threats, aggravated by the increasing coastal population and growing coastal economic activities [7.1]. Many coastal estuaries are polluted by wastewater discharged from industrial parks and urban areas [7.2-7.3]. Chemical and heavy metal pollution is one of the dangers to the habitats of invertebrates, fish and humans [7.4].

The Ha Long Bay, located in the north-west part of the Gulf of Tonkin, is a UNESCO World Heritage Site and a popular travel destination. The Bay has an area of about 43,400 ha and over 1600 islands and islets, most of which are uninhabited and unaffected by humans [7.5]. In recent years, the Bay is threated by high levels of pollution. Tourism growth and industry have resulted in water contamination and solid waste accumulation, damaging the ecosystem and affecting visitor experience [7.6]. The landscape of the Bay is devastated by many negative impacts, one of which is the shallowing of the Bay. Along the Bay's coast, there are large scale open-pit coal mines with large amount of waste soil and rock materials, most are dumped near the mining sites on the top of the watersheds. Due to large steep terrains on the dumping sites, the risk of landslides and mud flooding is very high, often threatening the neighbouring residential areas and directly affecting the Bay's coast [7.7].

Nuclear techniques and isotopic techniques have been widely used to study the accumulation and transport of pollutants in and through the marine environment. The excess ^{210}Pb technique has been recognized and applied to evaluate the sedimentation rate. In addition, the ^{137}Cs technique has also been used to verify the age of sediments determined by the ^{210}Pb technique. The combination of the excess ^{210}Pb with ^{137}Cs techniques has become popular and has played an important role in determining the sedimentation rate of modern sediments over the past 100 years. Based on the sedimentation rate over time, it is possible to reconstruct the history of heavy metal pollution and other pollutants [7.8-7.12].

Previous studies (Fig. 7.1) showed that tidal flats in the north of Vietnam were impacted by land-ocean interaction processes and human activity [7.13-7.15]. The flats were experiencing erosion in some parts and accretion in others, and the sedimentation rates were in the range from 0.14 to 3.04 cm yr^{-1} [7.13]. The sediment core taken in Ha Long Bay coast (core MC3 in Fig. 7.1) had low sedimentation rate, averaging 0.14±0.11 cm yr^{-1} [7.13]. In the central water area of the Ha Long Bay, two sediment cores (HP16 and HP 21 in Fig. 7.1) were taken to the depths of 27 cm (HP16) and 62 cm (HP21) by a piston corer [7.14]. Sediment rates were calculated using the ^{210}Pb and ^{137}Cs techniques, and were found to be in the range of 0.41 - 0.75 cm yr^{-1} [7.14]. In another study, sediment cores at the estuary of the Binh Huong River, which discharges to the Ha Long Bay, and one sediment core at the area near the harbour of Cam Pha (QL in Fig. 7.1), were collected [7.15]. Results showed that the sedimentation rates ranged from 0.3 to 1.2 cm yr^{-1} with an average of 1.0 cm yr^{-1} [7.15].

Nevertheless, the use of nuclear and isotopic techniques in environmental studies in Vietnam is still in the initial stage, and the advantages of these techniques for real environment problems in Vietnam are not yet fully exploited. Most previous studies in Vietnam were not paying due attention to influencing factors such as sedimentation process and dynamics of the study area, so the criteria for sampling location selection and sampling techniques as well as models for calculating sedimentation rate have not been fully studied.

The aim of this study was to find a suitable dating method using ^{210}Pb and ^{137}Cs techniques for sediment chronology, and to evaluate the temporal trends of heavy metal and nutrient inputs into the Bay during the last century.

7.2. MATERIAL AND METHODS

New approaches to sediment sampling have been adapted. The sampling site was chosen in the undisturbed by modern human development coastal area. Fig. 7.1 depicts a map of the study area in which the sampling locations of past research [7.13-7.15] are also shown. The sampling sites were designed with the aim to elucidate the influence of the openings to and out the Bay to the sedimentation process. Two sediment cores were taken in March 2019 using a 6 cm diameter transparent acrylic corer.

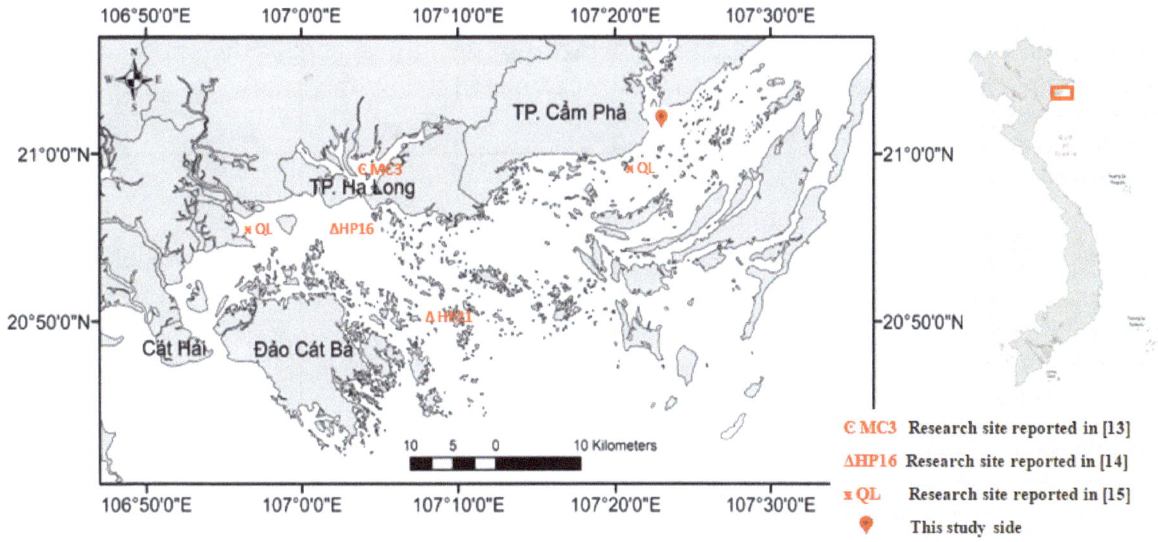

FIG. 7.1. The study site (base map source: 1:10000 scale Vietnam digital map).

The sediment cores were cut into slices of 2 cm in the field. Sliced samples were analysed for ^{210}Pb and ^{226}Ra to determine the excess ^{210}Pb ($^{210}Pb_{ex} = {}^{210}Pb - {}^{226}Ra$) and for ^{137}Cs. Sliced samples were also analysed for heavy metals, N, P and TOC as described in [7.15]. $^{210}Pb_{ex}$ values were used to determine the sediment age for each core by the following steps:

(a)Using the Constant Flux Constant Sedimentation Rate (CF:CS) model to determine the mean sedimentation rate. This model is based on two assumptions: the mass accumulation rate of each layer is constant and the $^{210}Pb_{ex}$ flux to the sediment surface is also constant. When these are met, $^{210}Pb_{ex}$ concentration would decrease exponentially with increasing mass depth by the following equation [7.12]:

$$\ln C_i = \ln C_0 - \frac{\lambda}{r} m_i \tag{7.1}$$

Where:

- C_i: $^{210}Pb_{ex}$ concentration in layer i (Bq kg^{-1});
- C_0: initial $^{210}Pb_{ex}$ concentration when layer i is formed (Bq kg^{-1});
- r: mass accumulation rate (kg m^{-2} yr^{-1});
- λ: ^{210}Pb disintegration constant (0.03114 yr^{-1});
- m_i: mean mass depth of layer i (kg m^{-2}).

By graphing in MS Excel the ln($^{210}Pb_{ex}$) versus accumulated mass depth, the trend line equation in the form of $y = a + bx$ gives $C_0 = e^a$, and $r = -\lambda/b$ [7.12].

(b)Applying CF:CS model for the segment of surface layers to estimate initial $^{210}Pb_{ex}$ concentration C_0 [7.12]. Using Constant Initial Activity (CIC) model that is based on the assumption that sediments have a constant initial $^{210}Pb_{ex}$ concentration C_0 regardless of accumulation rate. The age of each sediment slice is calculated the by the following equation [7.12, 7.16]:

$$t(i) = \frac{1}{\lambda} \ln \frac{C_o}{C_i} \tag{7.2}$$

Where $t(i)$ is time elapsed since formation of layer (i) (Bq m^{-2}).

(c)Applying CF:CS model for the segment of bottom layers to estimate the missing $^{210}Pb_{ex}$ inventory A(j) below the bottom layer [7.12]. The core $^{210}Pb_{ex}$ inventory (Bq m^{-2}) $A(0)$ is calculated by summing the inventory from the accumulated deposit to the incomplete core bottom and A(j) [7.12]. Using Constant Supply Rate (CRS) model that is based on the assumption that the rate of deposition of $^{210}Pb_{ex}$ from the atmosphere to the sediment surface is constant. The age of each sediment layer is calculated the by the following equation [7.12, 7.16]:

$$t(i) = \frac{1}{\lambda} \ln \frac{A(0)}{A(i)} \tag{7.3}$$

Where $A(i)$ is accumulated $^{210}Pb_{ex}$ deposit below layer (i) (Bq m^{-2});

(d)Using the Composite model [7.16] to correct ^{210}Pb dating by CRS model results if reference dates are known. In this study, the ^{137}Cs fallout peak has been tried to use as an independent reference date.

(e)Determining the mass accumulation rate and sediment accumulation rate for each sediment slice.

7.3. RESULTS AND DISCUSSIONS

Fig. 7.2 shows $^{210}Pb_{ex}$ and ^{137}Cs activities versus depth in the cores VD01 and VD02. The first core VD01 was taken from a shallow water depth of 1.6 m and had length of 60 cm. The VD02 core was taken from a deeper water depth of 3.5 m and had length of 38 cm. In both cores, $^{210}Pb_{ex}$ concentrations vary irregularly with depth. In VD01 they are relatively constant throughout the uppermost 20 cm of the core. Below that there is a significant non-monotonic feature between 24-34 cm. At greater depths the decline in $^{210}Pb_{ex}$ concentrations is more regular, though a steady increase in the gradient suggests a significantly lower sedimentation rate towards the base of the record.

It is evident from these observations that the CIC and CF:CS models are not applicable to either of the cores as a whole. The $^{210}Pb_{ex}$ concentration profiles do not decrease monotonically and the application of CIC model by equation (7.2) would give deeper layers with higher concentrations younger ages than the underlying layers [7.12]. The natural logarithmic graphs of $^{210}Pb_{ex}$ against accumulated mass are also not linear, and the CF:CS model would give results with large uncertainties. CF:CS model, however, could be applied for linear segments such as the segment of surface layers to estimate initial $^{210}Pb_{ex}$ concentration, and the segment of bottom layers to estimate the missing $^{210}Pb_{ex}$ inventory below the bottom layer [7.12]. Fig. 7.3 shows the results of applying the CF:CS model to the record below 50 cm of core VD01, which suggests a basal sedimentation rate (from the 1930s through to the early 1940s) of 0.31 ± 0.01 g cm^{-2} yr^{-1} (0.43 cm yr^{-1}). Similarly, the CF:CS model for the section below 25 cm of core VD02 gives a sedimentation rate of 0.21 ± 0.02 g cm^{-2} yr^{-1} (0.24 cm yr^{-1}). The missing $^{210}Pb_{ex}$ inventory A(j) below the bottom layer of each core was also derived from the regression equation (Fig. 7.3) and was used in CRS dating model by equation (7.3).

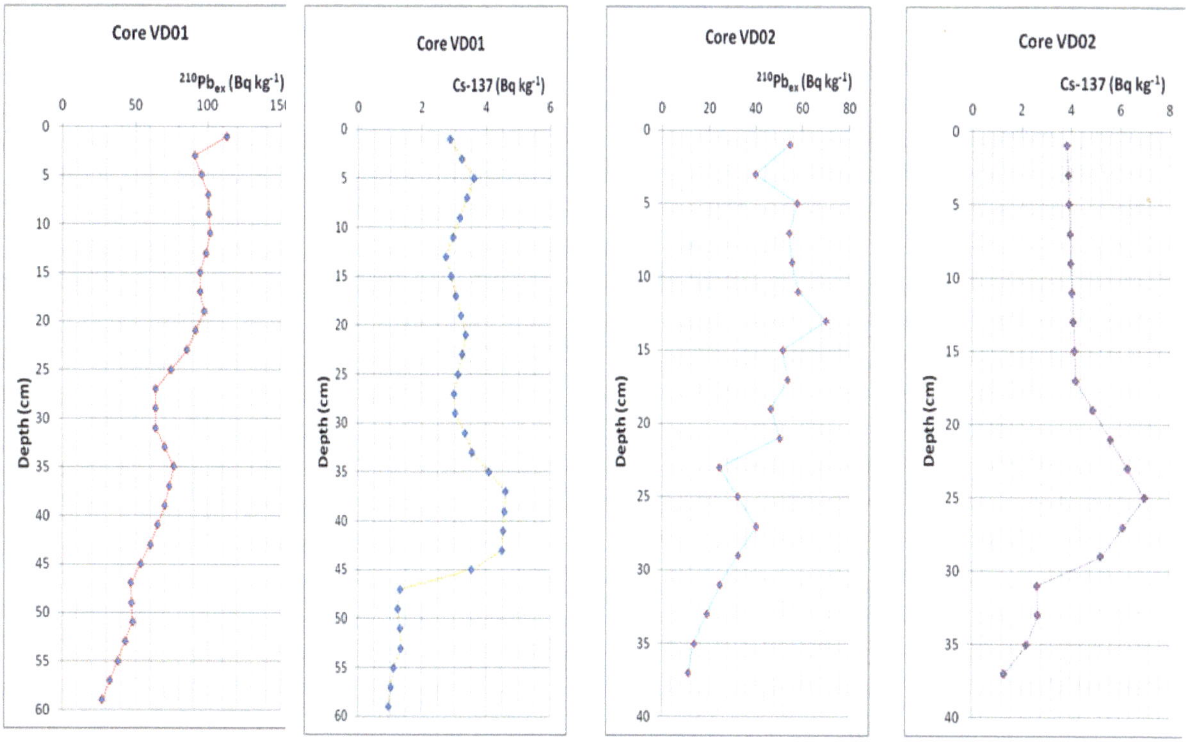

FIG. 7.2. Variation of ^{137}Cs and ^{210}Pb$_{ex}$ activity concentrations along the depth of sediment cores VD01 and VD02.

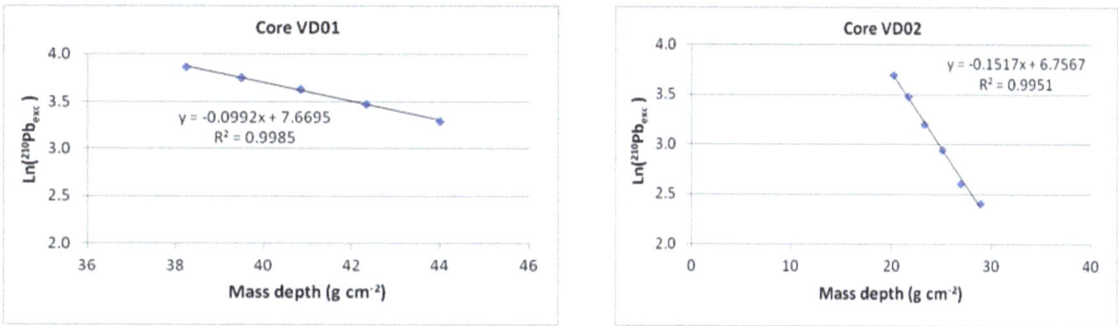

FIG. 7.3. Results of applying the CF:CS model to the record below 50 cm of core VD01 and below 25 cm of core VD02.

^{137}Cs dating technique has been used widely for independent validation of the ^{210}Pb chronology [7.16-7.18]. In this study, it is assumed that fallout ^{137}Cs in sediments was not redistributed by post-depositional processes such as human activities, diffusion within pore-waters, and only a small portion of ^{137}Cs concentration has been diffused downward. Fig. 7.2 demonstrates that the ^{137}Cs concentrations in VD01 have a broad but reasonably distinct peak between 36-44 cm that may record the early 1960s fallout peak from the atmospheric testing of nuclear weapons [7.17]. Assuming a date of 1964 for the ^{137}Cs fallout peak at depth 41 cm, this feature suggests a mean post-1964 sedimentation rate of 0.57 ± 0.10 g cm^{-2} yr^{-1} (0.74 cm yr^{-1}). The ^{137}Cs concentrations in VD02 also have a high peak at 25 cm, and by assigning a date of 1964 for the ^{137}Cs fallout peak, mean post-1964 sedimentation rate of VD02 could be estimated as 0.34 $\pm$ 0.08 g cm^{-2} yr^{-1} (0.45 cm yr^{-1}).

The ^{210}Pb inventories suggest ^{210}Pb supply rates of more than 1000 Bq m^{-2} yr^{-1} in VD01, and around 400 Bq m^{-2} yr^{-1} in VD02. Since these values are many times higher than the probable atmospheric flux, and it follows that the dominant processes controlling the ^{210}Pb supply are likely to be sediment transport via focussing or from catchment inputs, the Constant Rate of Supply (CRS) model needs to be applied with caution, particularly since the record is not complete. Nevertheless, the raw CRS model dates for VD01 are in relatively good agreement with the ^{137}Cs record in that they place 1964 at a depth of around 43 cm. The discrepancy is due to an apparent decline in the ^{210}Pb supply rate, from a pre-1960 value of ~1400 Bq m^{-2} yr^{-1} to a post-1960 value of ~1000 Bq m^{-2} yr^{-1}. This could be due to a change in the sedimentary processes delivering ^{210}Pb to the core site, or sediment mixing enriching older sediment with modern ^{210}Pb. However, the existence of the non-monotonic ^{210}Pb feature between 24-34 cm and maintenance of the ^{137}Cs peak between 36-44 cm together suggest that sediment mixing has had no more than a moderate impact on the radiometric record.

Assuming the ^{137}Cs date to be correct, composite model, or revised CRS model dates for the post-1964 period can be calculated using the ^{137}Cs date as a reference point [7.16]. The results suggest that during this period the sedimentation rate has increased from a value of around 0.41g cm^{-2} yr^{-1} (0.58 cm yr^{-1}) in the late 1960s (a value similar to the basal rate) to a contemporary value of around 1.08 g cm^{-2} yr^{-1} (1.75 cm yr^{-1}). Pre-1960 dates are more problematic, particularly in view of the fact that the core has not captured the complete ^{210}Pb record. A reasonable assumption is to suppose that the basal rate determined from the piecewise application of the CF:CS model to the lower part of the ^{210}Pb record persisted throughout the period spanned by this part of the core. The results of these calculations are shown in Fig. 7.4 and Table 7.1. Results of calculations for VD02 follow a similar pattern to VD01 though with a different timing for the brief episode of accelerated accumulation in the late 20th century (Fig. 7.4). Table 7.2 gives revised CRS model dates for the second core VD02. The sedimentation rate in the Bay has been increasing in recent years, possibly could be due to agricultural and coal mining activities as well as urbanization along the coast.

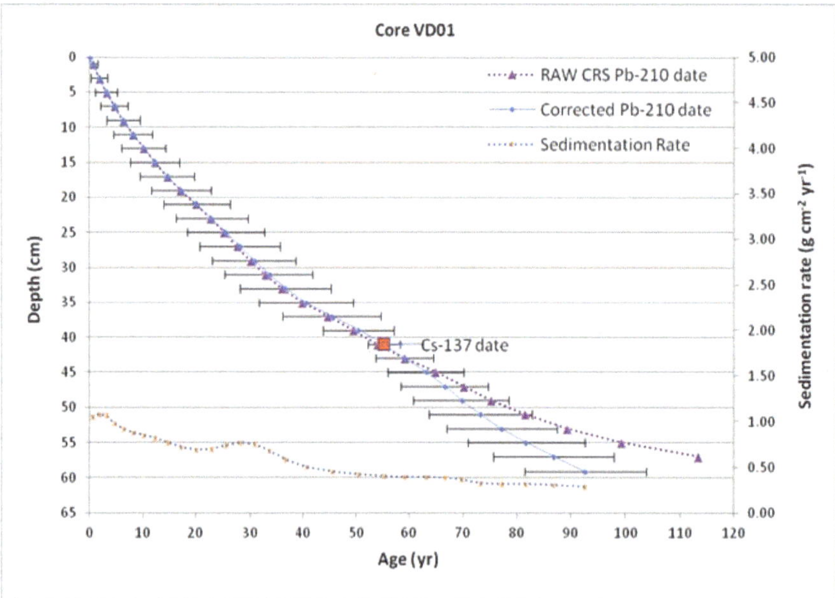

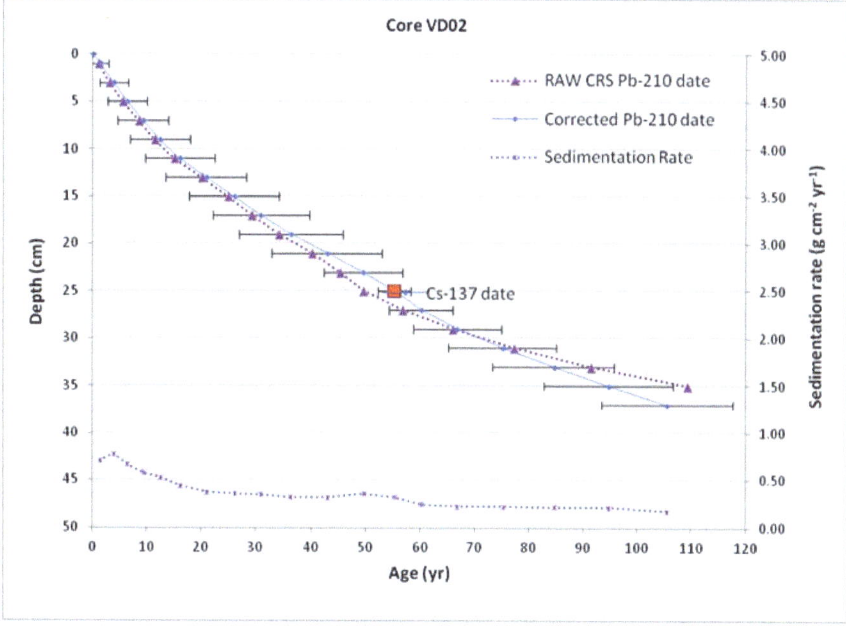

FIG. 7.4. Results of dating and calculations of sedimentation rates by CRS model.

TABLE 7.1. Pb-210 CHRONOLOGY OF CORE VD01

Depth		Chronology			Sedimentation Rate		
cm	g cm^{-2}	Date AD	Age y	±	g cm^{-2} yr^{-1}	cm yr^{-1}	± (%)
0		2019.3	0	0			
1	0.56	2018.7	0.6	0.9	1.05	1.74	9.99
3	1.74	2017.4	1.8	1.6	1.08	1.75	8.36
5	3.04	2016.1	3.2	2.1	1.07	1.58	8.37
7	4.44	2014.6	4.6	2.6	0.98	1.38	8.36
9	5.89	2013.0	6.3	3.1	0.92	1.25	8.37
11	7.39	2011.1	8.1	3.6	0.88	1.14	8.38
13	8.96	2009.1	10.2	4.2	0.85	1.07	8.39
15	10.55	2006.9	12.3	4.6	0.82	1.02	8.41
17	12.17	2004.6	14.6	5.1	0.77	0.94	8.44
19	13.85	2002.0	17.2	5.7	0.72	0.84	8.45
21	15.60	1999.1	20.1	6.2	0.69	0.82	8.71

23	17.22	1996.3	22.9	6.7	0.70	0.89	9.24
25	18.73	1993.7	25.6	7.1	0.74	0.96	9.47
27	20.29	1991.1	28.2	7.5	0.77	0.97	9.48
29	21.92	1988.4	30.8	7.8	0.75	0.92	9.49
31	23.54	1985.7	33.6	8.2	0.68	0.84	9.53
33	25.17	1982.6	36.7	8.5	0.59	0.71	9.59
35	26.86	1978.7	40.6	8.9	0.51	0.59	9.58
37	28.63	1973.8	45.4	9.2	0.46	0.55	9.54
39	30.21	1968.8	50.4	6.7	0.43	0.58	9.55
41	31.61	1964.0	55.3	3.0	0.41	0.58	9.57
43	33.06	1960.2	59.0	5.3	0.40	0.54	9.79
45	34.58	1956.2	63.1	7.0	0.40	0.56	10.33
47	35.91	1952.6	66.6	8.1	0.39	0.63	10.55
49	37.06	1949.6	69.7	8.9	0.36	0.62	10.56
51	38.26	1946.1	73.2	9.6	0.33	0.55	10.54
53	39.50	1942.1	77.1	10.3	0.32	0.49	10.52
55	40.85	1937.6	81.7	10.9	0.32	0.44	10.49
57	42.34	1932.4	86.8	11.1	0.31	0.40	10.40
59	44.01	1926.6	92.6	11.1	0.28	0.32	7.30

TABLE 7.2. Pb-210 CHRONOLOGY OF CORE VD02

Depth		Chronology			Sedimentation Rate		
cm	g cm^{-2}	Date AD	Age y	±	g cm^{-2} yr^{-1}	cm yr^{-1}	± (%)
0		2019.3	0	0			
1	0.72	2017.9	1.4	1.6	0.70	0.91	7.46
3	2.23	2015.4	3.9	2.6	0.78	0.98	6.56

5	3.91	2012.9	6.4	3.6	0.67	0.80	7.00
7	5.55	2010.0	9.3	4.7	0.58	0.73	6.75
9	7.09	2006.8	12.4	5.5	0.53	0.67	6.69
11	8.69	2003.1	16.1	6.4	0.44	0.54	7.01
13	10.36	1998.4	20.9	7.4	0.37	0.47	6.81
15	11.87	1993.2	26.0	8.2	0.36	0.51	6.08
17	13.22	1988.4	30.9	8.8	0.35	0.51	5.73
19	14.62	1982.9	36.3	9.4	0.32	0.45	5.66
21	16.08	1976.4	42.9	10.0	0.32	0.44	5.58
23	17.48	1969.8	49.5	7.2	0.36	0.52	5.37
25	18.82	1964.0	55.3	3.0	0.32	0.48	5.36
27	20.21	1959.1	60.1	5.9	0.24	0.34	5.52
29	21.66	1952.4	66.8	8.2	0.22	0.29	5.94
31	23.28	1944.1	75.1	9.9	0.22	0.26	6.63
33	25.08	1934.7	84.6	11.2	0.22	0.24	7.13
35	26.96	1924.6	94.6	12.0	0.21	0.22	7.32
37	28.91	1913.7	105.5	12.0	0.17	0.17	5.22

Table 7.3. Summary of analyses for heavy metals, TOC, N and P. It can be seen that the heavy metal concentrations in Bay's coastal sediments are lower than the limits set by the Vietnam National Technical Regulation on Sediment Quality [7.19].

TABLE 7.3. ANALYSIS RESULTS FOR HEAVY METALS, TOC, N AND P

	Cr (mg kg^{-1})	Co (mg kg^{-1})	Ni (mg kg^{-1})	Cu (mg kg^{-1})	Zn (mg kg^{-1})	As (mg kg^{-1})	Cd (mg kg^{-1})	Pb (mg kg^{-1})	TOC (%)	N (%)	P (%)
Average value	11.82	37.97	41.07	28.78	32.37	1.27	0.82	14.78	4.78	0.08	0.02
Min value	9.13	32.77	30.46	22.58	20.09	0.02	0.13	8.10	2.60	0.05	0.01
Max value	17.34	44.85	63.11	44.99	54.12	2.89	1.11	23.90	9.91	0.14	0.05

Safe limits set by [19]	160	-	-	108	271	41.6	4.2	112	-	-	-

Temporal trends of heavy metals inputs into the Ha Long Bay are depicted in Fig. 7.5 (at core VD01) and Fig. 7.6 (at core VD02). It is evident that the concentrations of Cr, Co, Ni, Zn, Pb, Cu, As and Cd in the Bay sediment have not varied much over the time. It could be concluded that all heavy metals input into the Bay are from the same source consistently over time.

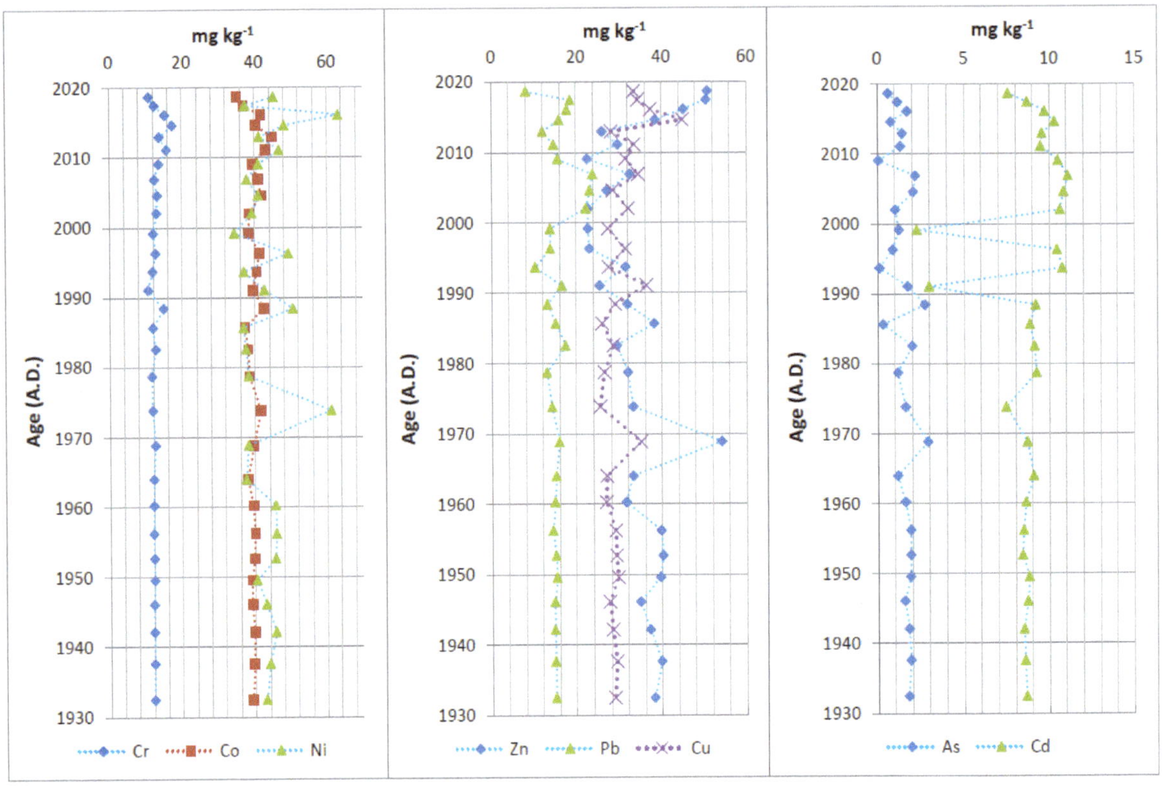

FIG. 7.5. Temporal trend of heavy metals input into the Ha Long Bay (at core VD01).

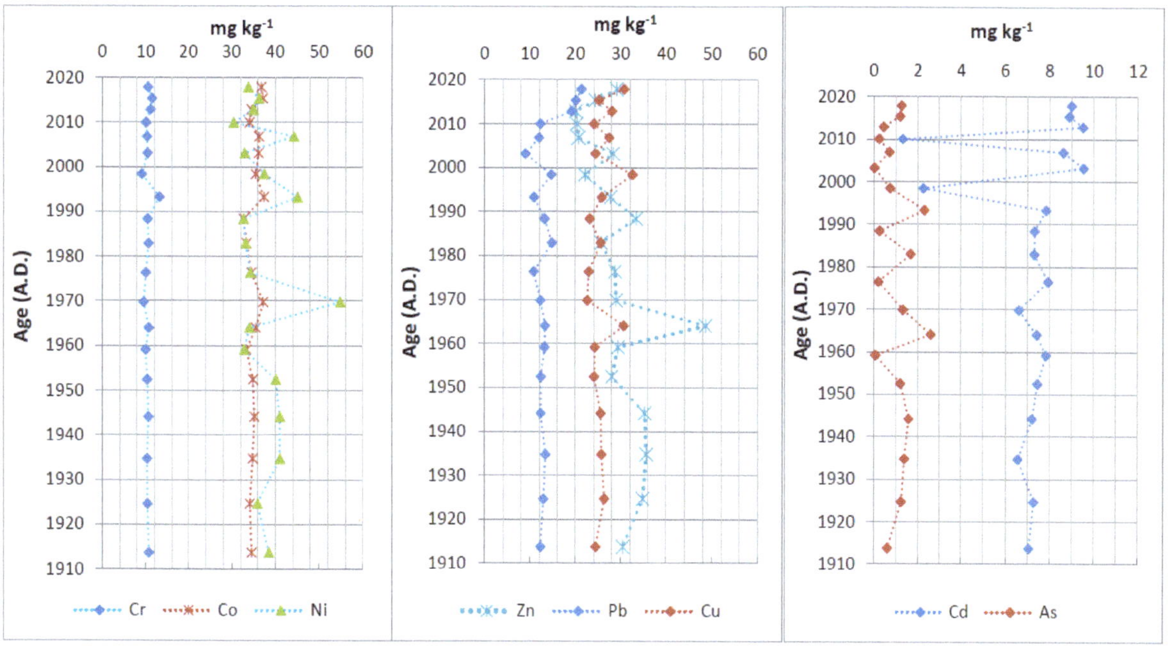

FIG. 7.6. Temporal trend of heavy metals input into the Ha Long Bay (at core VD02).

Temporal trends of nutrients (N, P andTOC) inputs into the Ha Long Bay are depicted in Fig. 7.7 and 7.8. It could be seen the tendency of low N, P, and TOC values from 1910s to about 1990 and increasing values from 1990s onward. TOC concentrations before 1990 were low, around 3%, and increased gradually from 1990s onward, reaching a value of about 8-10% in 2016. Trends of increasing concentrations of nutrients from 1990s onward could be the evidence of booming economic activities in Vietnam after "Renovation", the economic reforms starting in 1986 with the objective of creating a "socialist-oriented market economy" [7.20].

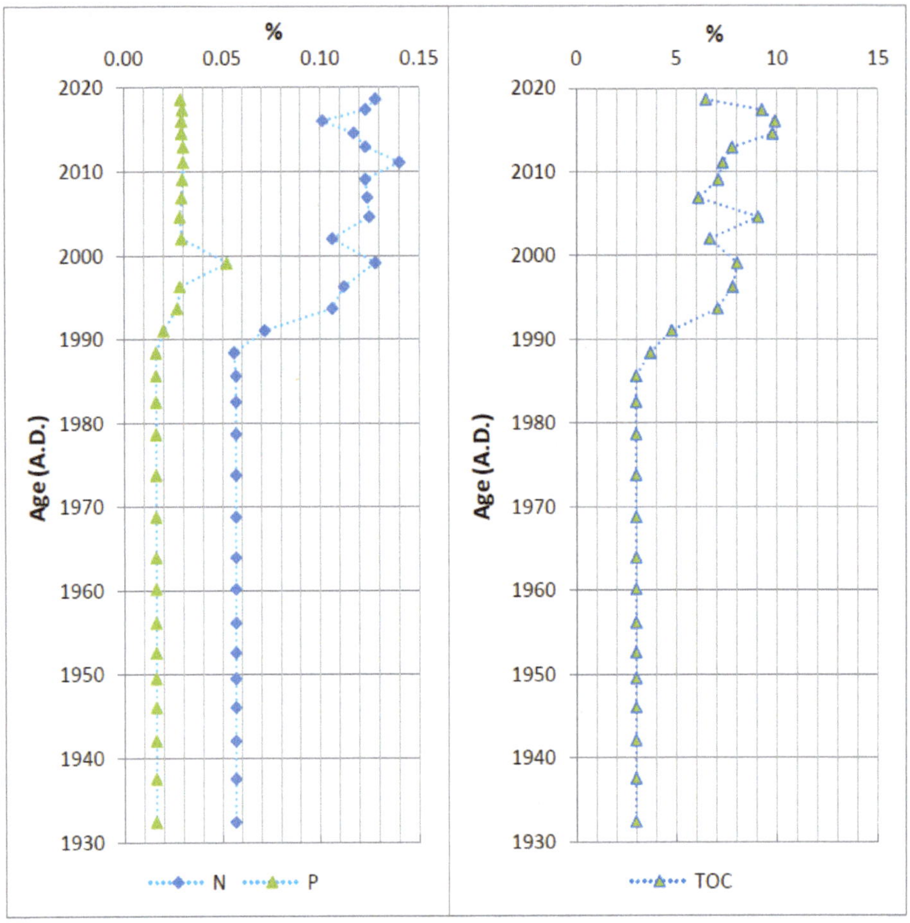

FIG. 7.7. Variation of N, P and TOC concentrations along the depth age of sediment core VD01.

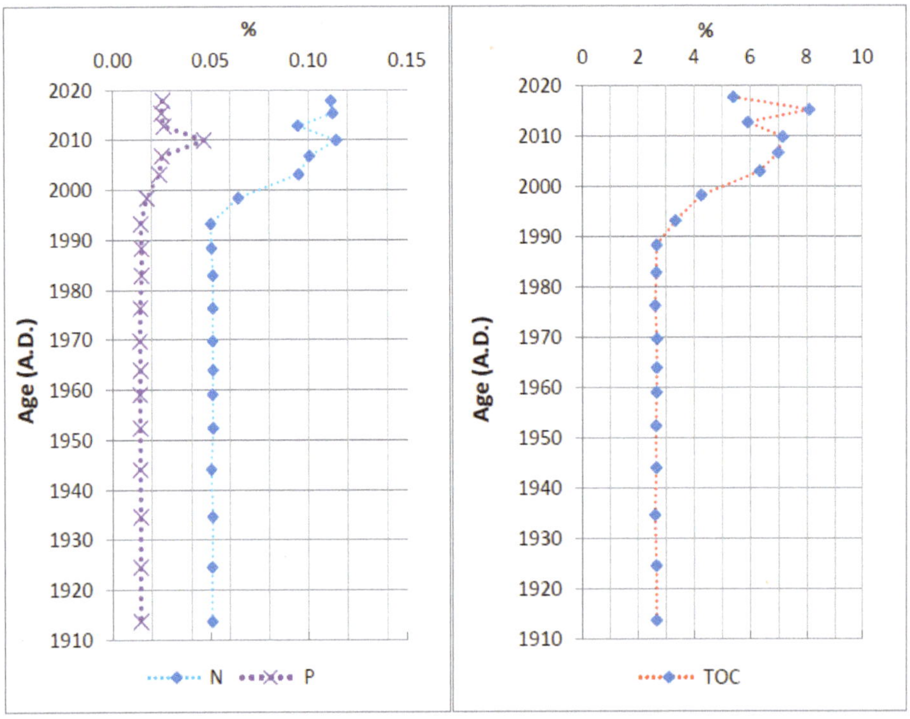

FIG. 7.8. Variation o N, P and TOC concentrations along the depth age of sediment core VD02.

7.4. CONCLUSION

Several ^{210}Pb dating models were applied for sediment chronology of the two cores taken in the coastal area of the Ha Long Bay. It was found that the CIC model would not be suitable for dating these cores due to the $^{210}Pb_{ex}$ concentration profiles do not decrease monotonically, and the application of CIC model would give deeper layers with higher concentrations younger ages than the underlying layers. CF:CS is also not applicable to either of the cores as a whole because the natural logarithmic graphs of $^{210}Pb_{ex}$ against accumulated mass are not linear, and the CF:CS model would give results with large uncertainties. CF:CS model, however, could be applied for linear segments such as the segment of surface layers to estimate initial $^{210}Pb_{ex}$ concentration, and the segment of bottom layers to estimate the missing $^{210}Pb_{ex}$ inventory below the bottom layer.

In this study, CRS model was used in combination with the use of ^{137}Cs fallout peak as an independent reference date. With the assumption that fallout ^{137}Cs in sediments was not redistributed by post-depositional processes, the use of the 1964 date for the ^{137}Cs fallout peak for correcting CRS model data could be the most suitable for dating of the two cores. Dating results of the two Ha Long Bay sediment cores showed that the sedimentation rates in each core vary, probably depending on sediment flux from the surrounding coastal watersheds and rivers, and were in the range from 0.17 to 1.08 g cm^{-2} y^{-1} (0.17 to 1.75 cm yr^{-1}). The sedimentation rate in the Bay has been increasing in recent years, possibly could be due to agricultural and coal mining activities as well as urbanization along the coast. Coastal sediments were found to be below regulatory standards for heavy metal concentrations. However, TOC, N, and P profiles show increasing concentrations trend since 1990 onward, which could be due to increased agricultural activities in the continent along the Bay's coast.

7.5. REFERENCES

[7.1] UNITED NATIONS, First Global Integrated Marine Assessment (First World Ocean Assessment) (2016), https://www.un.org/Depts/los/global_reporting/WOA_RegProcess.htm

[7.2] GREEN, M.O., COCO, G., Review of Wave-Driven Sediment Resuspension and Transport in Estuaries, Rev. Geophys. 52(1) (2014) 77-117.

[7.3] ROMERO, A.F., ASMUS, M.L., MILANELLI, J. C. C., BURUAEMA, L., ABESSA, D.M.S., Self-Diagnosis Method as an Assessment Tool for Environmental Management of Brazilian Ports, J. Integr. Coast. Zone Manage. 14(4) (2014) 637-644.

[7.4] ULUTURHAN, E., KUCUKSEZGIN, F., Heavy Metal Contaminants in Red Pandora (Pagellus Erythrinus) Tissues from the Eastern Aegean Sea, Turkey, Water Research 41 (2007) 1185-1192.

[7.5] *UNESCO, Ha Long Bay,* http://whc.unesco.org/en/list/672

[7.6] LE TUAN ANH, Stituation Analysis of the Water Quality of Ha Long Bay, Quang Ninh Province, Viet Nam: A Social Study from Tourism Businesses' Perspectives, Gland, Switzerland: IUCN, 2015.

[7.7] QUANG NINH PROVINCIAL DEPARTMENT OF NATURAL RESOURCES AND ENVIRONMENT, Environmental Planning of Halong Bay to 2020 Vision to 2030, Summary Report, Quang Ninh, 2014.

[7.8] FUKUE. M., NAKAMURA.T., KATO. Y., YAMASAKI. S., Degree of Pollution for Marine Sediments, Engineering Geology 53 (1999) 131-137.

[7.9] LIGEROA.R.A., BARRERAA. M., CASAS-RUIZA.M., SALESB. D., LÓPEZ-AGUAYOC.F., Dating of Marine Sediments and Time Evolution of Heavy Metal

Concentrations in the Bay of Ca´ Diz, Spain, Environmental Pollution 118 (2002) 97-108.

[7.10] ZOURARAH, B., MAANAN, M., CARRUESCO, C., AAJJANE, A., MEHDI. K., FREITAS, M.C., Fifty-year Sedimentary Record of Heavy Metal Pollution in the Lagoon of Oualidia (Moroccan Atlantic Coast), Eustuarine, Coastal and Shelf Science 72 (2007) 359-369.

[7.11] MIL-HOMENS, M., STEVENS, R.L., BOER, W., ABRANTES, F., CATO, I., Pollution History of Heavy Metals on the Portuguese Shelf Using ^{210}Pb-geochronology, Science of the Total Environment 367 (2006) 466-480.

[7.12] SANCHEZ-CABEZA, J.A., RUIZ-FERNANDEZ, A.C., ^{210}Pb Sediment Radiochronology: An Integrated Formulation and Classification of Dating Models, Geochimica et Cosmochimica Acta 82 (2012) 183–200.

[7.13] NHON DANG HOAI, THANH TRAN DUC, HUY DINH VAN, HAI PHAN SON, HA NGUYEN MANH, LAN TRAN DINH, DUAN XIAOYONG, VE NGUYEN DAC, Sedimentation Rates and Heavy Metal Concentrations in the Tidal Flats of North Vietnam, Pol. J. Environ. Stud. Vol. 28, No. 5 (2019) 3721-3733.

[7.14] *BUI VAN VUONG, ZHIFEI LIU, TRAN DUC THANH, CHIH-AN HUH, DANG HOAI NHON, NGUYEN DAC VE, DINH VAN HUY, Initial Results of Study on Sedimentation Rate, Sediment Source to The Ha Long Bay: Evidence from the ^{210}Pb and ^{137}Cs Radiotracer, Vietnam Journal of Marine Science and Technology Vol. 16, No. 1 (2016) 54-63.*

[7.15] NGUYEN QUANG LONG, TRAN THI TUYET MAI, PHAM NGOC KHAI, NGUYEN TRUNG THANH, NGUYEN VAN PHUC, DOAN THUY HAU, DUONG VAN THANG, HA LAN ANH, VO THI ANH AND NGUYEN THI THU HA (2013), Quantitative Research for Pollution Levels in Marine Sediments of Halong Bay by Nuclear Technique, https://inis.iaea.org/collection/NCLCollectionStore/_Public/45/058/45058910.pdf

[7.16] APPLEBY, P.G., Chronostratigraphic techniques in recent sediments, in *Tracking Environmental Change Using Lake Sediments Volume 1: Basin Analysis, Coring, and Chronological Techniques*, (eds W M Last & J P Smol), Kluwer Academic (2001) 171-203.

[7.17] RITCHIE, J.C., McHENRY, J.R., Application of radioactive fallout cesium-137 for measuring soil erosion and sediment accumulation rates and patterns: a review. J. Environ. Qual. 19 (1990) 215–223.

[7.18] KLAMINDER, J., APPLEBY, P., CROOK, P., RENBERG, I., Post-deposition diffusion of ^{137}Cs in lake sediment: Implications for radiocaesium dating. Sedimentology 59 (2012) 2259-2267.

[7.19] QCVN 43: 2012/BTNMT, National Technical Regulation on Sediment Quality, Hanoi (2012).

[7.20] https://en.wikipedia.org/wiki/Đổi_Mới.

8. PLUTONIUM IN THE SOUTHERN BALTIC SEA AND ITS BUDGET AND INFLOW TO THE GULF OF GDAŃSK AND THE GDAŃSK BASIN

D. STRUMIŃSKA-PARULSKA
University of Gdańsk, Faculty of Chemistry, Laboratory of Toxicology and Radiation Protection
Gdańsk, Poland.
Email: dagmara.struminska@ug.edu.pl

Abstract

This study presents the ^{238}Pu, $^{239+240}$Pu, and ^{241}Pu isotopes activity concentrations in different coastal compartments of the southern Baltic Sea. Also, $^{239+240}$Pu inflow, its inventory in the Gulf of Gdańsk and the Gdańsk Basin, as well as the most important sources, are given, allowing the establishment of the trend in Pu contamination assessment of these areas. The Baltic Sea organisms of different taxonomy or trophic levels did not accumulate plutonium to the same extent. The highest amount of $^{239+240}$Pu existed in the middle parts of sediments, came from global atmospheric fallout and occurred in the mobile, connected to carbonates, fraction. Opposite, the most considerable amount of ^{241}Pu appeared in the surface layers of sediments and came from the Chernobyl accident. The Vistula and the Odra River were significant sources of plutonium in the southern Baltic Sea. Annually in 2002 and 2003, the Vistula River enriched its estuary in 10.3 MBq ^{238}Pu, 89.0 MBq $^{239+240}$Pu and 1.65 GBq of ^{241}Pu. The annual inflow of ^{238}Pu, $^{239+240}$Pu, and ^{241}Pu with the Odra River in 2002 and 2003 to the Pomeranian Bay was 9.51 MBq, 45.86 MBq and 0.62 GBq, respectively. The total $^{239+240}$Pu activity deposited in the Gulf of Gdańsk and the Gdańsk Basin sediments was estimated at 1.18 TBq and 3.77 TBq. Based on the calculations and data available in the literature, the sediments of both regions were highly enriched with plutonium – the sediments of the Gulf of Gdańsk and the Gdańsk Basin contained 6.4% and 22.0% of $^{239+240}$Pu deposited in whole Baltic Sea sediments, whereas their areas constituted only 1.2% and 4.4% of its total area.

8.1. INTRODUCTION

The Baltic Sea is a small, shelf and mostly enclosed sea with limited water exchange, very sensitive to human activities. The large amount of radioactive substances produced by human use of nuclear power is a significant current and future concern, not only when related to the Baltic Sea. Plutonium is one of the most significant transuranic elements found in the environment as ^{238}Pu, ^{239}Pu, ^{240}Pu and ^{241}Pu [8.1, 8.2]. The key plutonium sources in the Baltic Sea have been: global atmospheric fallout from the high-yield nuclear weapon tests and the Chernobyl accident. At present, the contribution of the global fallout is much smaller and equals about 1 GBq per year, while the Chernobyl accident needs to be taken into consideration in risk assessment on the environment [8.1, 8.3, 8.4, 8.5, 8.6]. The other potential source of plutonium has been the releases from spent fuel facilities in Sellafield (UK) and Cap de la Hague (France). However, they seem less important – the North Sea inflows to the Baltic Sea through the Skagerrak and the Kattegat are short-term, every few years, and rarely reach the Słupsk Narrow [8.7, 8.8]. Atmospheric fallout after the Chernobyl accident appears to have more importance as it resulted in plutonium emission of around 30 TBq of ^{238}Pu, 61.5 TBq of ^{239}Pu+^{240}Pu and 5 PBq ^{241}Pu [8.9].

Marine plants and animals accumulate radionuclides from the aquatic environment [8.10, 8.11]. Thus, it is essential to determine radionuclides accumulation in living organisms and the potential transfer to the human body. The highest plutonium activity concentrations in marine ecosystems were found in sediments, but its complex biogeochemical cycle caused it to be present in all environment compartments [8.1, 8.3]. Many studies suggested the possible increase in both the plutonium concentration in the southern Baltic Sea and its bioavailability in the marine food chain as a result of desorption from sediments and subsequent accumulation in benthic organisms [8.8, 8.12, 8.13, 8.14, 8.15, 8.16].

The work summarizes the results of the studies on activity determination of the most important α-emitting plutonium isotopes (^{238}Pu and $^{239+240}$Pu) as well as β^{-}emitting ^{241}Pu in the southern Baltic Sea as well as inflowing river using the alpha spectrometry and accelerator mass spectrometry techniques as valuable tools to achieve comprehensive data on the presence of anthropogenic contaminants in the specific marine system, namely the southern Baltic Sea.

8.2. MATERIALS AND METHODS

8.2.1. Sampling

Seawater samples were collected in 1987 and 1999 throughout the southern Baltic Sea: Gulf of Gdańsk (Gdańsk seaport), Gdańsk Deep, Pomeranian Bay and open sea near Ustka. After collection of 100-300 L of seawater, the samples were filtered using pre-weighed membrane filters of 0.45 μm (suspended forms) and 0.23 μm (colloidal forms) pore diameter. The sediments cores were collected from the Vistula River confluence, the Gulf of Gdańsk, the Gdańsk Deep and the Bornholm Deep. Sediments were collected with Niemistö sampling tool to get clear, untouched cores. Surface sediments were cut: 1-10 cm layer every 1 cm, 10-30 cm every 2,5 cm, over 30 cm every 5 cm. Phytoplankton was sampled by vertical hauls and trawling using a Copenhagen-type net (20 μm mesh diameter). Zooplankton was collected by vertical hauls using a 505 μm mesh Bongo net, a 303 μm mesh Hensen net, and a Nansen net of 202 μm mesh diameter, and. Fish samples from the Gulf of Gdansk were caught by professional fishermen (Fig. 8.1).

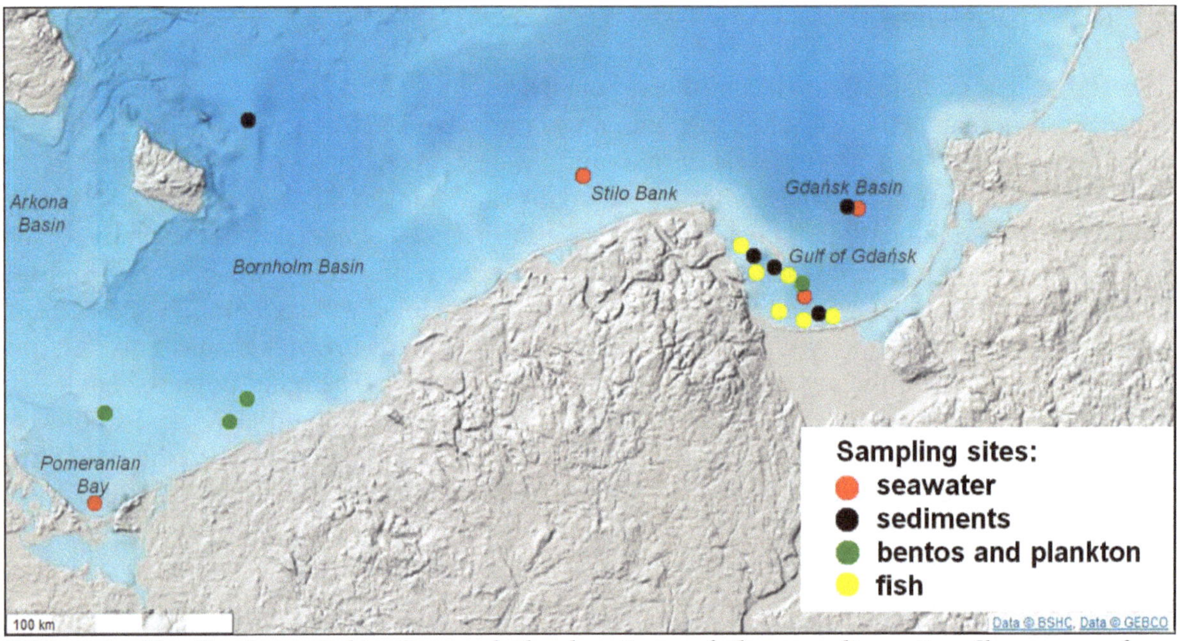

FIG. 8.1. The Southern Baltic Sea and the locations of the sampling sites [base map from http://data.bshc.pro/].

8.2.2. Radiochemical analysis and plutonium measurements

All samples were spiked with a known amount of ^{242}Pu. The separation and determination of plutonium in water samples were carried by methods that exploit the ability of radionuclide to co-precipitate with MnO_2. The sediments and biological samples were mineralized in HNO_3 and further separated and purified on Dowex anion exchange resins. Finally, the plutonium was electrolyzed on steel discs [8.17, 8.18]. The chemical yield varied from 50% to 90%.

Plutonium ^{238}Pu and $^{239+240}$Pu activity concentrations in analyzed Baltic samples were measured using an alpha spectrometer (Canberra, USA). The determination of ^{241}Pu in the samples was done indirectly by measurements of ^{241}Am ingrowth from the decay of β^--emitting ^{241}Pu. Previously prepared Pu targets that were a part of earlier work were re-measured after 10-15 years in alpha spectrometer again. All achieved spectra of plutonium alpha isotopes were compared with the same spectra obtained much earlier, and ^{241}Pu activities were calculated on the sampling date [8.18]. The values of ^{240}Pu/^{239}Pu atomic ratios in some samples were also measured using accelerator mass spectrometry (AMS). All AMS sources were prepared as Al/Cu cathode targets containing plutonium with Fe/Al. The targets were measured at CNA (Seville, Spain) using accelerator mass spectrometry (ETHZ/PSI, Switzerland) equipped with a 3x3 mm^2 silicon detector [8.18, 8.19].

8.3. RESULTS AND DISCUSSION

8.3.1. Plutonium isotopes (^{238}Pu, $^{239+240}$Pu, and ^{241}Pu) activity concentrations in the southern Baltic Sea ecosystem, as well as their activity and atomic ratios

8.3.1.1. Seawater

The seawater plutonium activity concentration in the southern Baltic changed from east to west. The greatest concentration was observed in the Pomeranian Bay (small inflows of saline water from the North Sea every 2-5 years, potentially richer in plutonium from Sellafield NRP discharges). For example, the total concentration of $^{239+240}$Pu in the total seawater sample (understood as unfiltered water) from the Pomeranian Bay was estimated at 150±4 $mBq{\cdot}m^{-3}$, while in the Gulf of Gdańsk it was 5.2±0.8 $mBq{\cdot}m^{-3}$ (Table 8.1, Fig. 8.2). A significant part of plutonium was noticed in filtered water (0.22 μm filters) as a dissolved fraction and ranged from 54 to 97 % of its total amount [8.20, 8.21]. Studies on ^{241}Pu in the southern Baltic seawater showed significant growth in a decade after the Chernobyl accident. In 1987 ^{241}Pu activity concentrations were 0.23±0.03 $Bq{\cdot}m^3$ in the water of the Gulf of Gdańsk and 0.11 ± 0.02 $Bq{\cdot}m^3$ in water from the Gdańsk Deep, while in 2000 its activities were 2.57±0.46 $Bq{\cdot}m^3$ in the water of the Gulf of Gdańsk and 2.48±0.04 $Bq{\cdot}m^3$ in the Gdańsk Deep (Fig. 8.2). The calculated values of the ^{238}Pu/$^{239+240}$Pu, and ^{241}Pu/$^{239+240}$Pu activity ratios indicated that the plutonium in southern Baltic water is mainly from the Chernobyl accident transported by river water [8.20-8.22].

TABLE 8.1. ^{238}Pu, $^{239+240}Pu$ AND ^{241}Pu ACTIVITY CONCENTRATION IN WATER FROM THE SOUTHERN BALTIC SEA; BDL – BELOW DETECTION LIMIT (*Adapted from [8.20-8.22]*)

Water Sample	Plutonium Activity Concentration		
	^{238}Pu (mBq·m^{-3} ± SD)	$^{239+240}Pu$ (mBq·m^{-3} ± SD)	^{241}Pu (mBq·m^{-3} ± σ)
Studies in 1987			
Gulf of Gdańsk	-	1.66 ± 0.32	0.23 ± 0.03
Gdańsk Deep	-	2.08 ± 0.33	0.11 ± 0.02
Studies in 2000			
Gulf of Gdańsk (seaport)			
dissolved	1.5 ± 0.5	2.8 ± 0.7	1.83 ± 0.09
suspended	0.8 ± 0.2	1.6 ± 0.4	0.59 ± 0.04
colloidal	-	0.8 ± 0.2	0.22 ± 0.02
total	2.3 ± 0.9	5.2 ± 0.8	2.57 ± 0.46
Gdańsk Deep			
dissolved	4.1 ± 0.5	7.9 ± 0.6	2.19 ± 0.04
suspended	1.2 ± 0.2	2.5 ± 0.2	0.16 ± 0.01
colloidal	0.4 ± 0.1	0.7 ± 0.1	0.16 ± 0.01
total	5.7 ± 1.0	11.1 ± 0.6	2.48 ± 0.04
Słupsk Bank			
dissolved	bdl	17.1 ± 2.3	1.56 ± 0.55
suspended	bdl	3.8 ± 0.9	1.94 ± 0.12
colloidal	bdl	1.1 ± 0.1	0.08 ± 0.01
total	bdl	20.9 ± 2.5	3.35 ± 0.17
Pomeranian Bay			
dissolved	70 ± 3	145 ± 4	1.86 ± 0.03
suspended	1.2 ± 0.2	4.0 ± 0.4	0.32 ± 0.01
colloidal	60.2 ± 0.2	1.0 ± 0.2	0.05 ± 0.01
total	71.8 ± 2.8	150 ± 4	2.21 ± 0.07

A historical trend of temporal anthropogenic contamination of the seacoast was observed. An increase of ^{241}Pu content would be noticed in other compartments of the Baltic ecosystem (Fig. 8.2). However, the late inflow of plutonium to the Baltic Sea from the catchment area would impact its presence in the sediments and distribution in organisms of different living habits.

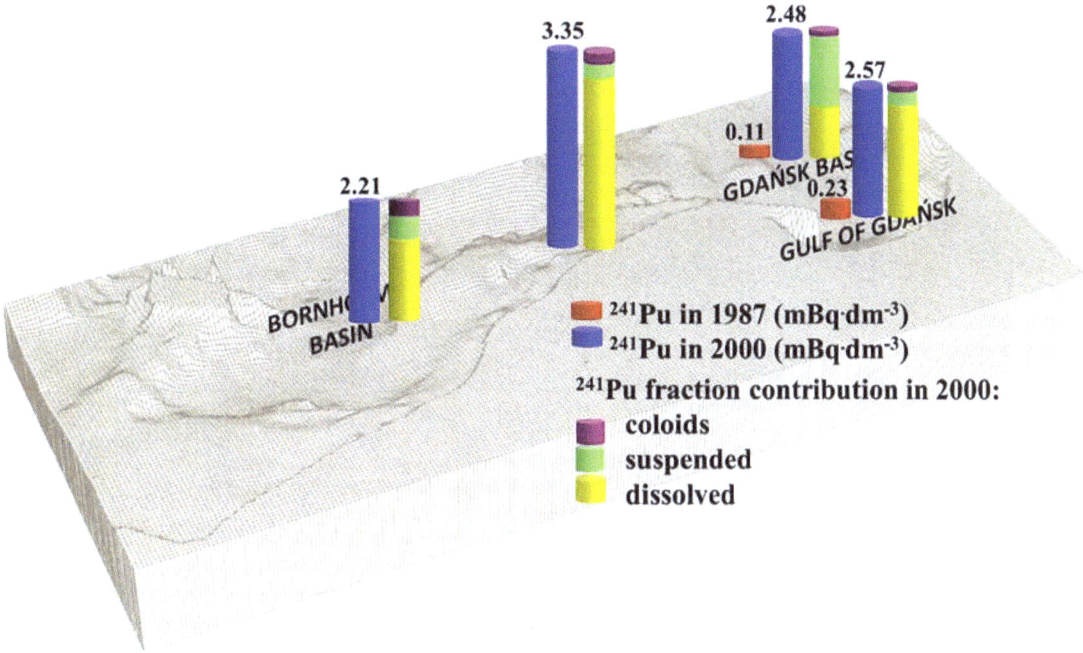

FIG. 8.2. ^{241}Pu in the southern Baltic Sea water and its activity concentration changes from 1987 to 2000.

8.3.1.2. Sediments

Some studies confirmed that plutonium particles settle out, and the aquatic sediments are its main reservoir [8.4, 8.5, 8.23, 8.24, 8.25, 8.26]. The southern Baltic Sea is not an exception, and plutonium activities in its sediments are much higher than in other Baltic ecosystem components. The research performed on the Baltic sediments showed the highest ^{238}Pu and $^{239+240}$Pu activity concentrations, which were determined in muddy sediments of the Gulf of Gdańsk. Opposite, ten times lower ^{238}Pu and $^{239+240}$Pu activity concentrations were noticed in sandy and poor in organic matter sediments of the Vistula River estuary. In all analyzed sediments, the total ^{238}Pu and $^{239+240}$Pu activities have decreased with sediment core depth. The study on ^{241}Pu content in the southern Baltic Sea sediments indicated its activity concentrations were not uniform and depended on the seafloor geomorphology and its depth and the location, similarly to alpha-emitting plutonium isotopes. In 2000, the highest ^{241}Pu activity concentration was found in the muddy sediments of the Gdańsk Deep (13.7±2.4 mBq·g^{-1} dw (dry weight)) and the value of the ^{241}Pu/$^{239+240}$Pu activity ratio was estimated at 37±8. The highest amount of ^{241}Pu has been present in the surface layer of all analyzed sediments. Compared to previous studies on ^{241}Pu, the effect of late inflow of plutonium to the Baltic Sea from the catchment area and ^{241}Pu activities increase in sediments when compared to water; e.g. from 0.90±0.14 Bq·kg^{-1} dw in surface sediments from the Gulf of Gdańsk in 1987 to 20.0±5.0 Bq·kg^{-1} dw in 2000 as well as from 14.2±2.5 Bq·kg^{-1} dw in surface sediments from the Gdańsk Deep in 1987 to 62±13 Bq·kg^{-1} dw in 2000. This situation could also be observed in the case of $^{239+240}$Pu (Table 8.2) [8.8, 8.23]. The analysis of the values of the ^{238}Pu/$^{239+240}$Pu and ^{241}Pu/$^{239+240}$Pu activity ratios allowed to recognize of plutonium sources and revealed the Chernobyl plutonium presence (including all radionuclides: ^{238}Pu, $^{239+240}$Pu, and ^{241}Pu) at 0-3 cm sediment layers in sediments from the Vistula River estuary and the Gulf of Gdańsk. In the other analyzed southern Baltic regions and its deeper sediment layers, the plutonium (mainly ^{238}Pu and $^{239+240}$Pu) came from the global atmospheric fallout as an effect of nuclear weapon tests in 1958-61 [8.8, 8.23].

TABLE 8.2. Pu-238, Pu-239+240 AND Pu-241 ACTIVITY CONCENTRATIONS IN SEDIMENTS FROM THE SOUTHERN BALTIC SEA; bdl – BELOW DETECTION LIMIT (*Adapted from [8.8, 8.22]*)

Sediment layer (cm)	Plutonium activity concentration		
	^{238}Pu (mBq·kg^{-1} ± SD)	$^{239+240}Pu$ (mBq·kg^{-1} ± SD)	^{241}Pu (Bq·kg^{-1} ± SD)
Studies in 1987			
Gulf of Gdańsk	-	0.12 ± 0.01	0.90 ± 0.14
Gdańsk Deep	-	0.37 ± 0.05	14.2 ± 2.5
Studies in 2000			
Vistula River estuary (54°22.00 N; 18°59.01 E), depth = 15.3 m			
0-1	47 ± 23	94 ± 33	16.5 ± 3.7
1-2	44 ± 19	131 ± 34	12.5 ± 2.6
2-3	60 ± 27	143 ± 41	11.4 ± 2.9
3-4	38 ± 22	141 ± 43	11.3 ± 4.0
4-5	42 ± 21	116 ± 35	8.8 ± 1.8
5-6	89 ± 45	423 ± 97	8.3 ± 1.8
6-7	80 ± 56	438 ± 132	12 ± 3
7-8	37 ± 26	93 ± 42	-
Gulf of Gdańsk (54°27.09 N; 19°02.09 E), depth = 70.5 m			
0-1	102 ± 39	234 ± 58	20.0 ± 5.0
1-2	32 ± 23	757 ± 110	8.1 ± 2.2
2-3	117 ± 24	793 ± 64	9.8 ± 2.0
8-9	850 ± 236	4576 ± 547	1.06 ± 0.12
17.5-20	3 ± 3	16 ± 7	0.02 ± 0.01
20-22.5	4 ± 4	18 ± 8	0.02 ± 0.01
Gdańsk Deep (54°49.91 N; 19°19.55 E), depth = 112 m			

0-1	27 ± 16	418 ± 74	62 ± 13
1-2	60 ± 23	1404 ± 110	21.0 ± 3.3
8-9	130 ± 130	391 ± 226	4.7 ± 2.0
10-12.5	1 ± 1	23 ± 3	1.5 ± 0.3
30-35	3 ± 1	15 ± 1	bdl
35-40	1 ± 1	11 ± 3	bdl
Bornholm Deep (55°09.00 N; 15°55.00 E), depth = 93 m			
0-1	11 ± 7	229 ± 30	6.0 ± 1.0
1-2	31 ± 9	747 ± 44	11.5 ± 1.0
10-12.5	175 ± 124	1491 ± 362	0.87 ± 0.25
12.5-15	22 ± 22	65 ± 38	0.74 ± 0.23
30-35	5 ± 4	27 ± 9	bdl

The experiment on plutonium fractionation in sediments collected in the southern Baltic indicated the highest amount of plutonium was connected with carbonates fraction: the Vistula River estuary (34%), the Gulf of Gdańsk (42%) and the Bornholm Deep (35%). In the Gdańsk Deep sediments, at the highest amount, the plutonium was associated with the fraction soluble in mineral acids (40%). A significant amount of plutonium in analyzed sediments was also complexed by organic matter (29%). As it occurred, about 40% of plutonium in the Gdańsk Deep sediments existed in low-mobile soluble in mineral acids fraction and potentially was not as dangerous to the ecosystem as carbonates fraction found in the other regions analyzed (Fig. 8.3) [8.8]. Although plutonium was released as a dioxide form mainly due to biological processes, it was transformed into bioavailable fractions; however, the amount was insignificant.

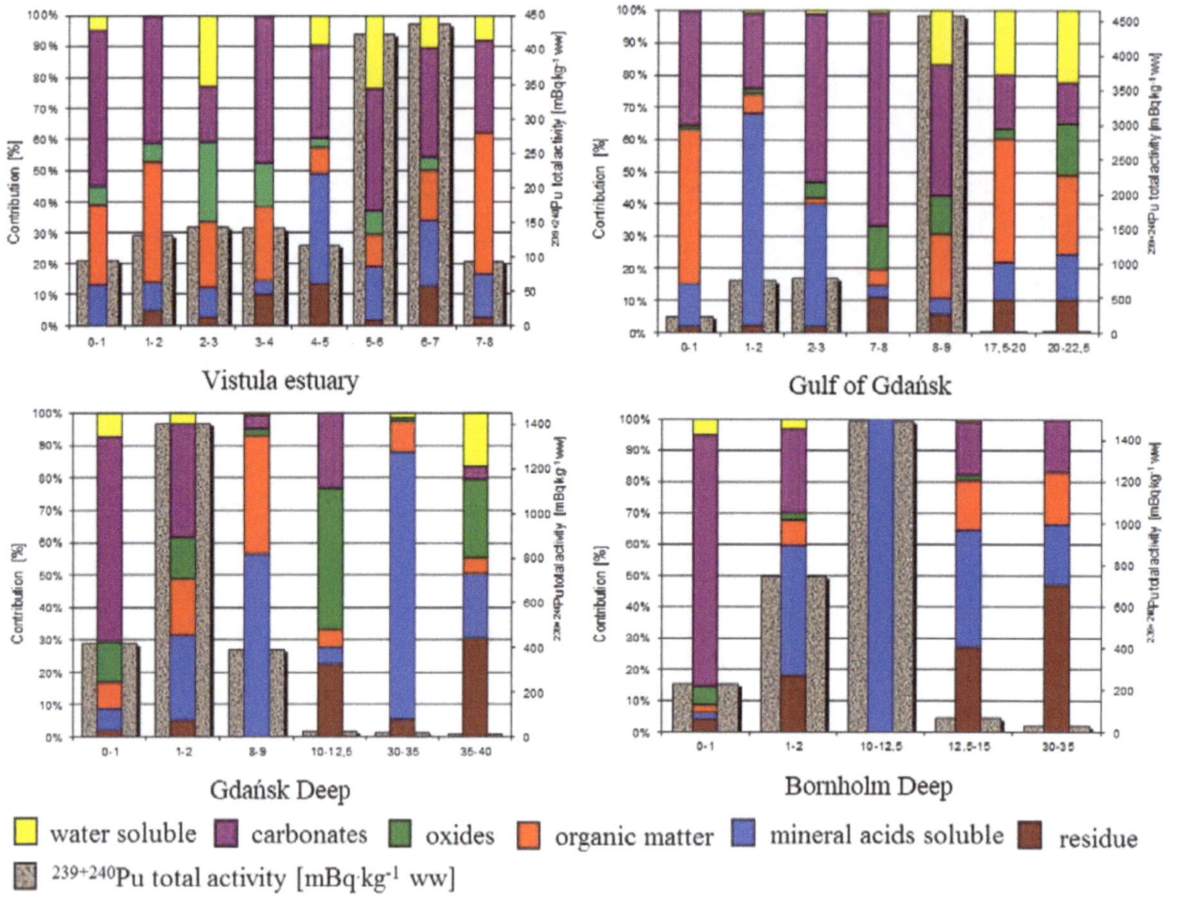

FIG. 8.3. $^{239+240}$Pu fractionation in the southern Baltic sediments.

8.3.1.3. Organisms

The Baltic Sea organisms of different taxonomy or trophic levels do not accumulate plutonium to the same levels (Table 3). The values of plutonium isotopes concentration activities in southern Baltic plants and animals ranged widely. $^{239+240}$Pu content in Baltic organisms varied from 0.94±0.05 mBq·kg^{-1} wet weight (ww) in fish (flounder *Platichthys flesus*) to 850±60 mBq·kg^{-1} ww in priapulid (*Halicryptus spinulosus*) [8.22, 8.27]. The average value of $^{239+240}$Pu concentration in phytoplankton was 6.31±0.89 mBq·kg^{-1} ww, while in zooplankton was estimated at 4.69±0.69 mBq·kg^{-1} ww (Table 3) [8.20, 8.22, 8.27]. The average values of ^{241}Pu activity concentration were calculated at 1.06±0.09 Bq·kg^{-1} dw in phytoplankton and 2.27±0.30 Bq·kg^{-1} dw in zooplankton, while in phytobenthos were lower (0.45±0.09 Bq·kg^{-1} dw). Among analyzed invertebrates, the highest values of ^{241}Pu activity concentration were in benthic organisms, namely Polychaeta (7.71±0.85 Bq·kg^{-1} dw) and Priapulida (9.20±1.20 Bq·kg^{-1} dw) [8.22].

The plutonium was also heterogeneously located among the analyzed fish tissues and organs and. Most of the $^{239+240}$Pu was located in their internal organs (predominantly alimentary tract (8-70 %) and skeleton (10-25 %)), and the lowest value was found in muscles (5-6 %). However, about 50% of the total $^{239+240}$Pu accumulated in herring's (*Clupea harengus*) gills [8.21, 8.27]. Similar observations were done according to ^{241}Pu – most of it was located in internal organs like the intestine (8-61 %), and a herring was an exception again – these fish contained 34% of total ^{241}Pu in gills [8.21]. It suggested that feeding habits had a significant impact on analyzed plutonium isotopes content and distribution in the body.

TABLE 8.3. ^{238}Pu, $^{239+240}Pu$ AND ^{241}Pu ACTIVITY CONCENTRATIONS IN ORGANISMS FROM THE SOUTHERN BALTIC SEA ECOSYSTEM; *RESEARCH DONE IN 1987 (BOTH ALGAE AND VASCULAR PLANTS) (*Adapted from [8.1,8.20,8.27,8.29]*)

Ecosystem component	Average plutonium activity concentration		
	^{238}Pu (mBq·kg^{-1} dw)	$^{239+240}Pu$ (mBq·kg^{-1} dw)	^{241}Pu (Bq·kg^{-1} dw)
Phytoplankton	3.09 ± 0.63	6.31 ± 0.89	1.06 ± 0.09
Zooplankton	-	4.69 ± 0.69	2.27 ± 0.30
Phytobenthos*	-	46.4 ± 4.0	0.45 ± 0.09
Zoobenthos*			
- Polychaeta	-	135 ± 15	7.71 ± 0.85
- Priapulida	-	850 ± 60	9.20 ± 1.20
- Crustaceans	-	32 ± 3.6	0.97 ± 0.14
- Bivalves (soft tissues)	-	73 ± 5	0.91 ± 0.07
Fish	^{238}Pu (mBq·kg^{-1} ww)	$^{239+240}Pu$ (mBq·kg^{-1} ww)	^{241}Pu (mBq·kg^{-1} ww)
- *Platichthys flesus*	0.17 ± 0.02	0.94 ± 0.05	104 ± 9
- *Gadus morhua*	0.45 ± 0.05	2.35 ± 0.11	254 ± 3
- *Clupea harengus*	0.86 ± 0.11	2.22 ± 0.11	317 ± 13
- *Perca fluviatilis*	0.49 ± 0.13	1.96 ± 0.03	666 ± 10
- *Neogobius melanostomus*	0.08 ± 0.02	6.50 ± 0.64	863 ± 66
Seabirds	^{238}Pu (mBq·kg^{-1} ww)	$^{239+240}Pu$ (mBq·kg^{-1} ww)	^{241}Pu (mBq·kg^{-1} ww)
- *Phalacrocorax carbo*	0.05 ± 0.01	0.22 ± 0.02	8.8 ± 0.4
- *Fulica atra*	0.13 ± 0.03	0.84 ± 0.09	19 ± 1
- *Alca torda*	0.41 ± 0.06	1.44 ± 0.12	46 ± 2
- *Somateria mollissima*	0.20 ± 0.06	1.16 ± 0.03	50 ± 3
- *Clangula hyemalis*	0.31 ± 0.09	2.10 ± 0.26	59 ± 2

- *Cepphus gryle*	0.20 ± 0.06	1.36 ± 0.15	38 ± 2
- *Govia stellata*	0.30 ± 0.06	1.16 ± 0.16	37 ± 2
- *Uria aalge*	0.18 ± 0.07	0.47 ± 0.10	22 ± 1

In the pelagic plankton-eating herring, most of the plutonium was present in the gills, while in the case of carnivorous pelagic cod (*Gadus morhua*), most of the plutonium was accumulated in its intestine. The different observation was done according to benthic, feeding on plankton and invertebrates flounder (*Platichthys flesus*) – both gills and intestine contained a significant comparable amount of $^{239+240}Pu$ and ^{241}Pu [8.21, 8.27]. These data might suggest a substantial amount of plutonium came directly from seawater, especially in the case of gills. The studies on plutonium $^{239+240}Pu$ and ^{241}Pu in carnivorous cod's gills and skin indicated that skin mucus contained the highest amount of plutonium (66% of total $^{239+240}Pu$ and 98% of total ^{241}Pu in the skin). The mechanism for uptake of $^{239+240}Pu$ and ^{241}Pu in the cod skin was passive adsorption onto the mucus surface. Only 13% of $^{239+240}Pu$ and 18% of ^{241}Pu in the gills of cod existed in adsorbed fraction what indicated plutonium content in gills of cod was involved in the metabolism of this radionuclide in fish. However, it could be the opposite in the case of pelagic herring that fed on plankton – its lamellas might contain a higher amount of plutonium due to the feeding style [8.20, 8.21, 8.28].

The calculated values of the $^{238}Pu/^{239+240}Pu$, and $^{241}Pu/^{239+240}Pu$ activity ratios in different components of the southern Baltic indicated that the plutonium has come as a consequence of the Chernobyl accident mostly [8.20-8.22, 8.27]. The significant amount of the Chernobyl plutonium in Baltic pelagic organisms has been confirmed by $^{240}Pu/^{239}Pu$ atomic ratio analysis. The obtained data have been differentiated; the highest value of the $^{240}Pu/^{239}Pu$ atomic ratio was found in the stomach of herring (0.44±0.03) and can be traced to its indirect contact with water and food water containing the Chernobyl plutonium. Lower $^{240}Pu/^{239}Pu$ atomic ratio has been measured in the skin of benthic flounder (0.26±0.04). It might indicate the global atmospheric fallout ($^{240}Pu/^{239}Pu$ 0.18) as the primary source of plutonium and the increasing impact of the Chernobyl-derived plutonium ($^{240}Pu/^{239}Pu$ 0.40). These calculations have been in good agreement with data determined by alpha spectrometry measurements. The analyses of $^{238}Pu/^{239+240}Pu$ and $^{241}Pu/^{239+240}Pu$ activity ratios revealed the smallest contribution of the Chernobyl-derived plutonium in benthic fish. In contrast, pelagic herring had a somewhat greater ratio (e.g. $^{238}Pu/^{239+240}Pu$ at 0.39±0.05) [8.18, 8.27, 8.39].

The reflection of the plutonium contamination level at the southern Baltic Sea coast might also occur in marine birds. Generally, the highest $^{239+240}Pu$ and ^{241}Pu activity concentrations were measured in the whole body of wintering seabirds such as black guillemot (*Cepphus grylle*), long-tailed duck (*Clangula hyemalis*), or common eider (*Somateria mollissima*), and in migratory seabirds such as common guillemot (*Uria aalge*), razorbill (*Alca torda*), or red-throated diver (*Gavia stellata*). These migratory birds contained the highest $^{239+240}Pu$ and ^{241}Pu concentrations in feathers, skeleton, and viscera, while the seabirds permanently habitating at the southern Baltic seacoast had the highest plutonium concentrations in the liver, e.g. great cormorant (*Phalacrocorax carbo*) (Table 3). The calculated values of the $^{238}Pu/^{239+240}Pu$, and $^{241}Pu/^{239+240}Pu$ activity ratios in individual parts of analyzed bird bodies indicated different plutonium sources, the Chernobyl accident, global atmospheric fallout or nuclear fuel reprocessing facilities [8.29, 8.30]. Some experiments were also conducted to estimate the $^{239+240}Pu$ sources in bird's feathers of the migratory bird; namely, razorbill was chosen, and about 82% of $^{239+240}$Pu present in feathers was adsorbed on their surface and came mainly from the external, namely the environment. The rest of $^{239+240}$Pu (18%), analogous to uranium or polonium, originated from metabolising the ingested isotope and was built in the feathers structure during their growth [8.29, 8.30, 8.31, 8.32].

8.3.2. Plutonium ^{238}Pu, $^{239+240}$Pu and ^{241}Pu influx to the southern Baltic Sea coastal basins

About 250 rivers enter the Baltic Sea with 500 km^3 of freshwater per year and in 2000-2003 contained about 1.5 GBq of $^{239+240}$Pu [8.1, 8.33]. Skwarzec et al. (2003) studied the plutonium isotopes content in Polish rivers, and they have shown rivers were the main source of plutonium in the Gulf of Gdańsk and the Gdańsk Basin (mainly the Vistula and the Neman River) [8.26].

The study on plutonium isotopes in the Vistula River catchment area indicated differences in their activities, and they depended on the season, weather conditions, geological structure. This research showed the highest annual surface entry of ^{238}Pu, $^{239+240}$Pu, and ^{241}Pu from the Vistula River drainage was observed for the mountain tributaries, like the Dunajec and the San, while the lowest was determined in lowland rivers like the Bug & the Narew or Bzura. Quarterly, the highest values of plutonium surface runoff were calculated for winter while the lowest for autumn and summer. In the study period 2002-2003, the southern Baltic Sea (the Gulf of Gdańsk and the Vistula Lagoon) was enriched with 10.3 MBq of ^{238}Pu, 89.0 MBq of $^{239+240}$Pu and 1.65 GBq of ^{241}Pu annually. The study showed rivers (mainly the Vistula and the Neman River) were the main source of plutonium in the Gulf of Gdańsk and the Gdańsk Basin and supplied these regions with 78% of its total content. According to the whole Baltic Sea, the Vistula brought about 7% of the total plutonium present in this sea (Table 8.4) [8.26, 8.34, 8.36].

TABLE 8.4. A SEASONAL AND ANNUAL FLUX OF ^{238}Pu, $^{239+240}$Pu, AND ^{241}Pu FROM THE VISTULA AND THE ODRA RIVER TO THE BALTIC SEA IN 2002-2003 (*Adapted from [8.34-8.36]*)

Season	Vistula			Odra		
	^{238}Pu (MBq)	$^{239+240}$Pu (MBq)	^{241}Pu (GBq)	^{238}Pu (MBq)	$^{239+240}$Pu (MBq)	^{241}Pu (GBq)
Winter	4.17	16.1	0.56	2.19	7.41	0.11
Spring	5.03	51.7	0.71	3.73	21.1	0.16
Summer	0.33	3.86	0.06	2.03	10.5	0.10
Autumn	0.81	17.3	0.32	1.57	6.89	0.25
Annual	10.3	89.0	1.65	9.51	45.9	0.62

The study on plutonium isotopes activities in the Odra River catchment area indicated that their differentiated concentrations depended on the season and weather conditions. The highest annual surface inflow of ^{238}Pu, $^{239+240}$Pu, as well as ^{241}Pu from the Odra River watershed, was observed for mountain tributaries from "the Opole anomaly" such as the Bóbr, the Nysa Kłodzka and the Nysa Łużycka [8.26, 8.35, 8.36].

Seasonally, the highest values of plutonium surface runoff were calculated for the winter, while the lowest was for the autumn and the summer, similarly to the Vistula River. Thus, the southern Baltic Sea (the Pomeranian Bay) could be, in period 2000-2003, enriched annually with 9.51 MBq of ^{238}Pu, 45.9 MBq of $^{239+240}$Pu, and 0.62 GBq of ^{241}Pu carried by the Odra. Therefore about 3% of the total plutonium came into the Baltic Sea with the Odra River water (Table 8.4, Fig. 8.5) [8.26, 8.35, 8.36]. The values of the ^{238}Pu/$^{239+240}$Pu, and ^{241}Pu/$^{239+240}$Pu activity ratios in the water of the Vistula and the Odra rivers indicated that the plutonium fraction that came from the Chernobyl accident was less than 40% [8.34-8.36].

The next crucial source of plutonium in the southern Baltic Sea was the atmospheric fallout, both the global from nuclear weapon tests and the Chernobyl accident. The study on dry atmospheric fallout collected monthly in 1986 in Gdynia (northern Poland, at the Gulf of Gdańsk) indicated a notable impact of the Chernobyl accident on plutonium presence. At the end of April 1986, there were an enormous increase of $^{239+240}$Pu and ^{241}Pu activities in the air over Gdynia and ^{240}Pu/^{239}Pu atomic ratio values changed significantly (Table 5) [8.7, 8.18, 8.22].

TABLE 8.5. $^{239+240}$Pu AND ^{241}Pu ACTIVITY CONCENTRATIONS, AS WELL AS THE VALUES OF ^{240}Pu/^{239}Pu ATOMIC RATIOS IN AIRBORNE DUST, COLLECTED OVER GDYNIA IN 1986 (*Adapted from [8.7, 8.18, 8.22]*)

Sampling Date	^{241}Pu (mBq·g^{-1} dw ± SD)	$^{239+240}$Pu (µBq·g^{-1} dw ± SD)	^{240}Pu/^{239}Pu
January	0.162 ± 0.034	4.14 ± 0.67	0.36 ± 0.06
February	1.008 ± 0.143	32.9 ± 3.2	0.23 ± 0.03
March	0.860 ± 0.160	26.2 ± 3.6	0.29 ± 0.04
April	3504 ± 32.9	65122 ± 723	0.47 ± 0.02
May	31.9 ± 4.1	662 ± 100	0.38 ± 0.02
June	2.45 ± 1.00	61.7 ± 25.2	0.41 ± 0.06
July	2.34 ± 0.49	58.2 ± 12.7	0.49 ± 0.10
September	2.85 ± 0.86	41.2 ± 13.0	0.47 ± 0.07
November	1.04 ± 0.35	18.4 ± 5.8	0.44 ± 0.15
December	0.99 ± 0.28	26.3 ± 5.9	0.37 ± 0.10

It was expected that the plutonium influx after the Chernobyl accident would be more prolonged and stretched over time. Still, the airborne dust plutonium determination revealed the opposite trend and showed its considerable dilution in 1986 over Gdynia at the Baltic seacoast (Fig. 8.4). The plutonium $^{239+240}$Pu and ^{241}Pu activity concentrations decreased considerably in a few months; consequently, the values of $^{241}Pu/^{239+240}Pu$ activity ratio and $^{240}Pu/^{239}Pu$ atomic ratio reached the pre-Chernobyl accident level.

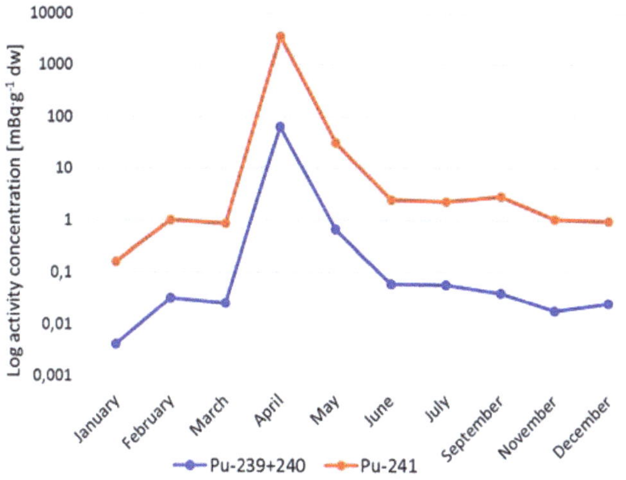

a)

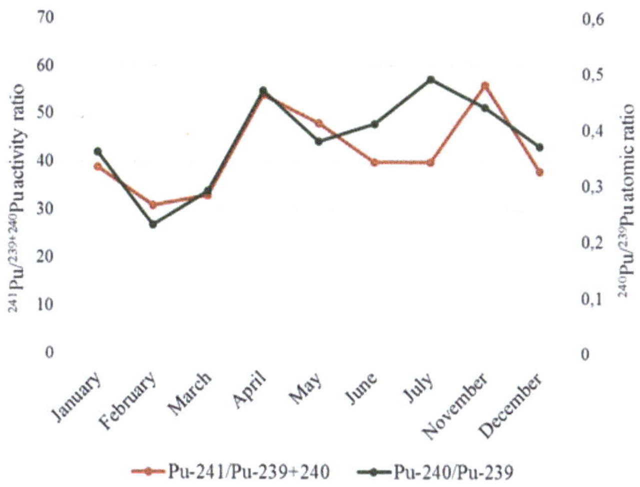

b)

FIG. 8.4. Trends in a) $^{239+240}Pu$ and ^{241}Pu activity concentrations, and b) the values of $^{241}Pu/^{239+240}Pu$ activity ratio and $^{240}Pu/^{239}Pu$ atomic in the airborne dust collected over Gdynia in 1986.

8.3.3. Plutonium $^{239+240}Pu$ budget in the Gulf of Gdańsk and the Gdańsk Basin

The mechanisms of water circulation in the Baltic Sea are different in comparison with other aquatic systems. The Gulf of Gdańsk, and to a less extent the Gdańsk Basin, are under the significant influence of the Vistula River water. This river provides about 32.1 km^3 of water to the Gulf of Gdańsk, constituting 7% of total river inflow to the Baltic Sea and 10% of water volume in the Gulf of Gdańsk [8.33]. At the mouth cone, there is no low muddy-loam fraction which constitutes about 60% of material transported with the water of the Vistula River. The intensive chemical and hydrodynamic processes have led to highly differentiated and unstable river-marine muddy-loam sediments with sandy layers [8.37]. These processes result in plutonium in the Vistula water being transported and deposited in deeper parts of the Gulf of Gdańsk [8.38]. The scheme of plutonium budget determination in the Gulf of Gdańsk and the Gdańsk Basin was presented in Fig. 8.5.

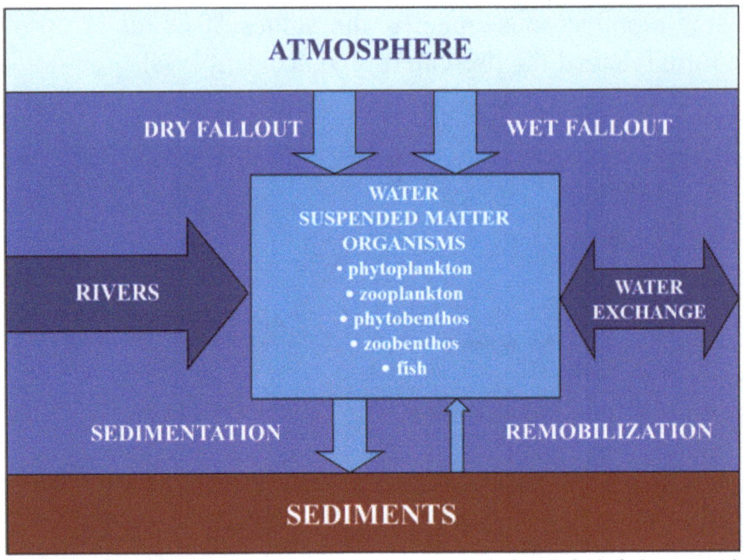

FIG. 8.5. Scheme of plutonium flux in the Gulf of Gdańsk and the Gdańsk Basin "(produced from Ref. [8.26] with permission courtesy of [Elsevier])".

In 2003 the Gulf of Gdańsk seawater (with suspended matter) contained about 2.33 GBq of $^{239+240}$Pu, and 9.92 GBq of $^{239+240}$Pu was present in the Gdańsk Basin. In both cases, 56% of $^{239+240}$Pu was associated with suspended matter. Living organisms of the Gulf of Gdańsk contained 3.81 MBq of $^{239+240}$Pu, while 7.45 MBq of $^{239+240}$Pu is in the Gdańsk Basin. From these values, in the Gulf of Gdańsk: 82.1% was deposited in zoobenthos, 1.6% in phytoplankton, 13.6% in phytobenthos, 1.2% in fish, and 1.5% in zooplankton. According to data from the Gdańsk Basin: 83.2% of plutonium was deposited in zoobenthos, 3.6% in phytoplankton, 2.5% in fish, and 7.5% in phytobenthos, [8.26].

The total $^{239+240}$Pu amount deposited in the Gulf of Gdańsk and the Gdańsk Basin in 2000-2003 was estimated at 1.18 TBq, and 3.76 TBq, respectively, and almost whole plutonium present in the Baltic Sea ecosystem (understood as biotic and abiotic components) was deposited in its sediments (about 99%). Based on the calculations and data available in the literature, the sediments of the Gulf of Gdańsk and Gdańsk Basin contained 6.4% and 22.0% of $^{239+240}$Pu deposited in sediments of the whole Baltic Sea. The areas of the Gulf of Gdańsk and the Gdańsk Basin constituted only 1.2% and 4.4%, respectively, of the total area of the Baltic Sea, and the capacities of the Gulf of Gdańsk and the Gdańsk Basin were 1.3% and 5.7%, respectively, of the total Sea. These results suggested that both regions contained sediments highly enriched with plutonium (Table 8.6) [8.26].

TABLE 8.6. $^{239+240}$Pu PLUTONIUM INVENTORY IN THE GULF OF GDAŃSK, GDAŃSK BASIN AND BALTIC SEA FOR 2003 (*Adapted from [8.3-8.5, 8.26]*)

Sample	Baltic Sea	Gdańsk Basin		Gulf of Gdańsk	
		Activity	Baltic part (%)	Activity	Baltic part (%)
$^{239+240}$Pu in:					
• water (GBq)	200	4.3	2.2	1	0.5

• water and suspended matter (GBq)	-	9.9	5.0	2.3	1.2
• sediments (TBq)	15.2-24.2	3.77	22.0	1.19	6.4
• organisms (MBq)	'a few GBq'	7.50	~0.2	3.81	~0.1
Area (km^2)	415266	18178	4.4	4940	1.2
Capacity (km^3)	21721	1239.5	5.7	291.2	1.3

8.4. CONCLUSIONS

A historical trend of temporal anthropogenic contamination of the southern Baltic seacoast with plutonium was observed. The analysis of the values of the $^{238}Pu/^{239+240}Pu$ and $^{241}Pu/^{239+240}Pu$ activity ratios allowed following plutonium sources and revealed the Chernobyl was a dominant source in 2000. Significant increase of ^{241}Pu content in seawater within 10-15 years after the Chernobyl accident occurred. Its growth could be expected in the other compartments of the Baltic ecosystem due to the effect of late inflow to the Baltic Sea from the catchment area. Plutonium fractions in the southern Baltic sediments also occurred as bioavailable fractions; however, the amount was insignificant. The Baltic Sea organisms did not accumulate plutonium to the same levels. The analysis of plutonium in airborne dust revealed the plutonium influx after the Chernobyl accident was not prolonged and stretched over time, but the trend was opposite and showed its considerable dilution in the air.

In 2000-2003, the total $^{239+240}Pu$ amount deposited in the Gulf of Gdańsk and the Gdańsk Basin was estimated at 1.18 TBq, and 3.76 TBq, respectively, almost whole plutonium present in the Baltic Sea ecosystem (understood as biotic and abiotic components) was deposited in its sediments (about 99%). Based on the calculations, the Gulf of Gdańsk and Gdańsk Basin sediments contained 6.4%, and 22.0% of $^{239+240}Pu$ deposited in the whole Baltic Sea's sediments suggested that the sediments of both regions were highly enriched with plutonium.

8.5. REFERENCES

[8.1] SKWARZEC, B., Polonium, Uranium and Plutonium in the Southern Baltic Sea, Ambio **26** 2 (1997) 113-117.

[8.2] DONARD, O.F.X., BRUNEAU, F., MOLDOVAN, M., GARRAUD, H., EPOV, V.N., BOUST, D., Multi-isotopic determination of plutonium (^{239}Pu, ^{240}Pu, ^{241}Pu and ^{242}Pu) in marine sediments using sector-field inductively coupled plasma mass spectrometry, Anal. Chim. Acta 587 (2007) 170-179.

[8.3] HOLM, E., Plutonium in the Baltic Sea, Appl. Radiat. Isot. 46 (11) (1995) 1225-1229.

[8.4] SALO, A., Strontium-90, cesium-137, plutonium-239 and plutonium-240 inventories in the Baltic Sea, Proc. 7th Nordic Society for radiation Protection Ordinary Meeting, Copenhagen, Denmark, 10-12 October, pp. 10-16 (1984).

[8.5] SALO, A., TUOMAINEN, K., VOIPIO, A., Inventories of certain long-lived radionuclides in the Baltic Sea, In Study of radioactive materials in the Baltic Sea, IAEA-TECDOC-362, Vienna, pp. 53-62. (1986).

[8.6] PAATERO, J., JAAKKOLA, T., IKAHEIMONEN, T.K., Regional distribution of Chernobyl derived plutonium deposition in Finland. J. Radioanal. Nucl. Chem. 252 (2002) 407-412.

[8.7] STRUMIŃSKA-PARULSKA, D.I., SKWARZEC, B., Plutonium isotopes ^{238}Pu, $^{239+240}$Pu, ^{241}Pu and ^{240}Pu/^{239}Pu atomic ratios in the southern Baltic Sea ecosystem, Oceanologia 52(3) (2010) 499-512.

[8.8] STRUMINSKA-PARULSKA, D.I., SKWARZEC, B., PAWLUKOWSKA, M., Plutonium fractionation in southern Baltic Sea sediments, Isot. Environ. Health Stud. 48 (2012) 526-542.

[8.9] IAEA – International Atomic Energy Agency, Safety series, 75, INSAG-1, IAEA, Vienna (1986).

[8.10] HODGE, V.F., HOFFMAN, F.L., FOLSOM, T.R., Rapid accumulation of plutonium and polonium on giant brown algae, Health Phys. 27 (1974) 29-35.

[8.11] PILLAI, K.C., MATTHEW, E., Plutonium in the aquatic environment; its behaviour, distribution and significance, Proc. of the Symposium on Transuranium Nuclides in the Environment, 25-45, IAEA-SM-199/27, USER and IAEA, 17-21 November 1975 (1976).

[8.12] PENTREATH, R.J., The biological availability to marine organisms of transuranium and the other long-lived nuclides. Proc. of an Int. Symp. On the Impacts of Radionuclide Releases into the Marine Environment, Vienna, 241, IAEA-SM-248/102, IAEA, Vienna (1981).

[8.13] MORSE, J.W., CHOPPIN, G.R., Laboratory studies of plutonium in marine systems, Mar. Chem. 20 (1986) 73.

[8.14] Suplińska, M., Plutonium in sediments of the Baltic Sea in the period of 1991–1993, Nukleonika 40 (1995) 33.

[8.15] FRANCIS, A.J., Microbial mobilization and immobilization of plutonium, J. Alloys. Comp. 444–445 (2007) 500.

[8.16] OLSZEWSKI, G., ANDERSSON, P., LINDAHL, P., ERIKSSON, M., On the distribution and inventories of radionuclides in dated sediments around the Swedish coast, J. Environ. Radioact. 186 (2018) 142-151.

[8.17] SKWARZEC B, Radiochemical methods for the determination of polonium, radiolead, uranium and plutonium in the environmental samples, Chem. Anal. (Warsaw), 42 (1997) 107.

[8.18] STRUMIŃSKA-PARULSKA, D.I., A Study on determination of potentially hazardous plutonium isotopes in environmental samples, J. Environ. Sci. Health C 31 (2013) 145–169.

[8.19] CHAMIZO, E., GARCIA-LEON, M., SYNAL, H.A., SUTER, M., WACKER L., Determination of the Pu-240/Pu-239 atomic ratio in soils from Palomares (Spain) by low-energy accelerator mass spectrometry, Nucl. Inst. Meths. Phys. Res., 249 (2006) 768.

[8.20] STRUMIŃSKA, D.I., SKWARZEC, B., Plutonium concentrations in waters from the southern Baltic Sea and their distribution in cod (*Gadus morhua*) skin and gills, J. Environ. Radioact. 72 (2004) 355-361.

[8.21] STRUMIŃSKA-PARULSKA, D.I., SKWARZEC, B., ^{241}Pu concentrations in water, plankton and fish from the southern Baltic Sea, Radiochim. Acta 101 (2013) 405-412.

[8.22] STRUMIŃSKA, D.I., SKWARZEC, B., ^{241}Pu concentration in southern Baltic Sea ecosystem, J. Radioanal. Nucl. Chem. 268(1) (2006) 59-63.

[8.23] STRUMIŃSKA-PARULSKA, D.I., Vertical distribution of ^{241}Pu in the southern Baltic Sea sediments, Mar. Poll. Bull. 89 (2014) 12–15.

[8.24] HOLM, E., Sources and distribution of anthropogenic radionuclides in the marine environment, In Radioecology, Lectures in Environmental Radioactivity, ed. E. Holm, World Scientific, Singapore-New Jersey-London-Hong Kong, 63-83 (1994).
[8.25] HELCOM, Baltic Sea environment proceedings, Radioactivity in the Baltic Sea 1984-1991, Helsinki Commission, 61, (1995) 69-113.
[8.26] SKWARZEC, B., STRUMIŃSKA, D.I., PRUCNAL, M., Estimates of $^{239+240}Pu$ inventories in Gdańsk bay and Gdańsk basin, J. Environ. Radioact. 70 (2003) 237-252.
[8.27] SKWARZEC, B., STRUMIŃSKA, D.I., BORYŁO, A., Bioaccumulation and distribution of plutonium in fish from Gdańsk Bay, J. Environ. Radioact. 55 (2001) 167.
[8.28] SKWARZEC, B., HOLM, E., ROOS, P., PEMPKOWIAK, J., Nickel-63 in Baltic fish and sediments, Int. J. Appl. Radiat. Isot. 45 (1994) 609-611.
[8.29] STRUMIŃSKA-PARULSKA, D.I., SKWARZEC, B., FABISIAK, J., Plutonium bioaccumulation in seabirds, J. Environ. Radioact. 102 (2011) 1105-1111.
[8.30] STRUMIŃSKA-PARULSKA, D.I., SKWARZEC, B., Characterization of ^{241}Pu occurrence, distribution, and bioaccumulation in seabirds from northern Eurasia, Environ. Sci. Pollut. Res. 22 (2015) 7821-7832.
[8.31] FURNESS, R.W., CAMPHUYSEN, C.J., Seabirds as monitors of the marine environment, J. Mar. Sci. 54 (1997) 726-737.
[8.32] SKWARZEC, B., FABISIAK, J., Bioaccumulation of polonium ^{210}Po in marine birds, J. Environ. Radioact. 93 (2007) 119-126.
[8.33] MIKULSKI, Z., Wody śródlądowe w strefie brzegowej południowego Bałtyku, PIHM, Oddział Morski (1970).
[8.34] SKWARZEC, B., JAHNZ-BIELAWSKA, A., STRUMIŃSKA-PARULSKA, D.I., The inflow of ^{238}Pu and $^{239+240}Pu$ from the Vistula River catchment area to the Baltic Sea, J. Environ. Radioact. 102 (2011) 728-734.
[8.35] STRUMIŃSKA-PARULSKA, D.I., SKWARZEC, B., TUSZKOWSKA, A., The inflow of ^{238}Pu and $^{239+240}Pu$ from the Odra and Pomeranian rivers catchments area to the Baltic Sea, J. Environ. Radioact. 113 (2012) 63-70.
[8.36] STRUMIŃSKA-PARULSKA, D.I., SKWARZEC, B., ^{241}Pu in the biggest Polish rivers, J. Radioanal. Nucl. Chem. 298 (2013) 1693-1703.
[8.37] UŚCINOWICZ, S., Geochemistry of Baltic sea surface sediments, PIG – PIB (2011).
[8.38] ANDRULEWICZ, E., Toxic substances in the water, bottom sediments and organisms of the Gdansk Basin, Oceanological Stud. **25** (1996) 77-85
[8.39] BUESSELER, K.O., The isotopic signature of fallout plutonium in the North Pacific, J. Environ. Radioact. 36(1) (1997) 69.
[8.40] *http://data.bshc.pro/* Baltic Sea Hydrographic Commission, 2013, Baltic Sea Bathymetry Database version 0.9.3. Downloaded from http://data.bshc.pro/ on 16.01.2022

9. SEDIMENTARY STUDIES TO RECONSTRUCT AND CALCULATE LEVELS AND TRENDS IN HISTORICAL NUCLEAR DISCHARGES ON THE SWEDISH COAST

M. Eriksson
Linköping University
Linköping Sweden
Email: mats.k.eriksson@liu.se

G.Olszewski
Linköping University
Linköping Sweden
Email: grzegorz.olszewski@liu.se

P.Törnquist
Linköping University
Linköping Sweden
Email: per.tornquist@liu.se

H.B.L. Pettersson
Linköping University
Linköping Sweden
Email: hakan.pettersson@regionostergotland.se

Abstract

Release of radionuclides into the marine environment takes place today from most of the nuclear facilities around the Baltic Sea. Usually these releases are well documented and data on the levels can be reviewed. Such data is valuable when sedimentary studies on the geochemical behavior of elements are conducted. This study focuses on actinides as well as other elements in a semi anoxic basin on the Swedish coast about 100 km south of Stockholm. Discharge records are compared with levels found in dated sediment depth profiles, past unknown releases from the nuclear facility Studsvik are reconstructed.

9.1. INTRODUCTION

The nuclear facility at Studsvik was built in the late 1950s on the eastern coast of Sweden, approximately 100 km south of Stockholm on the shore of the bay called Tvären. Tvären, which is semi-sealed, was created about 450 million years ago by a meteor impact. Tvären has shallow (10-20 m) inlets and a deep central area with a maximum depth of 80 meters (Fig 9.1). The surface area is 18 km^2 with thermocline laying on average at a depth of 11 meters. The water temperature at the bottom of the deep parts range between 0-4 °C. There are periods with anoxic conditions at the bottom that create unique conditions to study sedimentation processes for different radionuclides [9.1, 9.2]. In the deep part of Tvären, undisturbed laminated sediment is formed, with high time resolution in the sediment profiles making it possible to distinguish single years.

Aquatic releases of radionuclide from the nuclear facilities and the permissible discharge levels are regulated and controlled by the SSM (Swedish Radiation Safety Authority) in Sweden. Historically, contamination have also occurred through fallout from accidents such as the

Chernobyl accident [9.3] and fallout from atomic weapon tests [9.4]. Today, after decommissioning of their nuclear reactors, the facility is mainly used for processing radioactive waste. Studsvik has a wastewater treatment plant of its own and liquid waste is listed in eight different categories [9.5-9.7]. Categories 1 and 2 have never been released, while categories 3 and 4 after purification and analysis for the radioactive levels, release in Bergösundet (see Fig 9.1, point A). There are 10 upward pointing nozzles placed 10 meters apart, at the bottom of Bergösundet, at a depth of 10 m. On average, the wastewater releases have occurred on a weekly basis, emitting 200-300 m^3 in 3-4 hours, until 1985 when the amount decreased to 150 m^3 and the time of release changed to 10 hours. The interval between the wastewater releases have been approximately 5 days since 1985 [9.8]. Categories 5 to 8 and cooling water were released through another pipe at a steady flow rate of about 1 m^3/s. The outlet of that pipe is situated about 600 m from the shore, in the Tvären (see Fig. 9.1. discharge point "B"). According to the annual reports, the cooling water contained very small amounts of anthropogenic radionuclides [9.6] and the regulations state that the activity released in the Tvären pipe should not exceed 1/10 of the released activity at Bergösundet [9.5].

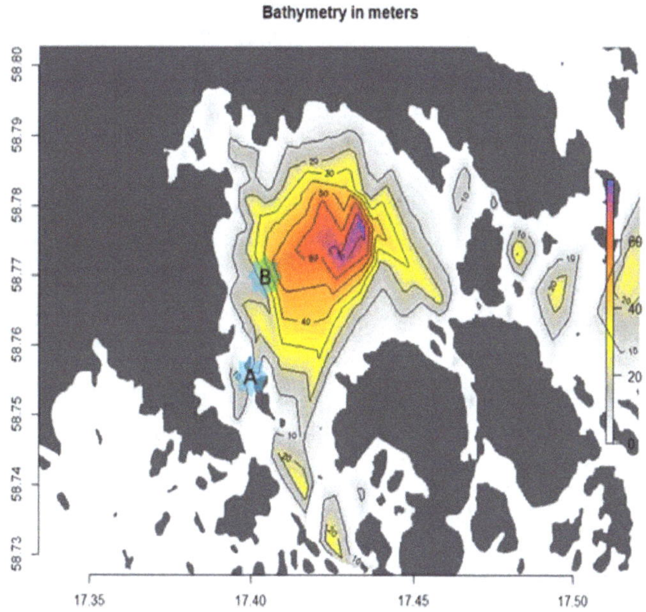

FIG. 9.1. The location and the bathymetry of the bay Tvären. The two aquatic discharge points from the Studsvik facility are indicated by "A" and "B" in the map.

9.1.1. Historical overview of Studsvik activities that might have led to discharges to the local marine environment.

At the beginning Studsvik was a research center concentrating on nuclear reactor technology. It also had the capacity to process enrichment of uranium to plutonium [9.9]. In the past the Studsvik facility used research reactors: a heavy water zero power reactor called R0, a pressurized water reactor called R2, a reactor called TZ (a pressurized Zero Energy Bare Reactor Assembly), a light water moderated reactor called KRITZ [9.10] and a neutron research laboratory. Experiments concerning different fuel symmetries and different fuel compositions under different conditions were conducted as well as experiments with neutrons, such as neutron capture therapy [9.11]. Additionally, a van de Graaff generator was used for various experiments relating to neutron physics [9.12]. The last operating reactor was shut down in 2005. Studsvik with is its spin-off companies, have also developed and incorporated several methods for reducing the volume of different kinds of waste containing radioactive materials, e.g. by incineration or metal melting.

In 2016 the Studsvik area housed the following nine radioactive facilities: The Hot cell laboratory, (HCL), The Active central laboratory, (ACL), the Incineration facility, (HA), the Melting facility, (SMA), the Treatment facility for intermediate waste, (HM), the Interim storage for low and intermediate waste, (AM), the Storage for solid intermediate waste, (AT), The Storage facility (FA) and Storage for radioactive waste, (AU) [9.13]. Work at all these facilities causes aquatic radioactive waste of which some amounts of low-activity radioactive waste have been released into the Baltic Sea, under strictly controlled conditions. One of the important projects managed by Studsvik was the Inter Ramp Project that started in 1975. It was an international fuel research project assigned to examine the failure propensity and characteristics of 20 short unpressurized fuel rods of a characteristic 8 x 8 Boiling Water Reactor (BWR) design. After having finished a prior long-term irradiation scheme containing a patterned cyclic variation in power levels, they were subjected to overpower ramp operation. The goal of the ramp operation is ito rapidly increase the power level of a fuel rod to a value substantially higher than those during pre-ramp operation. The ramp test's main objective was to show evidence of overpower failure limits for each fuel rod and, if failure occurs, to examine the characteristics and mechanisms of the failure. All irradiations, both long-term and power ramp, for the 20 fuel rods with average burnup of 10 to 20 GWd/MTU were performed in the R2 test reactor at Studsvik [9.14]. Many experiments were conducted using THE ISOTOPE-SEPARATOR-ON-LINE facility called OSIRIS for the study of strongly neutron-rich fission products [9.15].

9.1.2. Discharge data from Studsvik

Reports exist on the discharges of ^{3}H, ^{60}Co, ^{90}Sr and ^{137}Cs from the first experiment, 1959, and onwards, ^{54}Mn and ^{134}Cs from 1977 and ^{152}Eu and ^{154}Eu with regularity from 1989. Past emissions were not documented in detail and certainly not for the α- and β-emitting radionuclides. Annual reports of all α-emitting isotopes, called *Total α*, exist up until 2006 but since then there exist isotope-specific reports of ^{234}U, ^{238}U, ^{238}Pu, ^{239}Pu, ^{242}Cm and ^{244}Cm. β-emitting isotopes are referred to as *Total-β* until 2006. In table 9.1 the total amounts released can be seen and in Figs 9.2 and 9.3 the annual discharge as a function of time can be seen for some selected elements. In Fig. 9.2 a steady decline can be noticed in reported annual discharges from around the year 2000 and onwards. This is probably due to the decommissioning of the research reactors and less experiments resulting in less radioactive waste was produced at Studsvik. For comparison with other nuclear facilities around the Baltic Sea a figure published by HELCOM [9.16] is shown in Fig. 9.3. It can be noted that Studsvik is the facility releasing most activity in the period 1999 – 2010, with a total activity equal to the sum of releases of all the other facilities in the Baltic Sea (note that nuclear power plant Ringhals is situated on the Swedish west coast and doesn't release activity into the Baltic Sea area). It is especially Studsvik ^{90}Sr-releases that are pronounced in the figure. A similar discharge comparison is not available for the actinides.

TABLE 9.1 REPORTED AQUATIC RELEASES FROM THE STUDSVIK FACILITY FOR SELECTED RADIONUCLIDES

Nuclide	Accumulated annual reported discharge	Reported period	Accumulated reported annual discharge, decay corrected to 2017
^{137}Cs	2 800 GBq	1959-2017	1 200 GBq
^{134}Cs	570 GBq	1977-2017	0.39 GBq
^{54}Mn	310 GBq	1977-2017	0.06 GBq
^{60}Co	1 700 GBq	1964-2017	37 GBq
^{152}Eu	61 GBq	1988-2017	18 GBq
^{3}H	1 900 000 GBq	1959-2017	350 000 GBq
^{90}Sr	1 700 GBq	1962-2017	840 GBq
Total alpha	53 GBq	1962-2017	-
Pu-238	7,0 MBq	2006-2017	-
Pu-239+240	2,9 MBq	2006-2017	-
Am-241	14 MBq	2006-2017	-

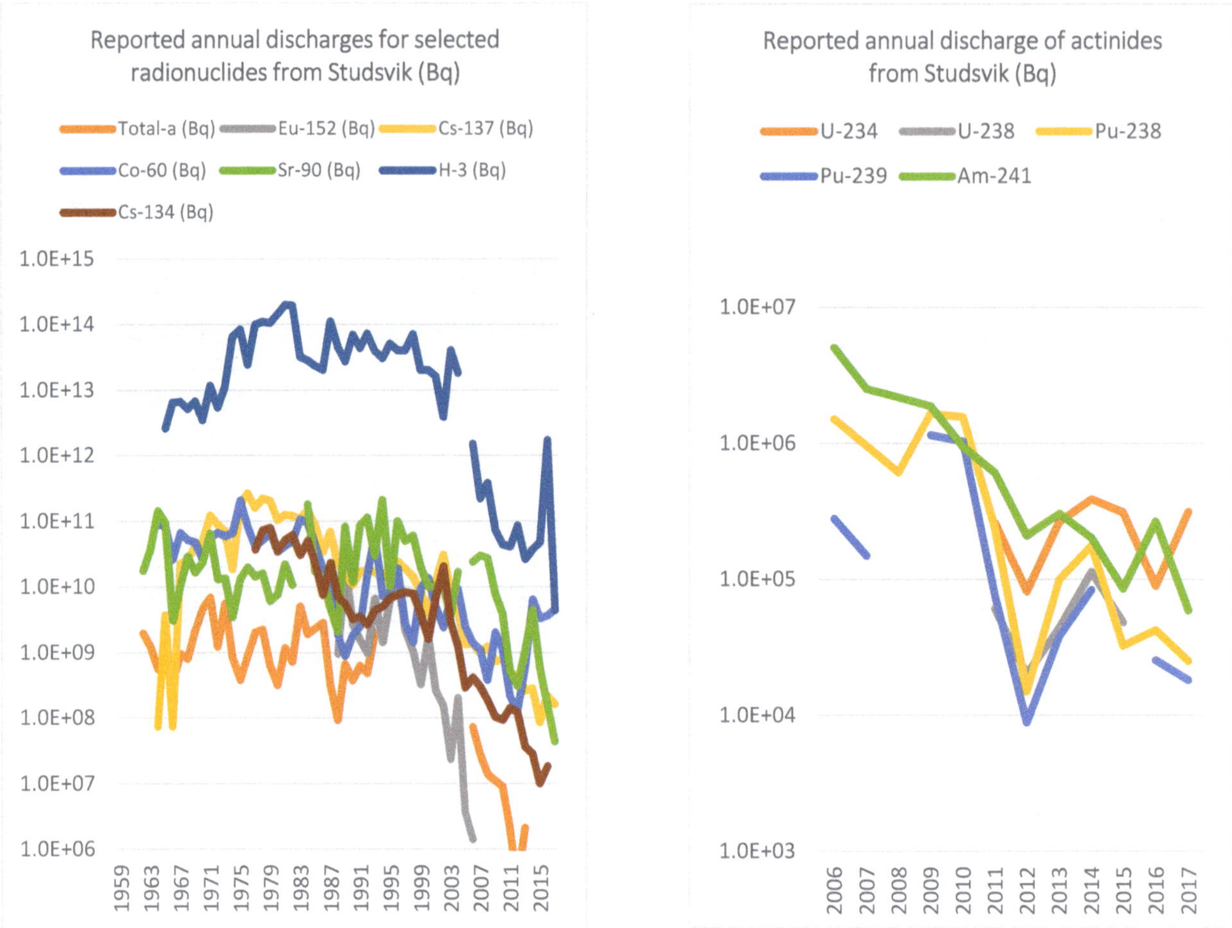

FIG. 9.2. Annual reported discharge to the Baltic Sea of selected radionuclides from Studsvik.

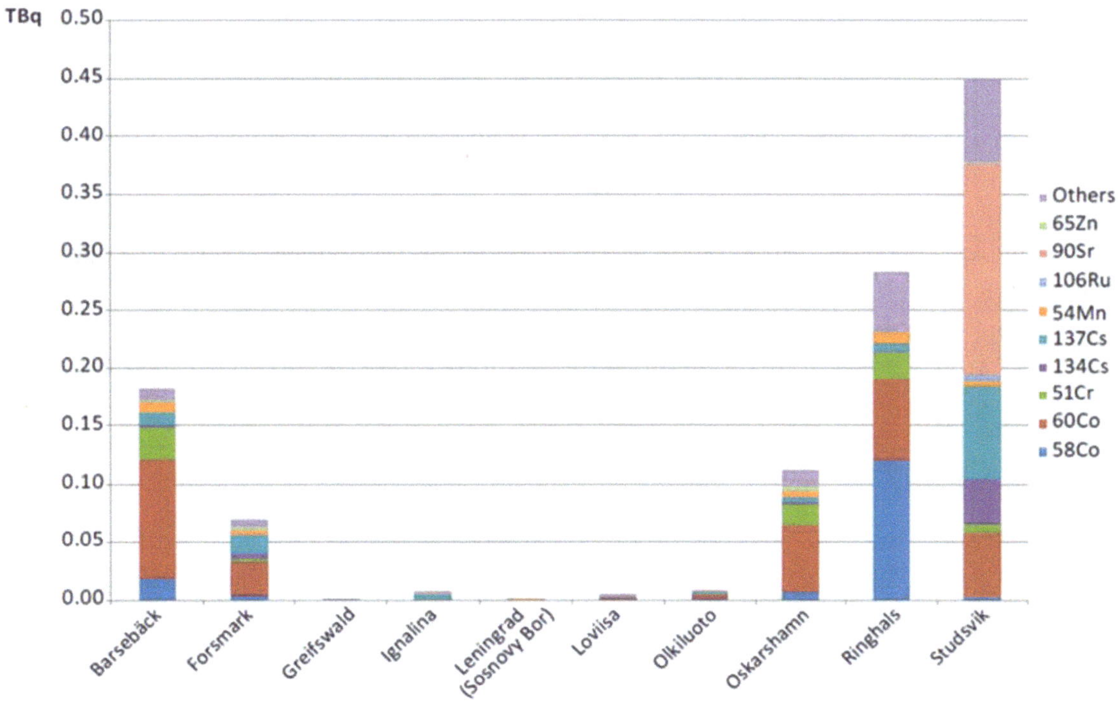

FIG. 9.3. Total aquatic discharges from local nuclear facilities into the Baltic Sea in 1999 - 2010 (excluding ^{3}H). The data were adapted from Ref. [9.16].

9.1.3. Other sources of radioactivity in the vicinity of Studsvik

In the Baltic Sea there are also other important sources of artificial radioactivity not originating from Studsvik operations. The special characteristics of the Baltic Sea (a semi-enclosed shallow sea) determine an increased accumulation of radioactive materials compared to other seas or oceans [9.17]. There were two major events that introduced artificial radioactivity, including plutonium and americium to the Baltic Sea ecosystem. First of all, the high-yield atmospheric nuclear weapons tests, with the maximum fallout dated to 1963 resulted in a $^{239+240}$Pu deposition of about 60 Bq·m^{-2} in the Northern hemisphere [9.18]. Around 40-50 Bq·m^{-2} of $^{239+240}$Pu deposited in the Baltic Sea region resulting with 16 to 18 TBq inventories [9.17]. The second main event was the Chernobyl accident in 1986, which introduced about 1.5 TBq of $^{239+240}$Pu to the Baltic Sea [9.17]. Additional smaller sources include rivers. It is estimated that riverine waters bring about 2 GBq of $^{239+240}$Pu per year [9.19]. In addition to actinides the fallout from nuclear test also deposited ^{90}Sr and ^{137}Cs to the Baltic Sea area and its catchment. The total global fallout of ^{90}Sr and ^{137}Cs has been estimated to 600 PBq and 960 PBq respectively [9.20, 9.21]. Most of the fallout was on the Northern hemisphere and around Studsvik latitude (~59 degrees north) this would mean an aerial deposition of around ~2 kBq·m^{-2} for ^{90}Sr and around 3.5 kBq·m^{-2} for ^{137}Cs. These activity concentrations have since then decreased, due to physical decay of the radionuclides. The Chernobyl accident released ^{90}Sr and ^{137}Cs at a ^{90}Sr/^{137}Cs activity ratio of about 1/10 and the deposition was heterogenous over Sweden. The Swedish radiation safety authority, SSM, has published a deposition map over land that are based on monitoring the deposition by airplane [9.22, 9.23]. The reported ^{137}Cs deposition is a sum of both Chernobyl and global fallout. Today about 1-2 kBq·m^{-2} can be found in the soil around Studsvik from these two sources.

Source signatures of the radionuclide emissions can be identified using activity and atomic rations between the radionuclides of interest. The ^{238}Pu/$^{239+240}$Pu activity ratio in global fallout has been estimated at 0.025-0.028 [9.17, 9.24], 0.3 for reprocessing plants like Sellafield [9.25], and 0.47 for Chernobyl fallout [9.17, 9.26]. In order to calculate the $^{241}Am/^{239+240}Pu$ activity ratio ^{241}Pu decay to ^{241}Am with a half-life of 14.35 years has to be accounted for. Assuming the $^{241}Am/^{239+240}Pu$ ratio in the Chernobyl fallout as 0.37 [9.27, 9.28], and $^{241}Pu/^{239+240}Pu$ activity ratio as 74.6 in 1986 [9.29], the $^{241}Am/^{239+240}Pu$ activity ratio of 2.3 would be present in 2017. Similar calculation led to the global fallout $^{241}Am/^{239+240}Pu$ activity ratio of around 0.54 in 2017 (assuming a global fallout $^{241}Pu/^{239+240}Pu$ activity ratio of 16 and a $^{241}Am/^{239+240}Pu$ activity ratio of 0.1 in 1963) [9.17, 9.30, 9.31]. ^{237}Np input with global fallout and Chernobyl accident are much lower compared to plutonium ($^{237}Np/^{239+240}Pu$ activity ratio of 0.0037 and 0.000125 in the global and the Chernobyl fallout, respectively) [9.26]. Atomic ratios of $^{240}Pu/^{239}Pu$ varies from 0.05 for weapons grade Pu to 0.18 for global fallout and 0.4 for the Chernobyl accident [9.29].

9.2. MATERIAL AND METHODS

A solid review of Studsvik activities and discharges was conducted, both by the access to Studsvik library and archive, personal interview with staffs (both active and retired) and from reported data to the SSM. This included archives of monthly discharges, both airborne and aquatic from 1959-2019, of which data from the aquatic releases were compiled.

9.2.1. Sediment sampling

To trace the mobility and to estimate the inventories of the discharged radionuclides several sediment cores were sampled during four sampling-campaigns. Sediment sampling was conducted with a boat with an electric derrick. The big advantage with that is that the core sampler reaches the bottom with a constant speed. After the first sampling campaign, it was observed that in order to penetrate deeper in the sediment, the sampler needed modification to allow for larger added weight. In total, 29 cores were collected using a gravitational piston core sampler, HTH Gravity Corer 90 mm, from Pylonex AB in Sweden. The sampling locations can be seen in Fig 9.4. The sample locations were selected to cover the whole Tvären area to gain the spatial distribution of the dischaged radionuclides. In order to get the chronology and to be able to reconstruct the discharge history, several cores was sampled in the centeral deep part of Tvären where the sedimentation rate is highest.

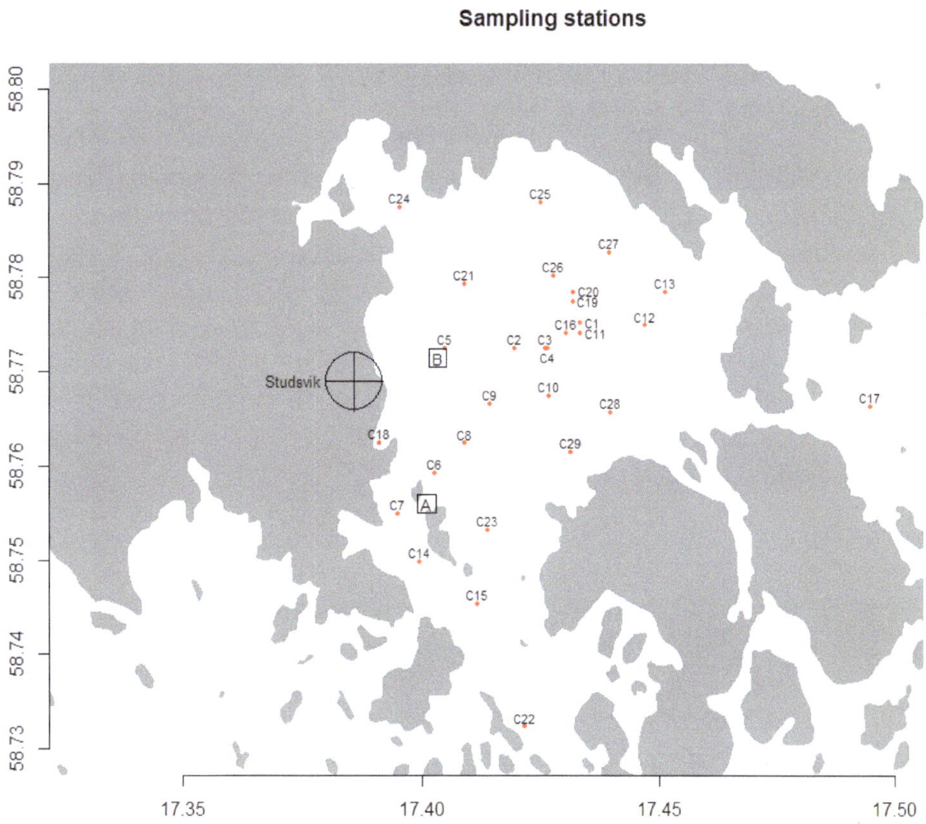

FIG. 9.4. Sediment stations around Tvären area.

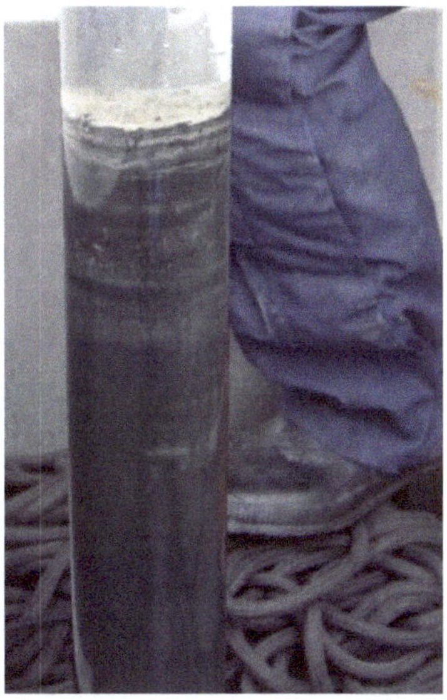

FIG. 9.5. A laminated sediment core taken in the central part of Tvären.

The sediment core, C3, that was measured in detail was taken in the part of Tvären at a depth of 76m. The sediment core was undisturbed with several visible layers, indicating a very low rate of bioturbation, see Fig 9.5.

9.2.2. Sample preparation

The cores were immediately transported to the port and sliced in 1 cm slices using an extruding device. The device is threaded so when turning the handles 360° a piston raises the core 5.0 mm. That is, 720° raises the core 1.0 cm. It is better to only take one or maximum two cores at a time. Decompression of gas was noted in one core that was left for about 15 minutes. Decompression of gas is an unwanted phenomenon due to the drastic change of pressure (compare with divers' disease), and to some extent the change of temperature, and can cause the sediment layers to mix. Therefore, it is important to process the cores as quickly as possible. If the ship is steady, immediate slicing onboard, is ideal. The slices were weighed the same day then put in a deep freezer. A vacuum freeze-dryer was used for the samples in order to get the dry weight of each slice. The samples were homogenized and placed in the calibrated sample containers. The sample containers were filled and made semi-tight for radon, using radon tight plastic bags and a conventional vacuum sealing system [9.32]. The samples were left for 3 weeks before measurement in the gamma spectrometer in order to gain secular equilibrium between ^{226}Ra and its short-lived daughters. Determination of ^{226}Ra concentration was evaluated as the weighted mean activity concentration of the daughters (^{214}Pb and ^{214}Bi) assuming no loss of radon from the sample containers. For the dating of the sediment cores ^{210}Pb was determined by its daughter ^{210}Po.

9.2.3. Measurements

For gamma spectrometry, a 43% HPGe-detector placed at the Department of Health, Medicine and Caring Sciences (HMV), Linkoping University was used. The detector efficiency calibration was done with a mixed gamma (^{60}Co, ^{134}Cs, ^{137}Cs, ^{152}Eu, ^{241}Am) certified reference standard. The samples were measured for about 48 h each. HMV uses a gas tight detector chamber that is flushed with nitrogen for each measurement in order to minimize ^{222}Rn-daughter background counts. Minimum detectable activity (MDA) for ^{137}Cs with the set up is 2.2 $Bq \cdot kg^{-1}$ using the Currie formulation for MDA.

Alpha emitting isotopes, ^{238}Pu, $^{239+240}Pu$, ^{241}Am, $^{243+244}Cm$ and ^{242}Cm were analysed at SSM using well-established methods [9.33]. Known amount of ^{242}Pu and ^{243}Am tracers were added to about 5 g sample. As Am and Cm have very similar chemical behaviour, they are following each other in the radiochemical separation method and deposition and are measured on the same disc using the ^{243}Am spike for both Am- and Cm-isotopes determination. The samples are incinerated over night at 450°C and leachedusing an *aqua regia* Next, Fe $(OH)_3$ co-precipitation was performed at pH=9, and Pu oxidation state was adjusted to (+4) followed by separation from U, Th, and Am, using anion exchange resin or TEVA resin (Triskem France). Before TRU resin (Triskem France) Am/Cm separation, CaC_2O_4 co-precipitation at pH around 4 was performed to remove Fe^{3+}. Finally, Am separation from U, Po, Pu, and Th residues, using TRU resin cartridges. In the last step, Am/Cm fraction was purified from rare earth elements using NH_4SCN with the aid of TEVA resinPrior to electrodeposition of the Pu and Am aliquots, the residues of NH_4SCN were decomposed using evaporation with mineral acids. Separated actinides fractions were electrodeposited on stainless steel discs from $Na_2(SO_4)$-H_2SO_4 solution (pH 2.1-2.5) [34]. Pu, Am, and Cm activities were measured using an Alpha Analyst (Canberra) alpha spectrometer. MDA values were evaluated as 0.09 mBq g^{-1}. according to Hurtgen et al. (2000) [35]. Pu activities were measured after two weeks allowing the ingrowth of ^{224}Ra from ^{228}Th what enabled corrections for ^{238}Pu count rates from ^{228}Th and ^{224}Ra spectral interferences. Obtained recoveries were between 70 to 95% for all analyzed radionuclides. To perform ^{210}Pb dating on sediments, ^{210}Po in secular equilibrium with ^{210}Pb was analyzed. For the determination of ^{210}Po in sediments a well-established deposition method was used where Po is self-deposited from weak HCl solution on noble metal disc i.e. Ag or Cu [36-38]

^{55}Fe, ^{63}Ni and ^{90}Sr were measured at DTU, Denmark, using radiochemical methods to separate each nuclide. ^{55}Fe and ^{63}Ni were measured using liquid scintillation counting. The separated ^{90}Sr was measured through its decay daughter ^{90}Y after 3 weeks ingrowth followed by separation of ^{90}Y using radiochemical method. The separated ^{90}Y was measured using ultra-low level beta counter (gas flow anti-coincidence GM counter).

During the first sampling cruise one test sediment core was taken, sliced into 5 cm section and analyzed in the laboratory to assess the levels of expected activities in the area. During further sampling cruises multiple sediment cores were collected from Tvären and based on the test core data sliced accordingly in-situ (1, 2 cm sections) and transported to the laboratory. All sediment slices were weighed, deep frozen and subsequently freeze dried. As a part of another project each slice was analyzed by HPGe gamma spectrometry to evaluate gamma emitting radionuclides. For dating and analysis of actinides 5 cores were selected that cover the area of Tvären and fulfil the intersect (Fig. 9.1). Selected cores were obtained from depths between 15 to 76 meters using HTH Gravity Corer (90 mm) from Pylonex AB in Sweden.

For stable elements X-ray fluorescence (XRF) analysis was conducted with a SPECTOR X-LAB PRO 2000 system at the IAEA environmental laboratories in Monaco. The calibration was made using sample cups with4 g samples and pre-mounted 4 μm thick Prolene thin-film. Measurement of IAEA Reference materials (IAEA-356, IAEA-405, IAEA-433) were used to validate the measurements. The sampling cups were filled with sediments that were homogenised and freeze-dried. Analysis was performed using a combination of measurements with three different targets, enabling the determination of elements from atomic number 13 (Al) to 92 (U).

9.3. RESULTS AND DISCUSSION

A comprehensive review of the Studsvik discharge history and past operations has been done and are presented in section 9.1.1. and 9.1.2. The review is based on published articles, internal and external Studsvik reports and interviews of former staff.

9.3.1. Dating of the sediment

Applying ^{210}Pb as a dating method is based on the disequilibrium between ^{210}Pb ($t_{1/2} = 22.3$ y) and its longer-lived parent isotope, ^{226}Ra ($t_{1/2} = 1602$ y), by diffusion of a short-lived gaseous intermediate isotope ^{222}Rn ($t_{1/2} = 3.82$ d). ^{226}Ra being a part of ^{238}U decay series is a natural radioisotope found in most rocks, soils, and sediments. When ^{226}Ra decays, ^{222}Rn is produced that diffuses into the atmosphere and further decays through a series of short-lived daughters to ^{210}Pb that is removed from the atmosphere with wet and dry precipitation. When reaching water bodies ^{210}Pb is attached to particulate matter and removed from the water column either by direct adhesion on the interface or by adsorption on inorganic particles or organic matter. ^{210}Pb is rather immobile so when adsorbed on sediment is a powerful marine tracer that enables dating up to 100 years [9.39, 9.40]. ^{210}Pb in sediments is measured as the sum of supported ^{210}Pb that is in secular equilibrium with ^{226}Ra and unsupported (excess) ^{210}Pb added to the sediment from water column. To quantify unsupported ^{210}Pb, the supported ^{210}Pb activity needs to be subtracted from total ^{210}Pb [9.40].

In literature there are many models developed to interpret unsupported ^{210}Pb profiles in the sediment core. If accumulation rate is steady and no sediment mixing is present, an ideal downcore ^{210}Pb activity profile has an exponential decrease. The most widely used ^{210}Pb models are: constant flux-constant sedimentation rate (CF:CS) model, constant rate of supply (CRS) model and constant initial concentration (CIC) model. All these models assume that ^{210}Pb when adsorbed on sediments, is immobile and that there are no interruptions in sedimentation. Otherwise different assumptions regarding specific activity, accumulation rates and flux of excess ^{210}Pb are applied [9.40]. It should be mentioned that it is difficult to assess both the total ^{210}Pb and supported ^{210}Pb (through ^{226}Ra daughters). Usually the activity concentrations typically have an uncertainty up to 10%. This is particularly true when the support level is determined by gamma-spectrometry due to possible ^{222}Rn leakage from the measured geometry in addition to the coincidence summing problem when determining ^{212}Bi and ^{212}Pb. Another challenge is application of self-attenuation corrections in the sample to correct the gamma spectra. All this will result in rather large uncertainties in the dating and the calculated sedimentation rates of the sediment core. In two interlaboratory sediment dating exercise performed by the IAEA this is clearly demonstrated and shows that even with the same set of data a wide range of results was obtained [9.41, 9.42].

Another widely used dating method, often employed as a validation process for ^{210}Pb dating, is establishing a ^{137}Cs concentration in time and relating the initial appearance or maximum ^{137}Cs activity to a particular year. In some cases, the use of $^{238,239,240}Pu$ is more advantageous as it is less prone to mobilization [9.43]. In the Northern Hemisphere the most significant ^{137}Cs fallouts from nuclear weapons testing occurred in 1953 and 1963. In some areas of the Northern Hemisphere 1986 fallout of ^{137}Cs from Chernobyl accident is also detectable. In the Southern Hemisphere this technique is practically not useful due to reduced activities although the same pattern is noticeable [9.40, 9.44, 9.45]. In the case of $^{239+240}Pu$, significant activities are related to the 1963 nuclear test, whereas ^{238}Pu is related to the Chernobyl accident [9.17]. In order to perform ^{137}Cs and $^{238,239,240}Pu$ dating correctly some assumptions have to be applied: a radionuclide, after entering the water column, is quickly adsorbed to suspended particle; once deposited the radionuclide is firmly adsorbed on the sediment particle; once deposited in sediment column, the radionuclide is immobile; physically or biologically mixed sediment surface layer needs to be included in calculations [9.40].

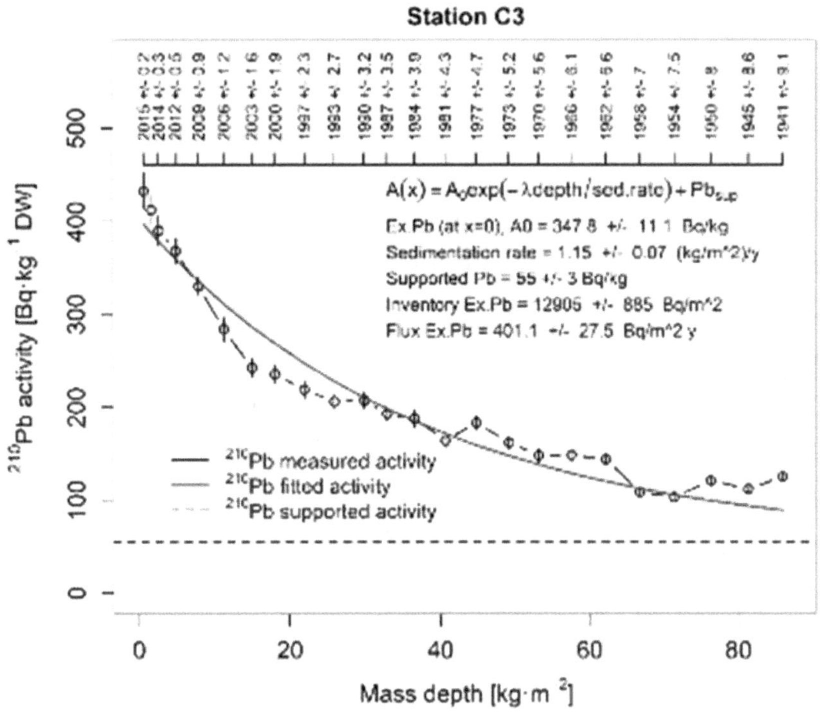
Station C3
A(x) = A0exp(−λdepth/sed.rate) + Pbsup
Ex.Pb (at x=0), A0 = 347.8 +/- 11.1 Bq/kg
Sedimentation rate = 1.15 +/- 0.07 (kg/m^2)/y
Supported Pb = 55 +/- 3 Bq/kg
Inventory Ex.Pb = 12905 +/- 885 Bq/m^2
Flux Ex.Pb = 401.1 +/- 27.5 Bq/m^2 y
210Pb measured activity
210Pb fitted activity
210Pb supported activity
210Pb activity [Bq·kg-1 DW]
Mass depth [kg·m-2]

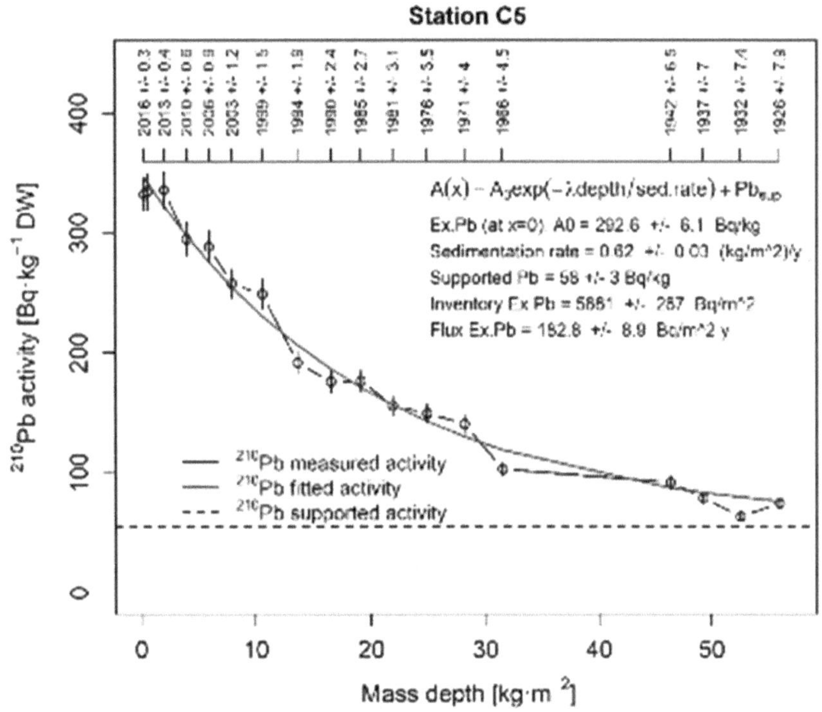
Station C5
A(x) = A0exp(−λdepth/sed.rate) + Pbsup
Ex.Pb (at x=0) A0 = 292.6 +/- 6.1 Bq/kg
Sedimentation rate = 0.62 +/- 0.03 (kg/m^2)/y
Supported Pb = 58 +/- 3 Bq/kg
Inventory Ex.Pb = 5881 +/- 287 Bq/m^2
Flux Ex.Pb = 182.8 +/- 8.9 Bq/m^2 y
210Pb measured activity
210Pb fitted activity
210Pb supported activity
210Pb activity [Bq·kg-1 DW]
Mass depth [kg·m-2]

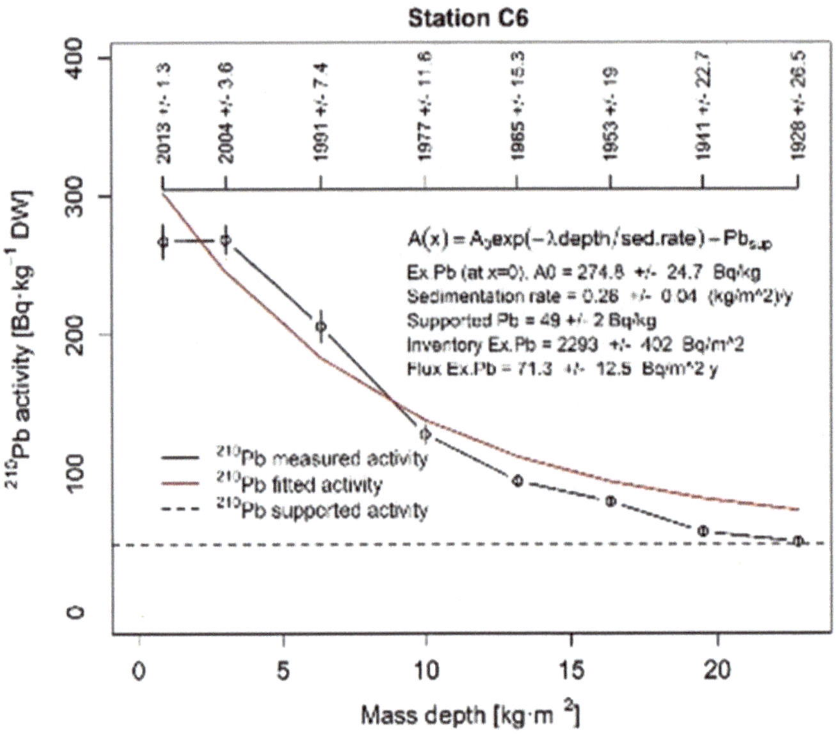
Station C6
2013 +/- 1.3
2004 +/- 3.6
1991 +/- 7.4
1977 +/- 11.6
1965 +/- 15.3
1953 +/- 19
1941 +/- 22.7
1928 +/- 26.5
A(x) = A0exp(−λdepth/sed.rate) − Pbsup
Ex.Pb (at x=0). A0 = 274.8 +/- 24.7 Bq/kg
Sedimentation rate = 0.26 +/- 0.04 (kg/m^2)/y
Supported Pb = 49 +/- 2 Bq/kg
Inventory Ex.Pb = 2293 +/- 402 Bq/m^2
Flux Ex.Pb = 71.3 +/- 12.5 Bq/m^2 y
210Pb measured activity
210Pb fitted activity
210Pb supported activity
210Pb activity [Bq·kg−1 DW]
Mass depth [kg·m−2]

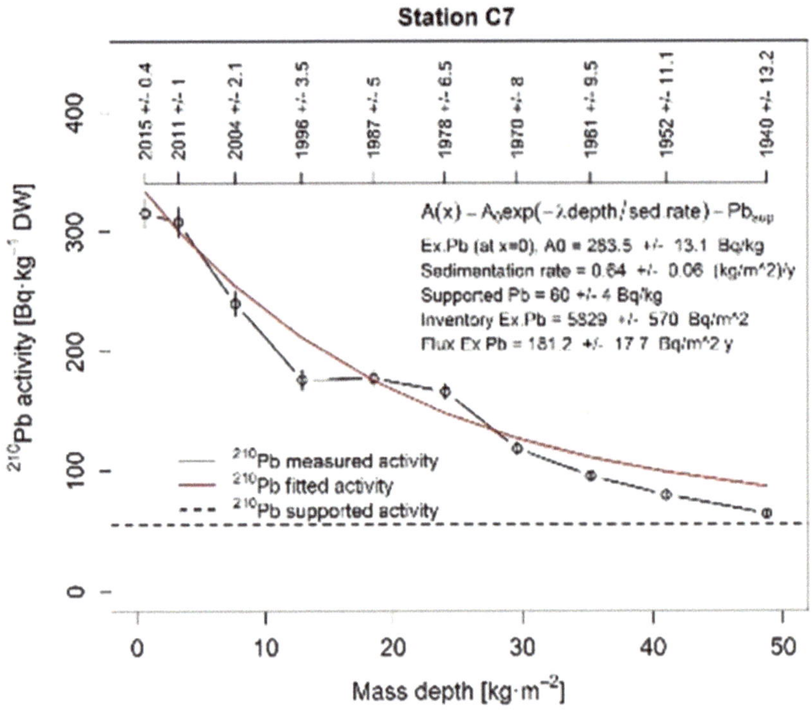
Station C7
2015 +/- 0.4
2011 +/- 1
2004 +/- 2.1
1996 +/- 3.5
1987 +/- 5
1978 +/- 6.5
1970 +/- 8
1961 +/- 9.5
1952 +/- 11.1
1940 +/- 13.2
A(x) = A0exp(−λdepth/sed.rate) − Pbsup
Ex.Pb (at x=0). A0 = 283.5 +/- 13.1 Bq/kg
Sedimentation rate = 0.64 +/- 0.06 (kg/m^2)/y
Supported Pb = 60 +/- 4 Bq/kg
Inventory Ex.Pb = 5829 +/- 570 Bq/m^2
Flux Ex.Pb = 181.2 +/- 17.7 Bq/m^2 y
210Pb measured activity
210Pb fitted activity
210Pb supported activity
210Pb activity [Bq·kg−1 DW]
Mass depth [kg·m−2]

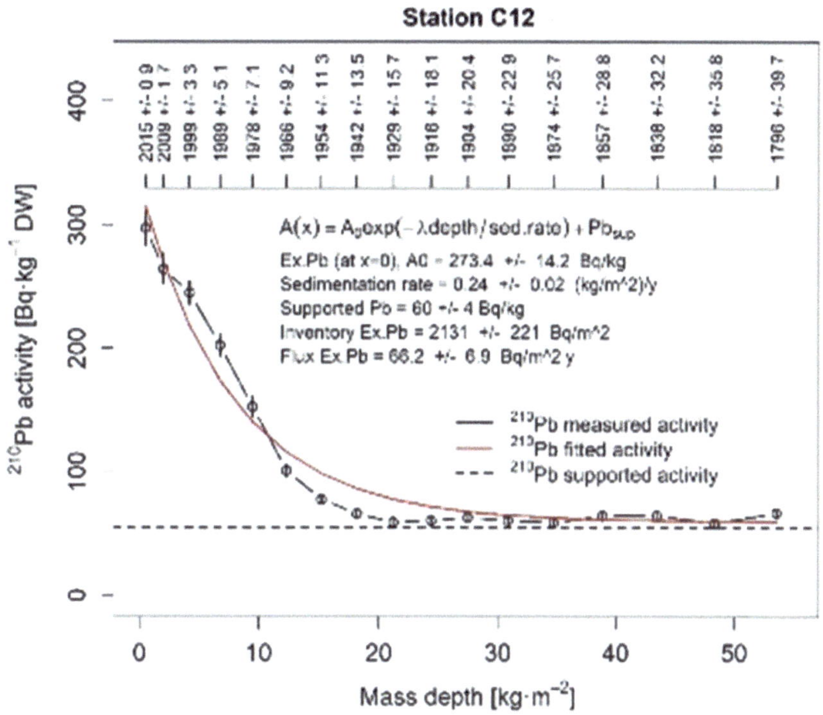

FIG. 9.6. Depth profiles of ^{210}Pb and the results of the CF:CS dating of selected sediment cores.

In the study area, ^{210}Pb dating is difficult to validate with the above-mentioned radionuclides, as all of these have been released into Tvären in relatively high quantities resulting in masking the expected fallout peaks in the sediment profiles. For that reason, stable Pb in the sediment was also analysed. It is well known that during the industrial age the emission of stable Pb to the atmosphere increased and in Sweden it reached its maximum in 1970 [9.46]. The emissions have since then decreased, due to better industrial filtering system and the ban of leaded petrol and are today almost back to historical levels before the industrialisation occurred. This stable Pb peak in the sediment profiles was used to calculate the sedimentation rates in 5 selected cores (Fig. 9.7.). The lead concentrations were measured by XRF. Maximum concentration of these sediments range between 60-70 µg/g of Pb with little variation between the stations.

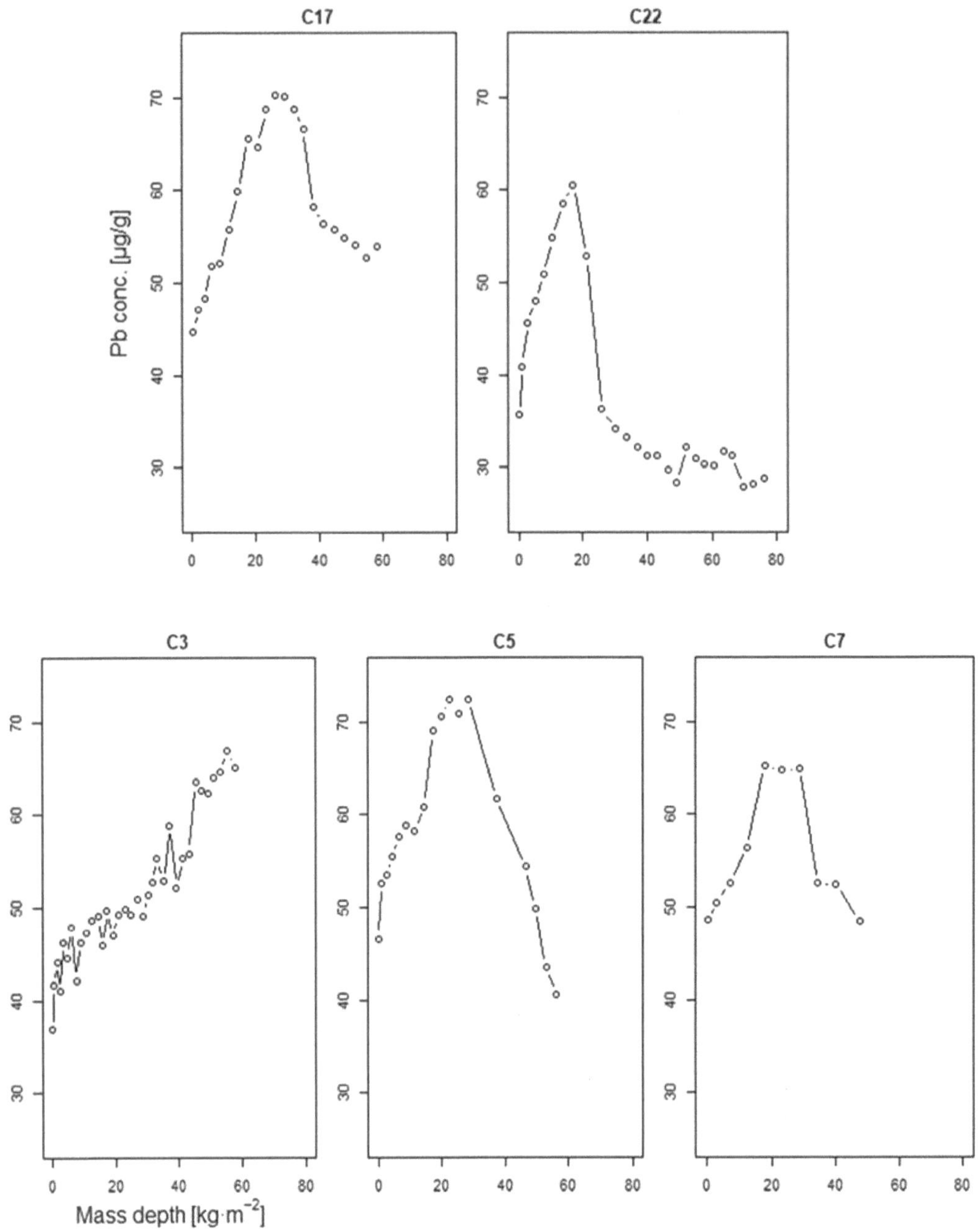

FIG. 9.7. Sediment depth profiles of stable Pb measured by XRF.

A comparison between the independent sedimentation rates can be seen in Table 9.2. The results agree rather well for most of the sediment cores with one exception, namely the sediment core from station 3, C3. This might relate to the determination of the ^{226}Ra concentration, i.e. the supported level of ^{210}Pb in the core. It seems that the level is rather high, but when the uncertainty is considered, the results can be seen as agreeing. However, this also affects the dating of the sediment slices and as the sediment core is rather long a larger bias occurs further down in the core and a misinterpretation of the release history can made. One has to always take such data with care and take this into account when using a dating model. Furthermore, the C3 core, taken in the central deep part of Tvären, has the highest sedimentation rate, more than double compared to the other stations from more shallow areas. This indicates some sediment focusing but also, as the sediment in the deepest station indicate anoxic condition (judged from the distinct H_2S smell from the sediment when slicing), that organic material does not decompose in the sediment. Station 6, core C6, indicates an erosion zone in Tvären and it is clear that the mass accumulation rates are not only dependent on the water depth but also other environmental conditions.

TABLE 9.2. MASS ACCUMULATION RATE) BY TWO INDEPENDENT PARAMETERS. I) USING STABLE LEAD PEAK AS A TIME MARKER IN THE SEDIMENT CORER, II) USING THE CF:CS ^{210}PB DATING MODEL

Site/ sediment core	Depth [m]	Sedimentation rate Stable-Pb (XRF) [kg m^{-2} y^{-1}]	Sedimentation rate ^{210}Pb [kg m^{-2} y^{-1}]
C3	76	≥1.21	1.10±0.07
C5	45	0.56±0.10	0.62±0.03
C6	28		0.26±0.04
C7	15	0.51±0.10	0.64±0.06
C12	33		0.24±0.02
C17	21	0.61±0.08	
C22	30	0.37±0.05	

9.3.2. Vertical distribution of radionuclides in Tvären sediment

The vertical distribution of selected gamma emitting radionuclides and selected sediment cores can be seen in Fig. 9.8. The highest concentrations are found in core C7 situated about 300 m to the west of the Bergösundet discharge point (marked "A" in Fig. 9.1.). It can be noticed that the activity concentrations in the cores, C6 and C8, are rather low, even though these sites are close to the discharge point "A". The reason for this is probably that it is a more dynamic environment where water is flowing in and out of Tvären. It was observed that the sediments at these sites consisted of more coarse-grained sediment compared to the other stations and several attempts were made before an appropriate a sediment core in this area of Tvären could be obtained. In the deepest part of Tvären, core C2 and C3, the longest core obtained consisting of laminated, very soft and fine-grained sediment (see Fig. 2.). The highest sedimentation rate and thereby also the highest time resolution was obtained in this part. For all these radionuclides, ^{137}Cs has the highest concentration in all cores, about ten times higher than the other measured radionuclides. While the differences between these radionuclides is that Studsvik's contribution of ^{137}Cs is rather small, compared to ^{60}Co and ^{152}Eu, where Studsvik is the only source to the sediments in the studied area. In Fig. 9.9. the reported annual discharges of these radionuclides can be seen. In cores C2 and C3 a rather good correlation between the vertical distribution of ^{152}Eu and ^{60}Co with the reported discharge pattern can be observed while for ^{137}Cs it is not agreeing. For ^{137}Cs there are mainly three different sources, i.e. global fallout from nuclear weapon tests, fallout from the Chernobyl accident and discharge from Studsvik. Even though the deposition from Chernobyl was not very high in the studied area (1-2k Bq·m^{-2}), about 300 km north of Studsvik the deposition was much higher (> 100 k Bq·m^{-2}). Considerable deposition also occurred over the Baltic Sea area, in Bothnia Sea. As the net Baltic Sea current flows anti clockwise from the north to the south relatively high contaminated sea water pass Studsvik with ^{137}Cs from the Chernobyl years after the accident. Relatively high levels that could be observed in the Baltic proper decays after the accident resulting in a rather steady input of Chernobyl ^{137}Cs by sea current to the Tvären, where some of the Cs is removed from the water by sedimentation. The sudden increase, followed with a flatter area in more recent sediments, of ^{137}Cs visible in the depth profiles is probably a result from this effect, and only limited activity originates from the Studsvik facility combined with Studsvik discharge. The sediment depth where the sudden increase of ^{137}Cs is dated to around 1986 in most cores. Activity concentrations of all analysed actinides (Pu, Cm, Am) were rather high compared to expected fallout levels in the area. Corresponding activity ratios are also unusually high which make it easy to separate actinides from fallout in the sediment. In Fig. 9.10 the vertical distribution of actinides and their activity ratios can be seen for four selected sediment cores.

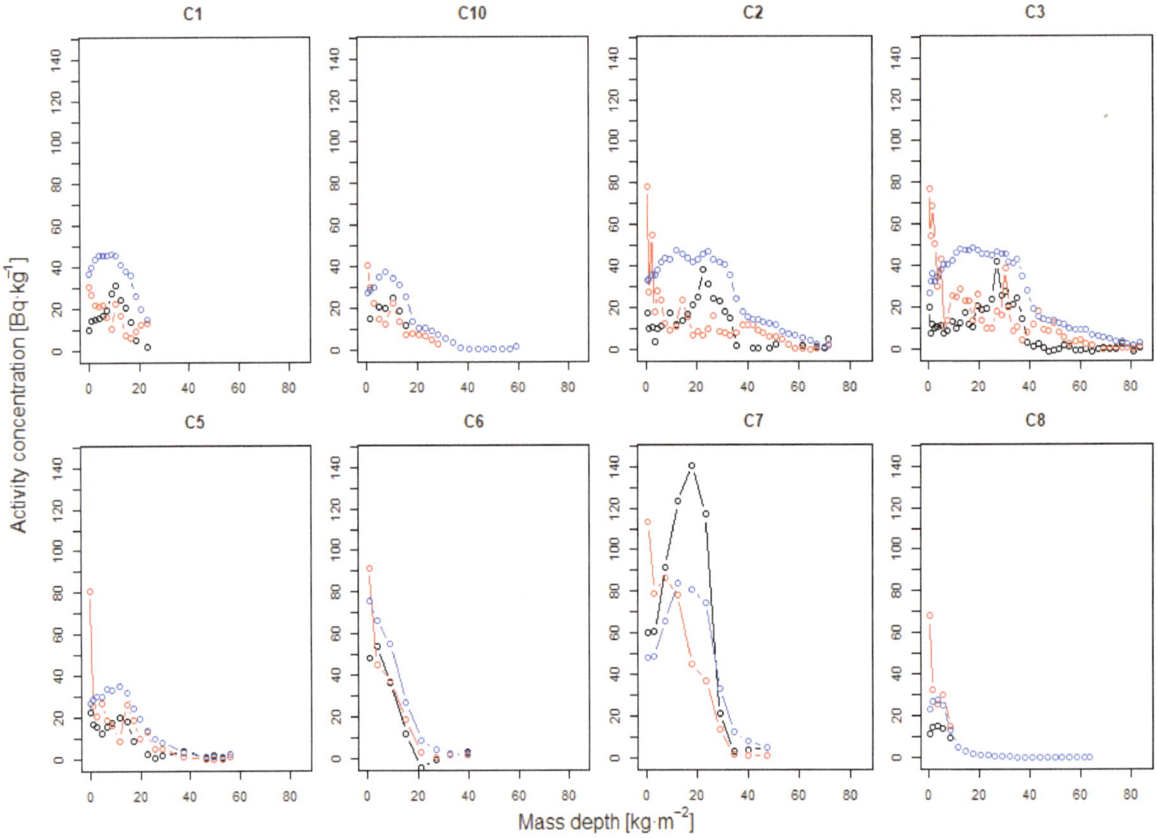

FIG 9.8. Vertical activity distribution in selected sediment cores. Black-^{152}Eu, Red-^{60}Co, Blue-^{137}Cs. Note that the ^{137}Cs activity concentations are divided by a factor of 10.

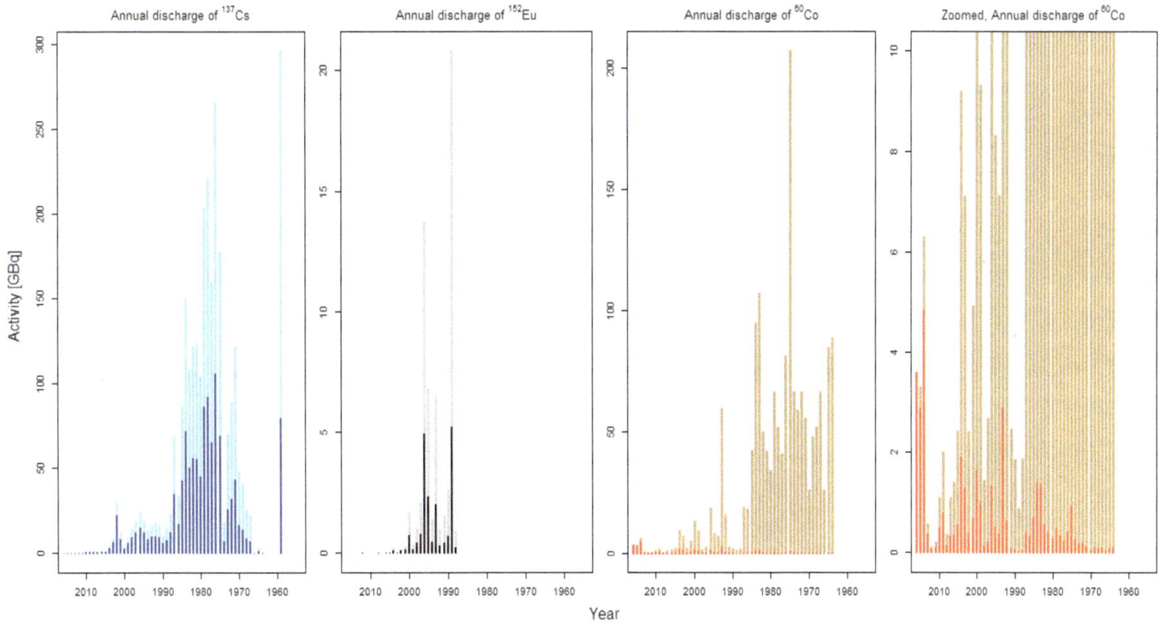

FIG. 9.9. Reported annual discharge activities of ^{137}Cs, ^{152}Eu and ^{60}Co. The smaller bars in each graph are decay corrected activities to 2016, enabling easier comparison with the vertical profiles in these radionuclides in the sediment cores shown in Fig 9.8. Last graph of ^{60}Co is zoomed to be able to see the decay corrected activities 2016.

Core, C3 Core, C5

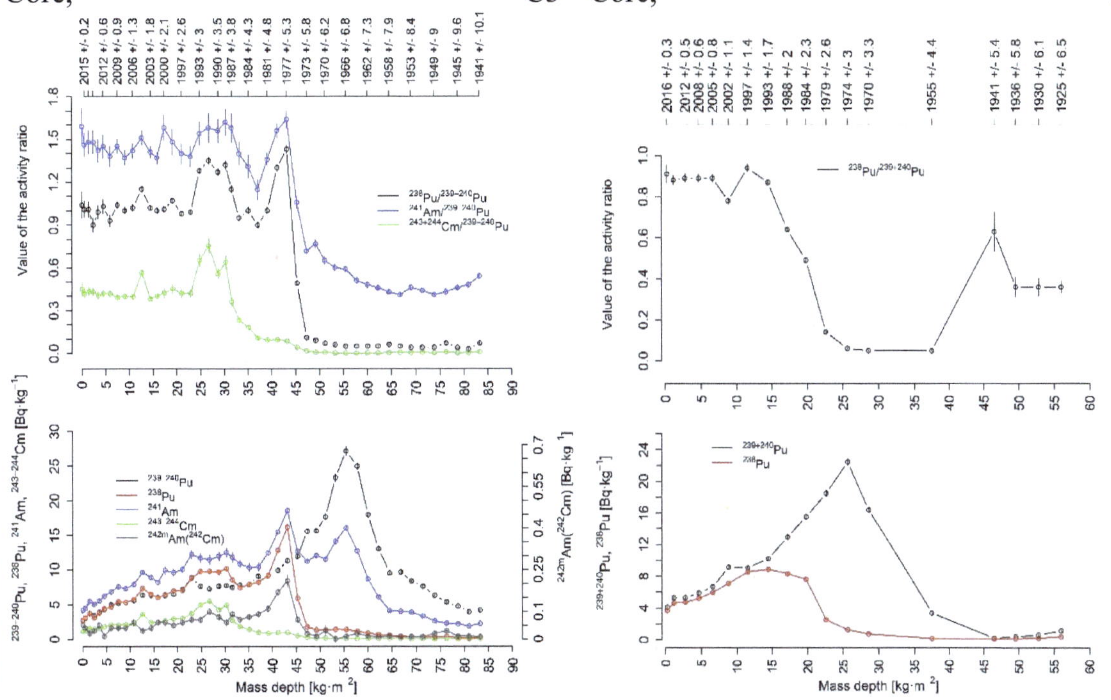

Core, C6 Core, C7

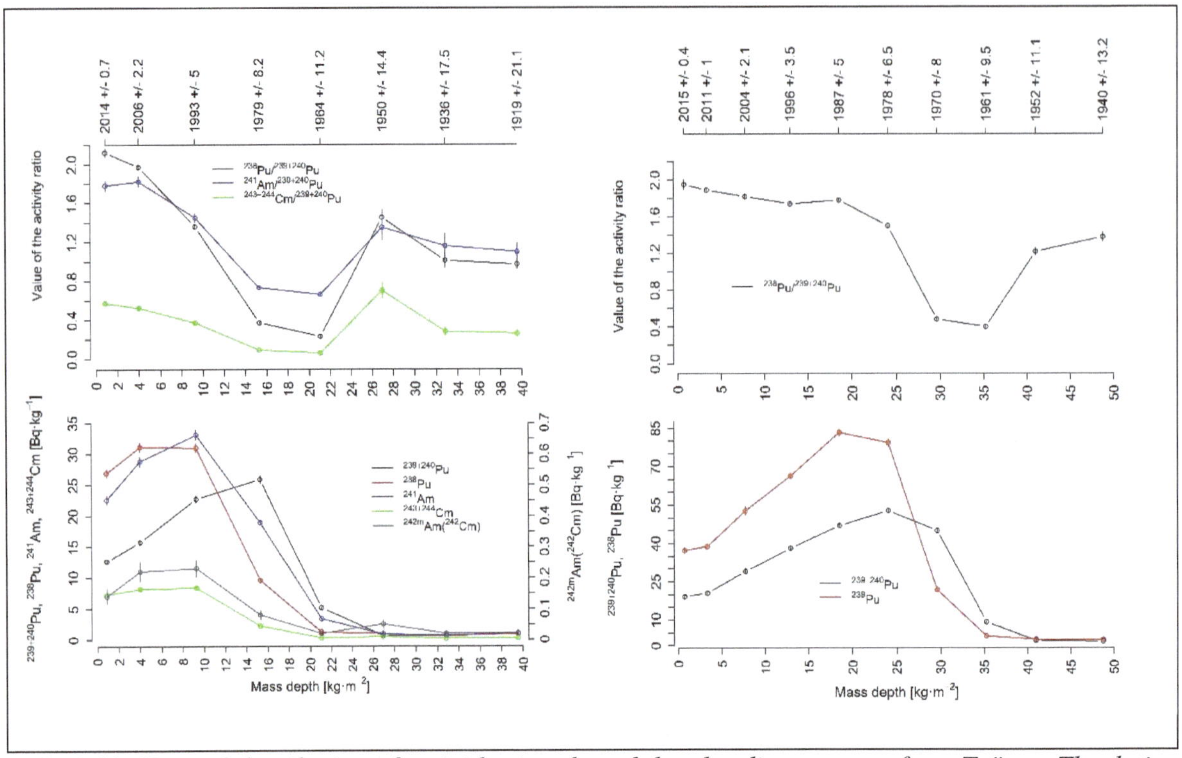

FIG. 9.10. Vertical distribution of actinides in selected dated sediment cores from Tvären. The dating axis is based on ages derived from the CF:CS ^{210}Pb model. For core C3 the presented age might be biased to an older age, see Table 9.2.

Sediment core, C3, was sampled in the deepest part of Tvären, where the highest mass accumulation rate was obtained (Table. 9.2), i.e. where the highest time resolution could be obtained in the sediments. In Fig. 9.10 the actinides vertical distribution can be seen. In core C3 clear visible actinides peaks correspond to aquatic discharges from the Studsvik facility. These peaks occur in three different time periods, i.e. around 1966 for $^{239+240}$Pu and ^{241}Am, around 1977 for ^{238}Pu, ^{241}Am, ^{242m}Am(^{242}Cm) and around 1993 for $^{243+244}$Cm. This probably reflects the material handling at the Studsvik facility. These discharges occurred before the nuclide specific information was mandatory and reported. Activity concentrations for all actinides gradually decrease in the top sediment layers from a maximum of 15.7±0.4 Bq·kg^{-1} for ^{238}Pu, 18.1±0.4 Bq·kg^{-1} for ^{241}Am and 0.20±0.03 Bq·kg^{-1} for ^{242m}Am(^{242}Cm) in layers dated to around 1977 to less than 5 Bq·kg^{-1} in the youngest layers suggesting that the impact of Studsvik releases diminish with time. At the same time the activity ratios remain rather constant with high averages for all three computed ratios (maximum of 1.62±0.09 for ^{241}Am/$^{239+240}$Pu and 1.43±0.03 for ^{238}Pu/$^{239+240}$Pu around 1977). This suggests that Studsvik is still the major contributor of actinides in this area. The maximum $^{243+244}$Cm/$^{239+240}$Pu activity ratio of 0.76±0.05 is seen around 1990 and is accompanied by an increased values of two other ratios suggesting that some works with high burn-up nuclear fuel in Studsvik were performed at that time. As mentioned in the historical overview (section 1.1.), in 1975 Studsvik Inter Ramp Project started and multiple unpressurized BWR fuel rods were tested. These experiments can maybe partially explain the increase of all activity ratios around 1980 indicating that a new type of material is handled in the Studsvik facilities.

In the deep sediment layers dated to around 1970 a maxima of $^{239+240}$Pu activity concentration are visible in all cores (Fig. 9.10). The activity ratios are lower than in the top layers, 0.074±0.04 for ^{238}Pu/$^{239+240}$Pu and 0.65±0.03 for ^{241}Am/$^{239+240}$Pu. This indicates that in the late 1960s and beginning of the 1970s Studsvik handled quite different material than in later years.

To get a better understanding of this discharge the $^{240}Pu/^{239}Pu$ isotopic ratios need to be studied. The ratios decrease even more in the deepest part of the cores where the activity concentrations are lower, 0.033±0.03 for $^{238}Pu/^{239+240}Pu$ and between 0.41±0.01 for $^{241}Am/^{239+240}Pu$. The latter ratios agree rather well with the global fallout from nuclear tests where the activity ratios should be around 0.03 and 0.54 in 2016. In core C3, the maximum $^{239+240}Pu$ activity concentration peak could also be a result of that dissolved $Pu^{+5,+6}$ in sea water might be reduced to $Pu^{+3,+4}$, in the anoxic environment in Tvären and get trapped, i.e. the sediments of Tvären will work like a sink for global fallout Pu. The uncertainty of this assumption is connected with the unknown Studsvik operations in 1960s. To verify whether $^{239+240}Pu$ increase in deeper layers of sediment is of Studsvik origin, additional $^{240}Pu/^{239}Pu$ atomic ratios measurements have to be performed. This ratio is considerably different for nuclear discharges and nuclear bombs tests.

Core C7, that was collected close to the pipeline 2 (Bergösundet, discharge point "A" in Fig. 9.1.), has the highest ^{238}Pu and $^{239+240}Pu$ activity concentrations measured among all analysed samples (83±2 and 53±1 $Bq \cdot kg^{-1}$, respectively). A similar range of activity concentrations was expected in core C6 that was sampled near core C7. However, measured activity concentrations in core C6 were twice as low (31±1 and 26±1 $Bq \cdot kg^{-1}$, respectively). One possible explanation might be that sedimentation conditions in core C6 (mass accumulation rate is twice as low for core C6 than for core C7, Table 9.2) do not favour radionuclide sorption that leads to much lower actinides activity concentrations comparable to areas more distant from pipeline 2 exit.

In general, Pu radioisotopes in stations C6 and C7 show similar patterns with significantly higher than unity ratio of $^{238}Pu/^{239+240}Pu$ activity ratios (Fig. 9.10). On the other hand, station C5 has similar $^{238}Pu/^{239+240}Pu$ isotopic signature as station C3 with ratios close to unity in top layers. The reason may be related to the location of the sampling points. Stations C3 and C5 are the most distant locations from pipeline 2 (discharge point "A" in Fig. 9.1). Cooling water released through pipeline 1 (discharge point "B" in Fig. 9.1.) historically contained not more than 1/10 of aquatic releases discharged through pipeline 2. It is also unknown how much $^{239+240}Pu$ was released with cooling water to Tvären through pipeline 1. Only by looking at $^{238}Pu/^{239+240}Pu$ activity ratios conclusions can be drawn that the closer to pipeline 1, the lower plutonium activity ratios in sediments are detected with maximum of 0.91±0.04 for station C5 and 1.43±0.03 for station C3. At the same time for stations C5 and C6, these ratios reach maximum of 1.95±0.05 and 2.12±0.04, respectively proving that aquatic releases from Studsvik are responsible for most of the ^{238}Pu activities in the sediments of Tvären. Moreover, this noticeable trend of decreasing Pu ratio with distance from the pipeline 2 can be related to dilution with global fallout Pu collected by the Tvären from the basin. Although, when comparing C3 and C6 there is no obvious decreasing trend for $^{241}Am/^{239+240}Pu$ (from 1.45±0.04 to 1.82±0.05 in core C6 and from 1.15±0.07 to 1.64±0.06 for core C3) and $^{243+244}Cm/^{239+240}Pu$ (from 0.37±0.01 to 0.57±0.02 for core C6 and from 0.38±0.01 to 0.76±0.05 for core C3) activity ratios in top sediment layers. Until now, Am and Cm radionuclides have not yet been measured for station C5 and C7 to confirm this pattern. Surely Am and Cm measurements for other cores will give additional insight into this problem.

An interesting phenomenon was noticed in the deeper sections of the sediment cores C5, C6 and C7 (Fig. 9.10). While activity concentrations of actinide decrease, the activity ratios increase reaching similar levels to those in top sediment layers. This pattern is not that obvious in core C3. The pattern indicates that Studsvik has worked with rather different nuclear material during the operation history and also released the actinides in different chemical forms, one possible explanation that the pattern was not that obvious in core C3 could be that the early releases contained more particles or particle reactive forms of the actinides. Another explanation could be slow actinide diffusion with pore water in anoxic conditions, especially as humate complexes after actinides adsorbed onto sedimentary Fe and Mn oxides and oxyhydroxides are released due to microbially-induced reductive dissolution [9.47-9.50]. This question cannot be answered without further analysis of $^{240}Pu/^{239}Pu$ atomic ratios that together with activity ratios will provide information of the origin of actinides in the deeper parts of the sediments.

Alpha spectrometry as a radiometric measurement technique enables detection of short-lived radionuclides at very low detections limits. In case of Studsvik sediments, when measuring Am/Cm fraction, it was possible to detect ^{242}Cm activities in almost all sediment layers of core C3 and C6 with maximum activity concentrations of 0.20±0.03 and 0.22±0.03 Bq·kg^{-1}, respectively. ^{242}Cm is an alpha emitter with relatively short half-life ($t_{1/2}$=163 d). Possible ^{242}Cm production paths are from ^{242}Am β-decay and ^{243}Cm (n,2n) reaction [9.51]. In fact, ^{242}Cm should be undetectable in environmental samples after a lapse of a few years, unless there is ingrowth from precursor ^{242m}Am ($t_{1/2}$=141 y) via its short-lived daughter ^{242}Am. One of the known sources of ^{242}Cm to the Baltic Sea environment was Chernobyl accident. Theoretical decay corrected activity ratios of supported ^{242}Cm to $^{243+244}Cm$ and $^{239+240}Pu$ that originated from Chernobyl fallout were around 0.03-0.07 and 0.001-0.004, respectively [9.52]. In Tvären maximum values of these ratios were detected in top sediments layers and are summarized in Table 9.3. Clearly these values differ from any other global sources of artificial radionuclides.

As seen from the vertical profile of actinides in core C3 and C6 (Fig. 9.10), higher activity concentrations of ^{242}Cm were rather associated with increased releases of ^{241}Am. This means that nowadays most ^{242}Cm is in equilibrium with ^{242m}Am.

TABLE 9.3. OBTAINED ACTIVITY RATIOS AT THE MAXIMUM ^{242}CM ACTIVITY CONCENTRATION IN THE DEPTH PROFILE.

Site/ sediment core	$^{242}Cm/^{243+244}Cm$ Activity ratio	$^{242}Cm/^{239+240}Pu$ rate Activity ratio	Sediment layer dating
C3	0.22±0.03	0.018±0.003	1977±5
C6	0.11±0.03	0.014±0.002	1993±4

In addition, uranium isotopes have been studied in one sediment core close to the discharge point at Bergösundet. As expected, no evidence of discharged U-isotopes (^{234}U, ^{235}U and ^{238}U) could be seen in the sediment profile. The concentration of ^{238}U was rather constant 83±6 Bq·kg^{-1} ($^{234}U/^{238}U$=1.06±0.04, $^{235}U/^{238}U$=0.042±0.009). These activity ratios are close to natural values suggesting limited impact from uranium in the aquatic releases on the Tvären sediments. This is possible due to two main reasons. Firstly, according to Studsvik releases reports, aquatic discharges containing uranium were not regular and average yearly released activities were rather low between 3 to 30 kBq for ^{234}U and ^{238}U with high $^{234}U/^{238}U$ activity ratios, between 2.5 to 8 which are typical for enriched uranium. Obviously considering quite high natural background of uranium in this area released uranium radioisotopes will be diluted enough to disappear in natural uranium content. Secondly, U(VI) in natural waters mainly exists as soluble carbonate complexes $[UO_2(CO_3)_3]^{4-}$ and $[UO_2(CO_3)_2]^{2-}$ that are not particle reactive and can be transported with currents to more distant locations, in this case, to the open Baltic Sea. Considering these results, this study will focus on ^{233}U and ^{236}U, as they do not exist in natural uranium but are rather a nuclear fingerprint of processed uranium, to evaluate the uranium discharge history.

9.3.3. Inventories and spatial distribution

Computed inventories for all analyzed radionuclides are presented in Table. 9.4. The highest inventories were computed for core C7 located close to the discharge point "A". In the deepest part of Tvären (core C3) the inventories of ^{152}Eu and ^{238}Pu are almost 5 times lower than in C7. The difference between $^{239+240}$Pu and ^{60}Co inventories are much lower for these two sites, where core C3 has about half the inventories of core C7. Both cores C6 and C7 are characterize by higher inventory of ^{238}Pu compared to $^{239+240}$Pu, while opposite proportion exists in cores C3 and C5 that can be associated with global fallout Pu impact or different releases characteristics from both discharge pipelines "A" and "B". On the other hand, inventories of Am and Cm are higher in core C3 compared to C6. One possible explanation was mentioned before, i.e. a low mass accumulation rate for core C6 that lead to much lower actinides activity concentrations. This trend needs to be confirmed with further analyses of Am and Cm in core C7 that is the closest to discharge point "A". In Fig. 11 and Fig. 12 the impact on the sediment inventory from discharge point "A" is seen. The highest inventories are seen around discharge point "A". For ^{152}Eu, ^{60}Co and to some extent ^{238}Pu, Studsvik is the exclusive source, while for ^{137}Cs, $^{239+240}$Pu and ^{241}Am a contribution to the inventory of fallout from Chernobyl and atmospheric nuclear tests cannot be excluded. It can be seen that ^{60}Co has higher inventories in the central part of Tvären compared to ^{152}Eu and ^{238}Pu. It might be due to Co having a lower k_d than Eu and Pu enabling longer transport from the discharge point "A". It can also be an effect of a greater amount of ^{60}Co released from discharge point "B" compared to ^{152}Eu and ^{238}Pu. This will be studied more in detail in future studies.

TABLE 9.4. COMPUTED INVENTORIES FOR SELECTED SEDIMENT CORES IN TVÄREN. NOTE THAT CORE C1 DOES NOT GIVE THE FULL INVENTORY DUE TO INSUFFICIENT CORE DEPTH. ALL VALUES ARE IN Bq·m^{2}

Radio-isotope	C1*	C2	C3	C5	C6	C7	C8
^{60}Co	381±40	720±120	1040±120	500±140	740±80	1830±150	280±20
^{137}Cs	6800±180	13950±180	16460±200	7800±160	7880±150	22010±200	2600±210
^{152}Eu	380±80	680±60	548±65	383±82	1270±130	3110±105	271±22
^{238}Pu	-	-	384±6	171±6	293±12	1904±24	-
$^{239+240}$Pu	-	-	808±9	433±9	259±9	1343±15	-
^{241}Am	-	-	739±12	-	325±12	2100±180	-
$^{243+244}$Cm	-	-	114±6	-	78±5	-	-
^{242}Cm (^{242m}Am)	-	-	4.0±0.3	-	2.0±0.6	-	-

*Not full inv.

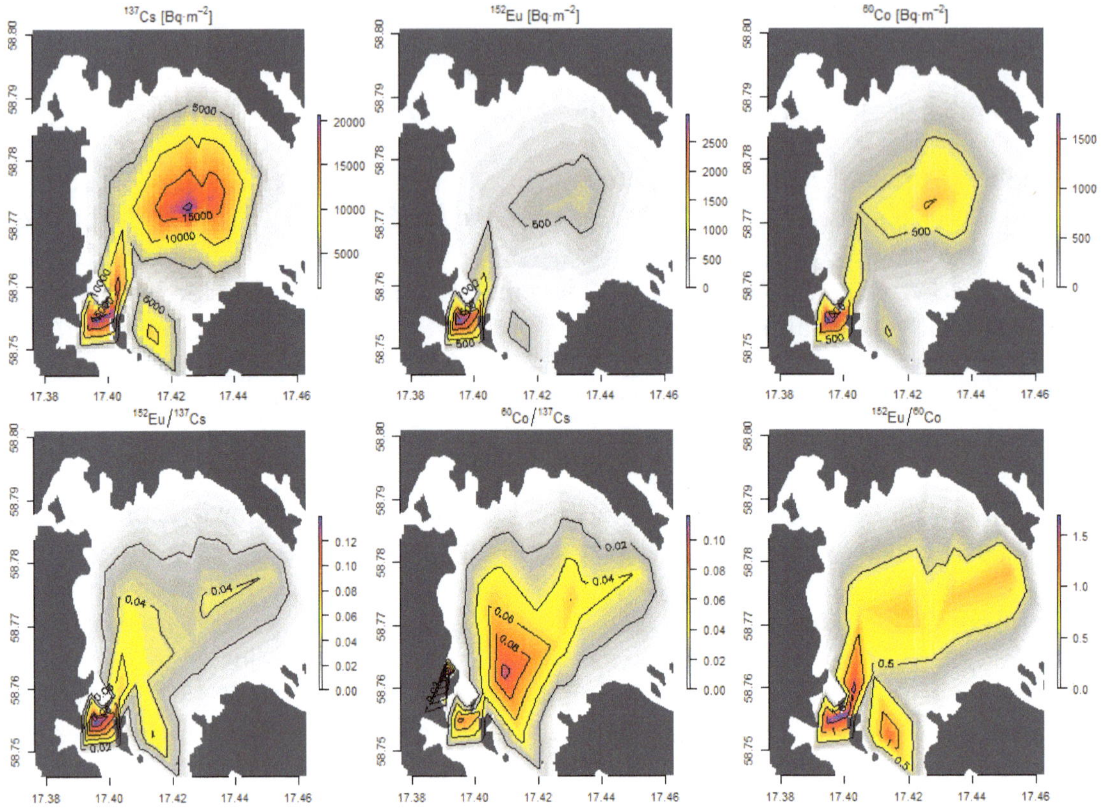

FIG. 9.11. Spatial distribution of inventories in Tvären sediments for ^{60}Co, ^{152}Eu and ^{137}Cs. The highest inventories are found close to discharge point "A" situated in the lower right of the map, see Fig 9.1 for the exact position. Highest inventories found were: ^{137}Cs about $20 kBq \cdot m^{-2}$, ^{152}Eu about $3 kBq \cdot m^{-2}$ and ^{60}Co about $2 kBq \cdot m^{-2}$. Lower row shows the activity ratio distribution and indicate that ^{137}Cs and ^{60}Co migrate longer from the discharge point "A" into Tvären. E.g. the $^{152}Eu/^{60}Co$ ratio is above 1.5 in core C7 and decreases to below 0.5 in the central part of Tvären.

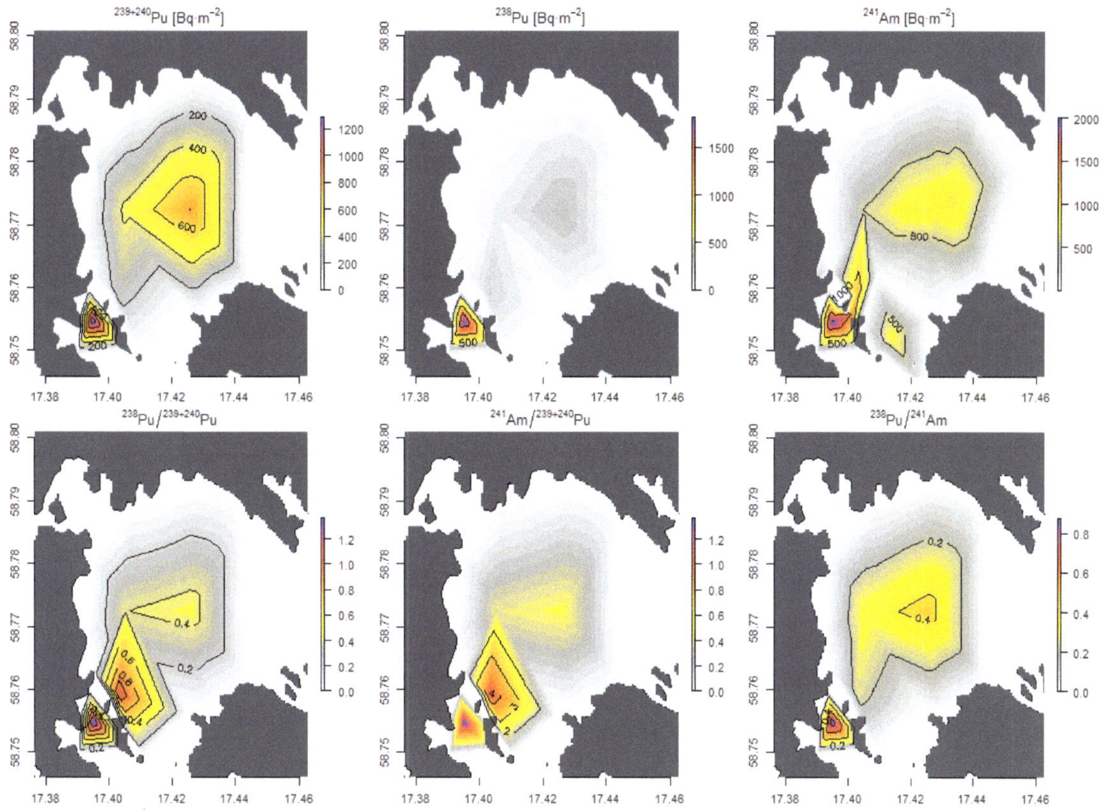

FIG. 9.12. Spatial distribution of inventories in Tvären sediments for $^{239+240}Pu$, ^{238}Pu and ^{241}Am. The highest inventories is found close to discharge point "A", see Fig 9.1 to see the exact position of the discharge points. The highest inventories are found for ^{238}Pu and ^{241}Am, about 2 kBq·m^{-2} and for $^{239+240}Pu$ about 1.2 kBq·m^{-2}. Lower row shows the activity ratios. The $^{238}Pu/^{239+240}Pu$ and $^{241}Am/^{239+240}Pu$ activity ratios are highest close to the discharge point "A", over 1.2. The $^{238}Pu/^{241}Am$ ratio is also highest in this area, indicating that ^{238}Pu is strongly associated to discharge from this pipeline.

In general, calculated inventories are very high when compared to reported levels in the Baltic Sea. The maximum inventories for $^{239+240}$Pu, ^{238}Pu and ^{241}Am reported to the Helsinki Commission in this area of the Baltic Sea are 150, 5 and 75 Bq·m^{-2}, respectively [9.16]. The inventories in Tvären show that additional sources of actinides to the Baltic Sea environment, like Studsvik facility, are important to control and investigate. This as some of the most 'exotic' artificial radionuclides can be used as environmental tracers and pose an important role in environmental radiological assessments.

9.3.4. Reconstruction of past unknown aquatic releases from Studsvik

Studsvik has regularly discharged effluents with different radionuclides. In the first discharge reports nuclide specific gamma emitters and total α and total β activity where reported. Since 2002 some reports are available containing released activities of ^{238}Pu. $^{239+240}$Pu, ^{241}Am, ^{244}Cm and ^{242}Cm. Summary of yearly aquatic releases for those radionuclides from both discharge points ("A" and "B") are given in Table 9.5 and Fig. 9.2.

For the reconstruction of Studsvik's releases using sediment cores, C3, C6 and C7 were chosen to study if reconstructed values are affected by the localization of the sediment core.

Reconstruction of the past Studsvik releases demands knowledge of the actinide inventories in the sediment as well as well documented release history. The uncertainty of the documented releases has not been stated nor how these values has been derived. By comparing, a specific radionuclide transfer factor, f, in the sediment cores can be derived and expressed for Pu as:

$$f = \frac{I}{A} \tag{9.1}$$

Where:

I – inventory in the sediment layers dated between 2002 and 2016 in $Bq{\cdot}m^{-2}$;

A – aquatic release between 2002 and 2016 in Bq.

The specific Pu transfer factor, f, calculated for C3 sediment layers dated between year 2002 and 2016 are:

for ^{238}Pu: 1.76 ± 0.02 $[1{\cdot}m^{-2} \cdot 10^{-6}]$

for $^{239+240}Pu$: 4.07 ± 0.05 $[1{\cdot}m^{-2} \cdot 10^{-6}]$

Based on those values, the "missing" releases were calculated based on the integrated ^{238}Pu and $^{239+240}Pu$ inventories for the sediment layers below 2002 where the aquatic release are unknown. Computed missing releases between years 1959 to 2002 are:

for ^{238}Pu: 216 ± 2 MBq

for $^{239+240}Pu$: 145 ± 2 MBq

From the reported aquatic discharge, shown in Table 9.5, the calculated missing release between years 1959 and 2002, the total aquatic release of the Pu radioisotopes to Tvären are:

for ^{238}Pu: 272 ± 4 MBq

for $^{239+240}Pu$: 160 ± 4 MBq

TABLE 9.5. AQUATIC ANNUAL RELEASE OF ACTINIDES FROM STUDSVIK TO TVÄREN BASED ON REPORTED VALUES TO THE SWEDISH RADIATION SAFETY AUTHORITY (SSM)

Release year	^{238}Pu [MBq]	$^{239+240}Pu$ [MBq]	^{241}Am [MBq]	^{244}Cm [MBq]	^{242}Cm [MBq]
2017	0.025	0.02	0.6	–	–
2016	0.042	0.03	0.27	–	–
2015	0.18	0.2	0.085	0.042	–
2014	0.033	0.08	0.02	–	–
2013	0.01	0.04	0.03	0.022	–
2012	0.015	0.009	0.02	–	–

2011	0.24	0.08	0.6	1.04	–
2010	1.6	1	0.94	–	–
2009	1.6	1.15	1.9	0.5	–
2008	0.64	1	2.2	1.8	0.3
2007	0.96	0.15	2.5	12.5	1.8
2006	1.5	0.3	5.1	4.5	4.5
2005	11	2.6	87	30	9.5
2004	6.6	1.3	10	39.1	2.8
2003	9.5	1.8	7.8	36.2	5.1
2002	21.5	5.6	8.3	45.4	15.3

The same calculations were performed for other cores and all computed results are presented in Table 9.6. Recalculation of the Studsvik releases for ^{238}Pu and $^{239+240}$Pu and was based on the assumption that the environmental conditions governing mobility and plutonium residence time was similar between 2002 to 2016 and 2002 to 1959.

In theory radionuclide specific transfer factor, f, in the sediment needs to be site specific and dependent on the particular radioisotope activity in the sediment and the mass accumulation rates. In the case of this study, the mass accumulation rates are constant, and the only variable is site specific Pu inventory. Differences in the calculated factors, f, are noticeable for each site but similar results were obtained for C3, C5 (collected from the center of Tvären) and C6, C7 (collected from the southern part of Tvären) pairs. Disagreements are probably due to uncertainties in sediment dating, especially in cores that have low resolution. In case of $^{239+240}$Pu it can be also attributed with an impact of the global fallout Pu that is difficult to assess for each sediment station. Even though, for the total releases reconstruction global fallout Pu was subtracted with the values of 150 Bq·m^{-2} for $^{239+240}$Pu and 5 Bq·m^{-2} for ^{238}Pu that are based on previous studies [9.33]. It is possible that they are under or overestimated for different locations of Tvären. Additionally, that the level of Pu retained in the sediment depends on site-specific environmental conditions including the amount of inorganic and organic colloids and water chemistry that govern Pu oxidation states and therefore its mobility. In general, total release reconstructions based on data from both locations are in good agreement and the total releases of both were around 200 to 300 MBq.

TABLE. 9.6. CALCULATED RADIONUCLIDE TRANSFER FACTOR (F) AND RECONSTRUCTION OF PAST UNKNOWN AQUATIC RELEASES FROM STUDSVIK CALCULATED FROM FOUR DIFFERENT SEDIMENT CORES.

Core	f [1·m^{-2} · 10^{-6}]		Missing release [MBq]		Total release [MBq]	
	$^{239+240}$Pu	^{238}Pu	$^{239+240}$Pu	^{238}Pu	$^{239+240}$Pu	^{238}Pu
C3	4.07±0.92	1.76±0.25	145±26	217±25	160±26	273±25
C5	1.00±0.07	1.22±0.08	229±16	106±12	162±26	244±22
C6	0.40±0.02	1.64±0.05	216±15	155±16	231±15	210±16
C7	5.03±0.28	0.10±0.03	181±14	164±17	196±14	219±17

Releases for other radionuclides cannot be reconstructed with present knowledge. Due to *in-situ* ^{241}Am activity in-growth from ^{241}Pu it is impossible to relate ^{241}Am release history to inventories in the sediment without the knowledge of ^{241}Pu activity concentrations in the sediment. And for ^{243}Cm and ^{244}Cm have relatively short half-lives of 29.1 and 18.1 years, respectively. In alpha spectrometry both radioisotopes are measured as the sum of both isotopes, due to similar α particles energies, and their deconvolution with acceptable precision is not always possible in low activity concentration samples. In this case decay correction for both Cm radioisotopes is not possible. As mentioned before, ^{242}Cm reported by Studsvik already decayed and is now in equilibrium with ^{242m}Am. In this case it is also impossible to relate the ^{242}Cm release history with inventories in sediment cores.

9.4. CONCLUSIONS

In general, the study revealed high level of environmental radioactivity in sediments. The levels are mostly related to the distance from the Studsvik aquatic discharge points and of the mass accumulation rates, especially in the deep center part of Tvären. Activity levels for analyzed radionuclides translated into high inventories in Tvären that are up to 300 times higher for ^{238}Pu, 9 times higher for $^{239+240}$Pu and 10 times higher for ^{241}Am than reported values for this area of the Baltic Sea. Other anthropogenic radionuclides were also characterized with high inventories. In general, the levels of radioactivity decreased in recent aquatic releases from Studsvik. Reported annual discharge was compared with the radioactivity in the dated sediment profiles. For most radionuclides and cores, there is fairly good agreement. The only exception is ^{137}Cs with high activities in top sediment layers and the distribution did not follow that expected from the reported discharges. The activity of ^{137}Cs in Tvären probably originates from Chernobyl fallout arriving to Tvären by the Baltic Sea water circulation currents. Therefore, it has rather steady input into bay's waters where a fraction is precipitated to the sediments. The profiles revealed increased activity ratios of actinides in the deepest layers of the analyzed sediments that can be due to slow actinide migration processes in the pore water of the sediments.

Activity ratios showed that the origin of the artificial radioactivity in Tvären is mainly due to Studsvik operations and, to some extent in the deeper sediment layers, due to global fallout. Additional ^{240}Pu/^{239}Pu and ^{241}Pu/^{239}Pu isotopic ratio measurements would also answer how much the global fallout impact on Tvären compared to activity from Studsvik releases.

Well established Pu profiles and known history of the discharges allow reconstruction of the total releases for both ^{238}Pu and $^{239+240}$Pu based on three Pu profiles from different locations.

The obtained results are in rather good agreement and indicating that Studsvik has discharged about 190MBq of $^{239+240}Pu$ and 240MBq of ^{238}Pu during its whole operation. These calculations assumed that activity of Studsvik aquatic releases are correct. Although according to the results presented here, the discharge of ^{238}Pu reported by Studsvik may be overestimated, especially the values reported between years 2002-2007.

The obtained results clearly show the impact of Studsvik discharges on the radionuclide inventories in Tvären area and for the ecosystem of the Baltic Sea.

9.5. ACKNOWLEDGEMENTS

The authors of this publication would like to acknowledge the Swedish Radiation Safety Authority for financial support to conduct parts of this study. Municipality of Linköping is acknowledged for partially funding the salary for Per Törnquist. The IAEA is acknowledged the use of the XRF machine and Isabelle Levy for helping in the XRF analysis. Marie Karlsson at Linköping University, Martin Hjellström, Nali Dilan, Karin Aquilonius, Charlotta Lager for assisting during sampling.

9.6. REFERENCES

[9.1] BYRDE, C. OCH ERIKSSON, B. A. VATTENOMSÄTTNING I TVÄREN, Aktiebolaget Atomenergi (AP-5130-15) (1963).

[9.2] KULLENBERG, M. Vattenomsättning i Tvären, Göteborg (1955).

[9.3] EVANS, S., LAMPE, S., Observations of the Chernobyl fallout in Tvaeren bay, Baltic Proper, Studsvik Nuclear, Report STUDSVIK-NP-90-4 (1990).

[9.4] AGNEDAHL, P., Measurements of Background and Fall-out Radioactivity in Samples from the Baltic Bay of Tvären, 1957–1963, Report AE-171, AB Atomenergi, Stockholm, Sweden (1965).

[9.5] LINDHE, S., LINDER, P., The handling of liquid waste at the research station of Studsvik, Sweden. Report AE-182, AB Atomenergi, Stockholm, Sweden (1965).

[9.6] LINDSKOG, A., A survey of the arrangements for waste management at Studsvik, Sweden, Report STUDSVIK/K1/3-79/11, Sweden, (1979).

[9.7] SWEDISH RADIATION SAFETY AUTHORITY. Decommissioning of the Nuclear Reactors R2 and R2-0 at Studsvik, Sweden: General Data as called for under Article 37 of the Euratom Treaty, Sweden, (2012).

[9.8] SUNDBLAD, B., Water release of tritium for model validation (1986).

[9.9] HULTGREN A., BERGGREN, G., BROWN, A., ENG, H. U., FORSYTH, R. S., The plutonium fuel laboratory at Studsvik and its activities. Report IAEA-SM-88/28, Brussels (1967).

[9.10] HULTGREN, A., OESTERLUND, C.G. Reprocessing in Sweden: History and perspective. National Board for Spent Nuclear Fuel (1990).

[9.11] LUND, E. Personal communication with Eva Lund 2017-05-11, in: Törnquist, P. (Ed.), p. 1.

[9.12] FOGELBERG, B. The Studsvik Science Research Laboratory Report NFL-31, (1982).

[9.13] ENVIRONMENT, M.O.T. Sweden's third national report under the Joint Convention on the safety of spent fuel management and on the safety of radioactive waste management, in: Environment, M.o.t. (Ed.), (2008).

[9.14] THOMAS, G. R., The Studsvik inter ramp project: An international power ramp experimental program. J. Nucl. Mater. **87** (1979) 215–226.

[9.15] BORG, S., BERGSTRÖM, I., HOLM, G. B., RYDBERG, B., DE GEER, L. E., RUDSTAM, G., GRAPENGIESSER, B., LUND, E., WESTGAARD, L. On-line separation of isotopes at a reactor in Studsvik (OSIRIS). Nucl. Instrum. Methods, **91** 1 (1971) 109–116.
[9.16] HELCOM. Report of the Sediment Baseline Study of HELCOM MORS-PRO in 2000–2005 and Baltic Sea Environment Fact Sheet. Helsinki Commission, Helsinki, Finland (2014).
[9.17] HOLM, E. Plutonium in the Baltic Sea. Appl. Radiat. Isot. **46** 11 (1995) 1225–1229.
[9.18] HARDY, E. P., KREY P. W., VOLCHOK H.L., Global Inventory and Distribution of Fallout Plutonium, Nat. **241** (1973) 444–445.
[9.19] BERGSTRÖM, S., CARLSSON, B., River runoff to the Baltic Sea: 1950–1990, Ambio **23** (4–5) (1994) 280–287.
[9.20] UNSCEAR, Ionizing radiation: Sources and biological effects, UNSCEAR Report (1982).
[9.21] UNSCEAR, Sources and Effects of Ionizing Radiation UNSCEAR Report (1993).
[9.22] ANDERSSON, P., et al., Strålningsmiljö i Svergie, SSI Rapport 2007:02 (2007).
[9.23] SWEDISH RADIATION SAFETY AUTHORITY. Map of fallout of cesium-137 on the ground. https://ssm-kartor.maps.arcgis.com/apps/webappviewer/index.html?id=b0b123c9e99549f58d5d23a2436016ba).
[9.24] SMITH, J.N., ELLIS, K.M., POLYAK, L., IVANOV, G., FORMAN, S.L., MORAN S.B. Pu-239,240 transport into the Arctic Ocean from underwater nuclear tests in Chernaya Bay, Novaya Zemlya, Cont. Shelf Res. **20** (2000) 255-279.
[9.25] VINTRÓ, L., SMITH, K.J., LUCEY, J.A., MITCHELL, P.I. "The environmental impact of the Sellafield discharges", SCOPE-RADSITE Workshop Proceedings, Brussels **4–6** (2000).
[9.26] KETTERER M.E., HAFER H.M., MIETELSKI J.W. Resolving Chernobyl vs. global fallout contributions in soils from Poland using Plutonium atom ratios measured by inductively coupled plasma mass spectrometry. J. Environ. Radioact. **73** (2004) 183–201.
[9.27] IRLWECK K., WICKE J. Isotopic composition of plutonium emissions in Austria after the Chernobyl accident. J. Radioanal. Nucl. Chem. **227** (1998) 133–136.
[9.28] SALMINEN S., PAATERO J., JAAKKOLA T., LEHTO J. Americium and curium deposition in inland from the Chernobyl accident. Radiochim. Acta **93** (2005) 771–779.
[9.29] SALMINEN S., PAATERO J., ROOS P., HELARIUTTA K. ^{237}Np in peat and lichen in Finland. J. Radioanal. Nucl. Chem. **281** (2009)405–413.
[9.30] STRUMIŃSKA-PARULSKA, D.I., SKWARZEC, B. Pu-241 concentration in southern Baltic Sea ecosystem. J. Radioanal. Nucl. Chem. **268** 1 (2006) 59–63.
[9.31] SALMINEN-PAATERO, S., JUSSI PAATERO, J., JAAKKOLA, T. ^{241}Pu and ^{241}Pu/$^{239+240}$Pu activity ratio in environmental samples from Finland as evaluated by the ingrowth of ^{241}Am. Boreal Environ. Res. **19** (2014) 51–65.
[9.32] MAURING A, GÄFVERT T. Radon tightness of different sample sealing methods for gamma spectrometric measurements of ^{226}Ra. Appl. Radiat. Isot. **81** (2013) 92-95.
[9.33] OLSZEWSKI, G., ANDERSSON, P., LINDAHL, P., ERIKSSON, M. On the distribution and inventories of radionuclides in dated sediments around the Swedish coast. J. Environ. Radioact. **186** (2018) 142-151.
[9.34] HALLSTADIUS, L. A., A method for the electrodeposition of actinides. Nucl. Instrum. Meth. Phys. Res. **223** (1984) 382–385.
[9.35] HURTGEN, C., JEROME, S., WOODS, M., Revisiting Currie - How low can you go? Appl. Radiat. Isot. **53** 1–2 (2000) 45-50.

[9.36] FLYNN, W.W. The determination of low levels of polonium-210 in environmental materials. Anal. Chim. Acta **43** (1968) 221–227.
[9.37] EHINGER, S.C., PACER, R.A., ROMINES, F.L. Separation of the radioelements 210Pb-210Bi-210Po by spontaneous deposition on noble metals and verification by Cherenkov and Liquid scintillation counting. J. Radioanal. Nucl. Chem. **98** (1986) 39–48
[9.38] HENRICSSON, F., RANEBO, Y., HOLM, E., ROOS, P. Aspects on the analysis of ^{210}Po. J. Environ. Radioact. **102** 5 (2011) 415–419.
[9.39] AKIE, H., SATO, I., SUZUKI, M., SERIZAWA, H., ARAI, Y. Simple formula to evaluate helium production amount in fast reactor MA-containing MOX fuel and its accuracy. J. Nucl. Sci. Tech. **50** 1 (2013) 107–121.
[9.40] REIDE CORBETT, D., WALSH, J. P. "^{210}Lead and ^{137}Cesium: establishing a chronology for the last century", Handbook of Sea-Level Research (SHENNAN, I., LONG, A. J., HORTON, B. P., Eds.), John Wiley & Sons, Ltd, Chichester (2015).
[9.41] BARSANTI. M., ET AL. Challenges and limitations of the ^{210}Pb sediment dating method: Results from an IAEA modelling interlaboratory comparison exercise, Quat. Geochronol. **59** (2020) 101093.
[9.42] GARCIA-TENORIO. R., ET AL., From radiometry to chronology of a marine sediment core: A ^{210}Pb dating interlaboratory comparison exercise organised by the IAEA, Mar. Pollut. Bull. **159** (2020) 111490.
[9.43] CRUSIUS, J., ANDERSON, R.F. Evaluating the mobility of ^{137}Cs, $^{239+240}$Pu and ^{210}Pb from their distributions in laminated lake sediments. J. Paleolimnol. **13** (1995) 119–141.
[9.44] AAKROG. A., The radiological impact of the Chernobyl debris compared with that from nuclear weapons fallout, J. Environ. Radioact. **6** 2 (1988) 151-162.
[9.45] UNSCEAR. Sources and effects of ionizing radiation UNSCEAR Report Vol. 1 (2000).
[9.46] BERGBÄCK, B., ANDERBERG, S., LOHM, U., Lead Load: Historical Pattern of Lead Use in Sweden, Ambio **21** 2 (1992) 159-165.
[9.47] CHOPPIN, G. R., ROBERTS, R. A., MORSE, J. W. Effects of Humic Substances on Plutonium Speciation in Marine Systems. Organic Marine Geochemistry, ACS Symposium Series (1986) 382–388.
[9.48] FRANCIS, A. J. Microbial transformations of plutonium and implications for its mobility. Radioact. Environ. **1** C (2001) 201–219.
[9.49] MAHARA, Y., KUDO, A. Plutonium mobility and its fate in soil and sediment environments. Radioact. Environ. **1** C (2001) 347–362.
[9.50] KERSTING, A. B. Plutonium transport in the environment. Inorg. Chem. **52** 7 (2013) 3533–3546.
[9.51] AKIE, H., SATO, I., SUZUKI, M., SERIZAWA, H., ARAI, Y. Simple formula to evaluate helium production amount in fast reactor MA-containing MOX fuel and its accuracy. J. Nucl. Sci. Technol. **50** 1 (2013) 107–121.
[9.52] HOLM, E., ROOS, P., AARKROG, A., MITCHELL, A., VINTRO, L. L. Curium isotopes in Chernobyl fallout. J. Radioanal. Nucl. Chem. **252** 2 (2002) 211–214.

10. BRAZIL ALTERNATIVES TO ^{137}CS FOR ^{210}PB DATING VALIDATION IN GUANABARA BAY SEDIMENTS: POINT AND DIFFUSE SOURCES

J.M. GODOY
Pontifical Catholic University of Rio de Janeiro

Rio de Janeiro, Brazil
Email: jmgodoy@puc-rio.br

C.G MASSONE
Pontifical Catholic University of Rio de Janeiro
Rio de Janeiro, Brazil

R.S. CARREIRA
Pontifical Catholic University of Rio de Janeiro
Rio de Janeiro, Brazil

A.R. WAGENER
Pontifical Catholic University of Rio de Janeiro
Rio de Janeiro, Brazil

O.G.P.G SANTOS
Pontifical Catholic University of Rio de Janeiro
Rio de Janeiro, Brazil

P.C.N. GUIMARÃES
Pontifical Catholic University of Rio de Janeiro
Rio de Janeiro, Brazil

V. HATJE
Federal University of Bahia
Salvador, Bahia, Brazil

F. CARVALHO
Baktron Microbiologia Ltda.
Polo de Meio-Ambiente, Saúde e Biotecnologia do Rio de Janeiro (MASBIO-RJ)
Rio de Janeiro, RJ, Brazil

Abstract

Alternatives to the use of ^{137}Cs as a ^{210}Pb sediment dating validation tool are proposed. In highly industrialized bays, such as Guanabara Bay in the Rio de Janeiro metropolitan region, several chemical/compounds from the surrounding industry releases may be applied as validation tools. Chromium and copper profiles in a sediment core adequately fit the expected pattern due to the implementation of a chemical plant in 1958, reaching their maximum discharge in 1982 and subsequently decreasing due to the operation of a new wastewater treatment plant. A diffuse source such, as the PCB-based mixture Askarel, was also applied as an alternative validation tool, and the observed concentration profile reproduced the expected behavior, with increasing concentration after the Second World War and a decrease after the ban of this product in 1981. The observed Aroclor 1254 and 1260 mixture chlorination rates fit the most widely distributed PCB-based products in the country.

10.1. INTRODUCTION

Guanabara Bay (22°24-22° 57' S and 43°33-43°19' W) is a 346 km^2 estuarine system located in the most industrialized and urbanized Brazilian coastal region, in the state of Rio de Janeiro [10.1]. Its drainage basin is formed by 91 rivers and canals which contribute to a mean water flow of 100 $m^3 s^{-1}$, ranging from 33 $m^3 s^{-1}$ in July (dry season) to 186 $m^3 s^{-1}$ in January (rainy season), and with 25 $m^3 s^{-1}$ of domestic wastewater delivered by the State Company for Water and Seawater (CEDAE), the total freshwater flow to the bay is approximately 125 $m^3 s^{-1}$ [10.2].

The hydrology of Guanabara Bay is described in 10.2 and 10.3. Briefly, the bay extends 30 km N-S and 28 km E-W, with a maximum width of 29 km and greatest continuous extent of 36 km. The total drainage basin volume is approximately $2.2 \cdot 10^9$ m^3, about 10% of which is exchanged each tidal cycle, resulting in a residence time of 20±5 days.

The sediments in the region range from sandy and silty-clay at Guanabara Bay to silt-sand on the coast of Caxias, and clay-silt-sand on the coast of São Gonçalo and the Catalão inlet (Fundão Island) Sands are dispersed at the bay's entrance and central channel and continue to Governor Island [10.3]. Particulate matter and flocculated clays in the northern part of the bay result in mud and sand deposits. Mangroves also heavily influence sedimentation in the northeastern region [10.3]. Unlike many regions where the sea is the main sediment source, the rivers, which contribute to around 4,000,000 tons $year^{-1}$ to the siltation of the Bay, are considered the main supply of sediment owing to heavy summer rains [10.3].

There is a substantial input of sediment into both the northwestern and western areas of the study region, reflecting the heavy influence of anthropogenic activities including sewage discharge, river channel modification, agriculture, and deforestation [10.3]. These historical changes have drastically increased the sedimentation rate, especially in the region between the mainland and Governor Island, which experienced an increase of 0.73 cm $year^{-1}$ between the 1850s and 1990s. Even higher rates were observed close to the mainland. Sedimentation rates in Northwestern Guanabara Bay, calculated using ^{210}Pb [10.4], were reported as to base 2.0 cm $year^{-1}$Wilken *et al.* [10.4] determined. Previous studies [10.5, 10.6] indicate that the sedimentation rates of different Guanabara Bay regions are similar, with values ranging between 1 and 2 cm $year^{-1}$. Therefore, Guanabara Bay sedimentation rates have increased 10-fold compared to the original rate. Higher sedimentation rates have resulted from either i) factors that led to wide-scale spatial increase (e.g., industrialization and urbanization), or ii) factors that increased rates locally (e.g., embankment construction, enlargement of the international airport, and the rapid growth of transportation routes).

Guanabara Bay sediments are highly enriched in the trace metals Pb, Zn, Cu and Cr compared to the average shale concentration and other regions worldwide [10.1, 10.7]. Several studies have suggested that the predominant heavy metal burden come from rivers in the northwest region, especially the Sarapui, Meriti, Iguaçu, Estrela and Irajá rivers [10.8-10.11]. The high metal concentrations in the northwest bay result from discharge of the most polluted rivers and a large oil refinery in the region [10.1]. However, studies show that the rivers that cross the cities of Niterói and São Gonçalo are also heavy metal sources for Guanabara Bay, especially to the east. The metal load in these rivers to the result of urban development and the large amounts of untreated domestic sewage inputs [10.12]. Two other areas exhibiting high heavy metal concentrations are the Rio de Janeiro port and Jurujuba Bay, which is heavily polluted by domestic sewage [10.13]. The lowest heavy metal concentrations occur in the southern region of the Bay, near the inlet. The northeastern and northwestern regions have similar sediments, although lower heavy metal concentrations are found in this region owing to the existence of an environmental protected area in the northeastern portion of the bay (Guapimirim Environmental Protected Area) [10.1].

Any ^{210}Pb sediment dating needs to be validated based on an independent time marker [10.14]. In the Northern Hemisphere, this is carried out based on the ^{137}Cs profile, the peak of which corresponds to the maximum of the atmospheric nuclear tests fallout in 1963 [10.15]. However, the nuclear test fallout in the Southern Hemisphere was 1/3 of that in the Northern Hemisphere and the actual activity is about 25% of the original values [10.15]. Nonetheless, ^{137}Cs application in the Southern Hemisphere is quite critical and alternatives need to be identified [10.16]. Considering that one of the main applications of ^{210}Pb dating comprises retrospective pollution studies, the depth profile of pollutants with known sources at each specific site can be used as an alternative to ^{137}Cs.

The present study presents the application of Cr and Cu as examples of alternative independent time markers for ^{210}Pb sediment dating, whose inputs to Guanabara Bay are related to one specific industrial plant with well-established operation data, and PCBs, a diffuse source with available information concerning their introduction and consumption in Brazil.

10.2. MATERIAL AND METHODS

The core assessed was collected from the location exhibited in Fig. 10.1, using a UWITEC gravity corer, with 6 cm internal diameter tubes, sliced at two centimeters. This region is one of the most environmentally impacted Guanabara Bay areas, receiving inputs from the Duque de Caxias oil refinery (REDUC), a waste dump site (Jardim Gramacho), and industries located along the Iguaçu and Sarapui river drainage basins.

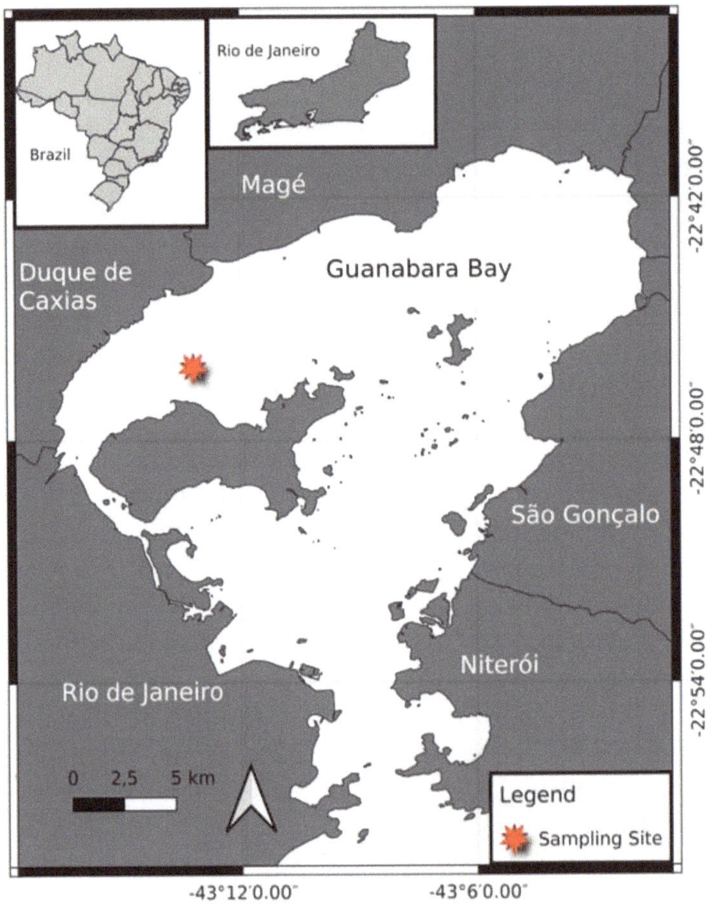

FIG. 10.1. Guanabara Bay map exhibiting the sediment core sampling site.

^{210}Pb concentrations were determined according to Godoy *et al.* [10.17]. To summarize, 5 g dry sediment aliquots were leached at 80 °C with 50 mL 0.5 mol L^{-1} HBr for 2 h. The resulting solution was isolated and the residue leached with 50 mL 0.5 mol L^{-1} HBr and 1.0 g hydroxylamine hydrochloride at 80 °C for 2 h. After addition of a lead carrier, the resulting solution was transferred to an ion-exchange column containing a Dowex 1X8 50-100 mesh. Subsequently, a cleaning step with 0.5 mol L^{-1} HBr and 1.0 g hydroxylamine hydrochloride was applied, followed by further elution with 1 mol L^{-1} HNO_3. Lead was precipitated as chromate, and the chemical yield was gravimetrically obtained. ^{210}Pb concentrations were determined as a function of the daughter decay product, ^{210}Bi, after a two-week ingrowth period by beta counting on a ten-channel low-level proportional counter (Perkin-Elmer Prof. Berthold LB-750). The detection limit of this technique for a 1,000-min counting time is 1 Bq kg^{-1}. Twenty-seven slices were analyzed for ^{210}Pb, one at each two slices, and one aliquot of each slice. The certified reference material IAEA-412 (Pacific Ocean Sediment) was applied for quality control.

A pseudo-total concentration method, similar to the EPA 3050-B method, was applied for the elemental analysis, in which 250 mg sediment aliquots were digested with aqua regia at 95 °C, [10.18] for ICP-AES but without the use of H_2O_2 to reduce blank values. Metal concentration was determined by applying the EPA 6020b method [10.19] using an Agilent 7500CX inductively coupled plasma-mass spectrometer (ICP-MS). The isotopes determined were ^{53}Cr and ^{63}Cu. Blank and reference samples (IAEA-405 sediment sample) were included in every sample batch. Based on the reference sample analysis, the mean uncertainty is lower than 5% for the determined elements. The limits of detection were several orders of magnitude lower than the elemental concentrations found in the sediment samples.

Samples were extracted (2 g) by the EPA 3550C ultrasound technique [10.20] after the addition of the surrogate standards in a dichloromethane/methanol solution (9:1 v/v) for 15 min, repeated three times. Sample extracts were eluted in a glass column (0.3 cm i.d. and 8 cm height) packed with anhydrous Na_2SO_4, alumina and silica before adding the internal standard.

The EPA 8270D method [10.21] in the scan ion mode (SIM) was adopted for PCB quantification. The method was developed based on the standard Quebec Ministry of Environment Congener Mix (C-QME-01 certified by the ISO Guide 34), from Accustandard®. 2,2′,4,5′,6- Pentachlorobiphenyl (PCB 103; CAS number 60145-21-3) and 2,2′,3,3′,4,5,5′,6-octachlorobiphenyl (PCB 198; CAS number 68194-17-2) were used as the surrogate standards. 50 μL of a 100 ng mL^{-1} internal standard (IS) made from 4,4′-Dibromobiphenyl (CAS number 92-86-4/MRC from SPEX CertiPrep) prepared in methanol at a concentration of 2 μg mL^{-1} were added to nine 1 mL standard solutions (500, 200,100, 50, 20, 10, 5, 2, and 1 ng mL^{-1}). The optimized chromatographic and mass spectrometry conditions are displayed in Table 10.1.

TABLE 10.1. OPTIMIZED CHROMATOGRAPHIC AND MASS SPECTROMETRY CONDITIONS FOR PCB DETERMINATIONS.

Chromatography Conditions	
Injector	250 °C Splitless injection

Gas	He, constant pressure (180 kPa)
	1.0 mL min^{-1}
	90 °C, 4 min
Temperature	15 °C min^{-1}, 160 °C
	4 °C min^{-1}, 225 °C
	7 °C min^{-1}, 305 °C
Transferline	290 °C
TriPlus AutoSampler	
Injection	1.5 μL
Mass Spectrometer Conditions	
Source temperature	250 °C
Ionization	EI, 40 eV
Emission Current	100 μA

10.3. RESULTS AND DISCUSSION

10.3.1. ^{210}Pb dating

The total ^{210}Pb profile according to mass depth is presented in Fig. 10.2. No mixing zone was observed, and the activity decreases with depth to the supported ^{210}Pb concentration (14.5±1.0 Bq kg^{-1}). The constant flux model (CF) [10.22] was chosen, based on previous publications [10.5, 10.23]. The plot calculated ages vs. cumulative mass depth shows two different mass accumulation rates (MAR), one up to 1982 and a lower one for the deeper layers (Fig. 10.3). Figure 10.4 presents the sediment accumulation rates (SAR), where three different rates were observed. One additional SAR in relation to MAR can be an effect of compaction. Godoy et al [10.23] observed the same time trend for a core taken from the same region, albeit with actual higher SARs, which can be explained by the fact that the sampling location in that study was closer to the shoreline than the one assessed herein, calculated by using [10.3]

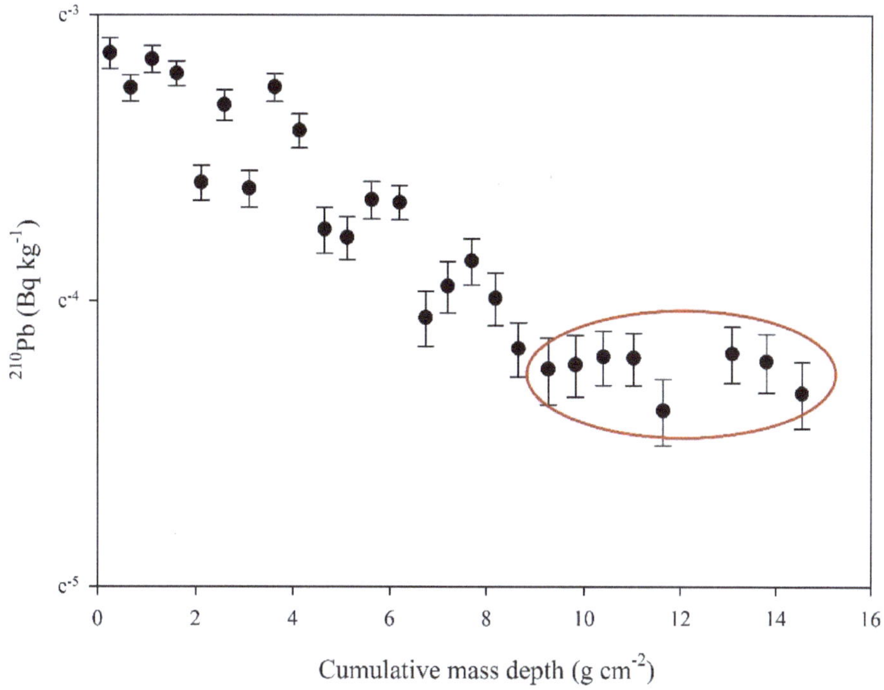

FIG. 10.2. Total ^{210}Pb activity profile according to mass depth. The red oval indicates the ^{210}Pb supported activity for a sediment core from Guanabara Bay.

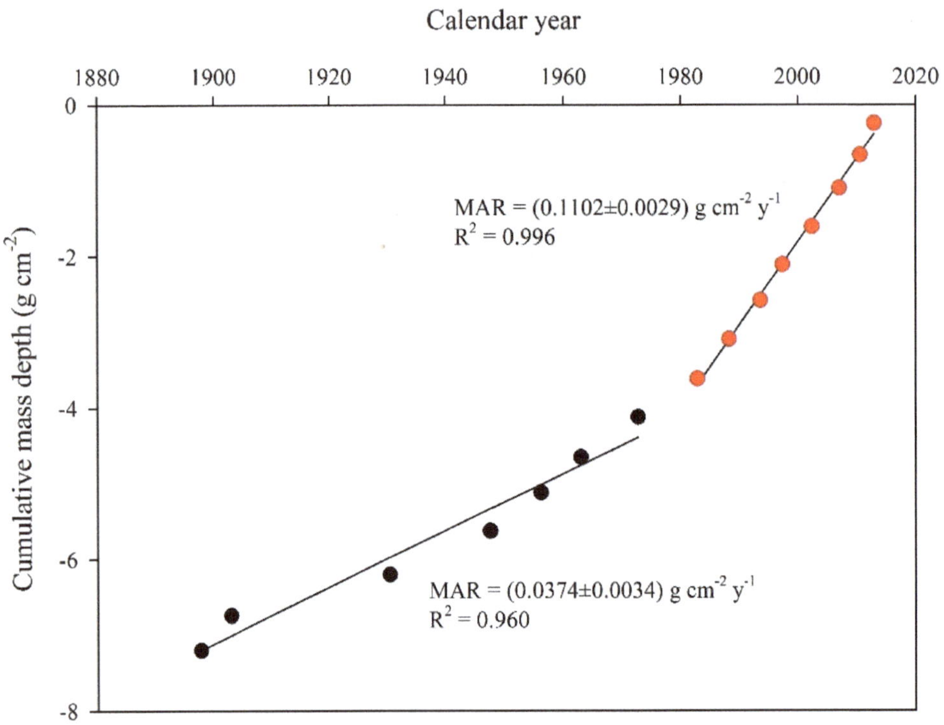

FIG. 10.3. Temporal variation of the mass accumulation rate (MAR) trends for a sediment core from Guanabara Bay.

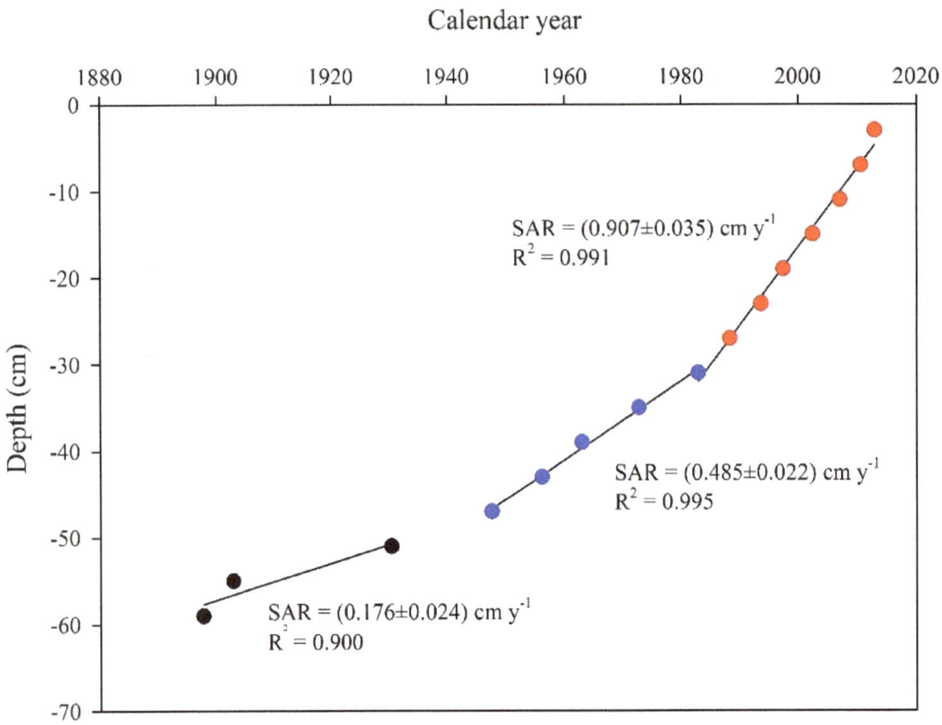

FIG. 10.4. Temporal variation of the sediment accumulation rate (SAR) for a sediment core from Guanabara Bay.

10.3.2. Chromium and copper profiles

In the past, the main Cr and Cu source to Guanabara Bay region was a chromium chemical facility belonging to the Bayer industrial complex, located at the Sarapuí River, which began operating in 1958 and released metal-rich effluents until 1982, when a new treatment plant was implemented [10.16]. Fig. 10.5 presents the Cr and Cu temporal profiles, where a clear increase of the concentration of these metals after the start of the operation of the Bayer chemical plant is noted, followed by a reduction with the introduction of the effluent treatment plant, validating the ^{210}Pb age-model.

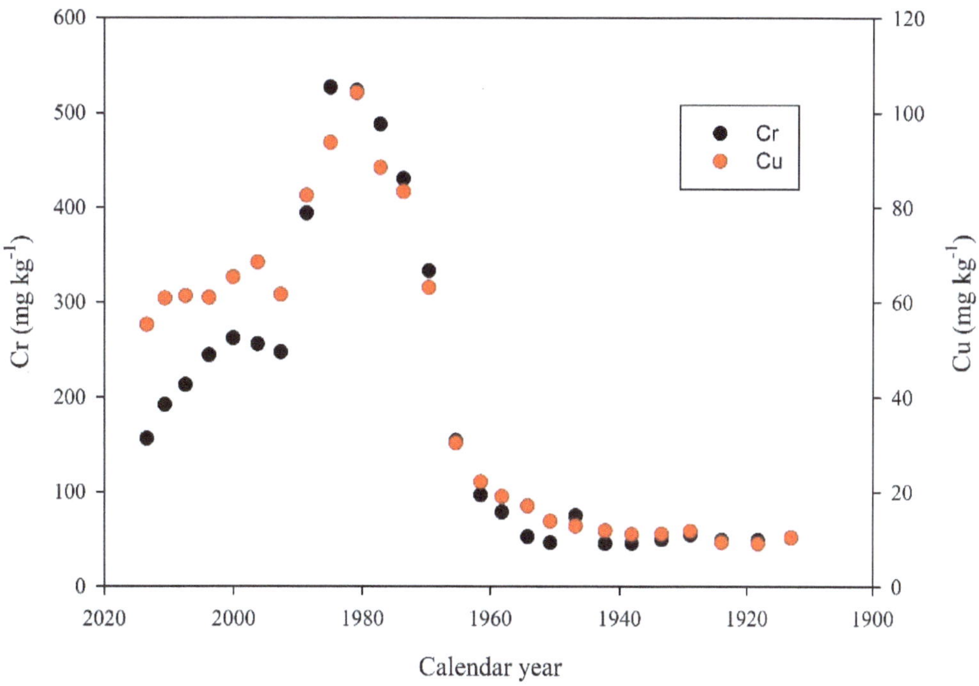

FIG. 10.5. Temporal profile of chromium and copper concentrations applied to validate the ^{210}Pb dating for a sediment core from Guanabara Bay.

The correlation between both metals during the period corresponding to the Bayer releases is displayed in Fig. 10.6. It can be observed that the natural relationship between Cr and Cu is interrupted after the Bayer plant has come into operation.

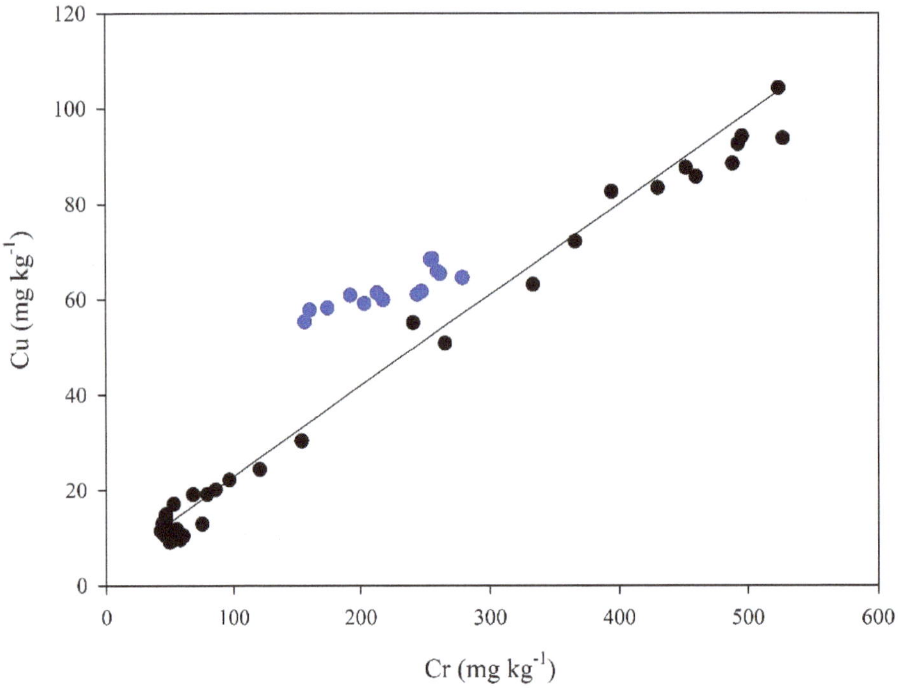

FIG. 10.6. Correlation of chromium and copper concentrations in a sediment core from Guanabara Bay. The black dots correspond to the period before to the metal concentration peak, and the blue dots to the data after the peak.

10.3.3. Polychlorinated biphenyl (PCBs) concentration profile

Polychlorinated biphenyls are synthetic organochlorine compounds, derived from the catalytic reaction of the biphenyl group with anhydrous chlorine. Although 130 PCBs congeners have been marketed, it is theoretically possible to synthesize up to 209 compounds, since PCB molecules can display different arrangements, according to the number of chlorine atoms (ranging from 1 to 10) present in their composition.

PCB-based synthetic products have many applications. Their use in fluid mixes is the most recurrent, although they can also be used in gas transmission turbines, hydraulic fluids, plasticizing resins, adhesives, heat transfer systems, and as additive flame retardant, cutting oils and lubricants. The production of PCB began in 1929 in the United States and subsequently extended worldwide, sold in several countries under different names, such as "Phenoclor®", in France; "Kanechlor®" in Japan "Clophen®"; in Germany and "Fenclor®, in Italy. In the United States it was produced by Monsanto, the main world producer, and the brand "Aroclor®" was marketed in Brazil under the name "Askarel®" 6,7,8. [10.27]. The peak of PCB production took place in 1970, in the electronics industry sector, totaling 50,000 tons [10.27]. In Brazil, the most applied PCB-based compound was Askarel, which led to the use of the term Askarel as a synonym for PCB in the country. The askarel were commercialized in Brazil as mixtures of tetra-, penta- and hexa-chloride isomers, corresponding to the Aroclor 1254 and 1260 mixtures sold in the USA, although heptaclorides PCBs were also found [10.24-10.26].

With the verification of their bioaccumulation potential and high toxicity to humans, PCB production and commercialization became limited or even prohibited in several countries, first in Japan and Sweden in 1972, followed by the USA in 1978, Norway in 1980, Finland in 1985 and Denmark in 1986 [10.28]. In Brazil, Ordinance 019 (MIC, MI, MME) was enacted in 1981. This legislation prohibited the commercialization and use of PCBs throughout the national territory, although existing equipment containing these compounds were permitted until the end of their useful lives. Recently, Stoll et al [10.29] reported PCB concentration above Brazilian regulations in river waters flowing to Guanabara Bay.

The PCB concentration temporal profile in the Guanabara Bay core is presented in Fig. 10.7, alongside their chlorination degree. Before the Second World War (1945), the observed concentrations were close to the limit of detection (<0.2ng g^{-1}). However, with the increasing industrialization in the country, during the 1970s, the so-called Brazilian miracle period, PCB concentrations and the presence of PCBs related to Askarel increased. After their restriction in 1981, PCB concentrations steadily decreased up to 1992, with a slight concentration increase observed after this period. This new increase is consistent with the recent findings of PCBs in the river waters of some Guanabara Bay tributaries [10.29]. The sources of this recent PCB input to the bay's sediment may be ascribed to diffuse inputs by leakages from illegal Askarel deposits or from the inappropriate disposal of electric equipment at the end of their useful life after decades of use. The relative concentration of PCBs by the degree of chlorination through sediment core (Fig. 10.7B) shows the prevalence of the tetra, hexa and hepta substituents, typical of the most marketed product in Brazil (Fig. 10.7C).

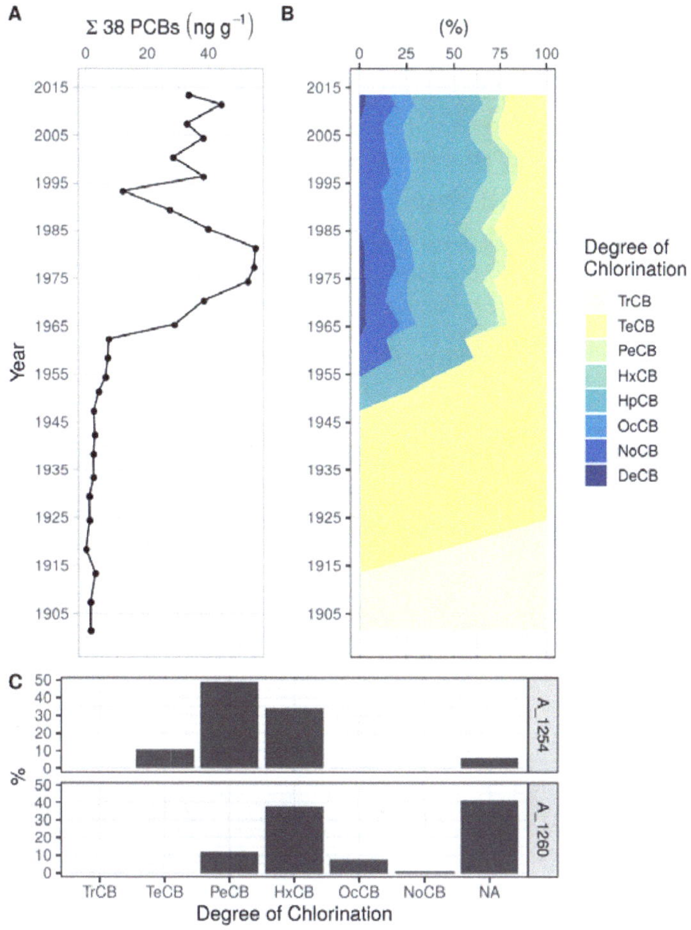

FIG. 10.7. Temporal profile of PCB concentrations (A) and the corresponding compound distribution (B) in in a sediment core from Guanabara Bay. Relative Aroclor distribution of the most marketed product in Brazil (C). The data were adapted from Ref. [10.27].

10.4. CONCLUSIONS

The present study has demonstrated that particle sinking pollutant originated from point or diffuse sources with well-known application dates may be used as a validation tool for ^{210}Pb sediment dating as an alternative to ^{137}Cs in the Southern Hemisphere. As an example, data from release records from a chromium chemical facility belonging to Bayer Chemical near Guanabara Bay were applied to explain the observed temporal concentration changes, with increasing values after facility operations began in 1958, reaching a maximum in 1982 and subsequently decreasing due to a new wastewater treatment plant, showing the coherency of the calculated ages. In addition, the PCB core-depth profile found for Askarel, the most common PCB product marketed and used in Brazil, is consistent with the increased use of this compound accompanying the 1970s Brazilian economic boom, the ban in the 1980s and residual inputs in recent years caused by illegal practices and inappropriate disposal of PCB-based electric products.

10.5. REFERENCES

[10.1] BAPTISTA NETO, J. A., GINGELE, F. X., LEIPE, T., BREHME, I. Spatial distribution of heavy metals in surficial sediments from Guanabara Bay: Rio de Janeiro, Brazil, Environ. Geol. **49** (2006) 1051-1063.

[10.2] KJERFVE, B., RIBEIRO, C. H. A., DIAS, G. T. M., FILIPPO, A. M., QUARESMA, V. S., Oceanographic characteristics of an impacted coastal bay: Baia de Guanabara, Rio de Janeiro, Brazil, Cont. Shelf Res. **17** (1997) 1609-1643.

[10.3] AMADOR, E. S.; Baía de Guanabara e Ecossistemas Periféricos: Homem e Natureza, ISBN 85.900432-1-5, Rio de Janeiro (1997).

[10.4] WILKEN, R. D., MOREIRA, I., REBELLO, A., ^{210}Pb and ^{137}Cs fluxes in a sediment core from Guanabara Bay, Brazil, Sci. Total Environ. **58** (1986) 195-198.

[10.5] GODOY, J. M., MOREIRA, I., BRAGANÇA, M. J., WANDERLEY, C., MENDES, L. B., A study of Guanabara Bay sedimentation rates, J. Radioanal. Nucl. Chem. **227** (1998) 157-169.

[10.6] JAPAN INTERNATIONAL COOPERATION AGENCY (JICA), "The Study on the Recuperation of the Guanabara Bay Ecosystem", Kokusai Kogyo Co. Ltd.: Tokyo, SSR-JE-94-035 (1994).

[10.7] ALMEIDA, A. C. M.; Geocronologia de Compostos Orgânicos de Estanho e de Alguns Metais de Relevância Ambiental. Estudo de Caso: Baía de Guanabara. *Ph.D.Thesis,* Pontifícia Universidade Católica do Rio de Janeiro, Chemistry Department, Rio de Janeiro, Brasil (2003).

[10.8] CARVALHO, C. N., SCHORCHER, H. D., Geochemical studies of the Sarapui River (tributary of Guanabara Bay) – Rio de Janeiro, Brazil, Environ. Tech. Lett. **3** (1982) 425-432.

[10.9] DE LUCA REBELLO, A., HAEKEL, W., MOREIRA, I., SANTELLI, R., SCHROEDER, F., The fate of heavy metals in a tropical estuarine system, Mar. Chem. **18** (1986) 215-225.

[10.10] PFEIFFER, W. C., FISZMAN, M., CARBONELL, N., Fate of chromium in a tributary of the Irajá River, Rio de Janeiro. Environ. Pollut. **1** (1980) 117-126.

[10.11] PFEIFFER, W. C., FISZMAN, M., ROCHA, N. C. C., Chromium in water, suspended particles, sediments and biota in the Irajá River estuary, Environ. Pollut. **4** (1982) 193-205.

[10.12] BAPTISTA NETO, J. A., CRAPEZ, M., MCALLISTER, J. J., VILELA, C. G., Concentration and Bioavailability of Heavy Metals in Sediments from Niterói Harbour (Guanabara Bay/S.E. Brazil), J. Coastal Res. **21** (2005) 811-817.

[10.13] BAPTISTA NETO, J. A., SMITH, B. J., MCALLISTER, J. J., Heavy metal concentrations in surface sediments in a nearshore environment, Jurujuba Sound, SE, Brazil, Environ. Pollut. **109** (2000) 1-9.

[10.14] SMITH, J. M., Why should we believe ^{210}Pb sediment geochronologies? J. Environ. Rad. **55** (2001) 121-123.

[10.15] MONETTI, M. A., Worldwide Deposition of Strontium-90 through 1990. Environmental Measurements Laboratory, US Department of Energy, NY, EML-579, (1996).

[10.16] GOMES, F. C., GODOY, J. M., GODOY, M. L. D. P., CARVALHO, Z. L., LOPES, R. T., SANCHEZ-CABEZA, J. A., OSVATH, I., LACERDA, L. D., Geochronology of anthropogenic radionuclides in Ribeira Bay sediments, Rio de Janeiro, Brazil. J. Environ. Rad. **102** (2011) 871-876.

[10.17] GODOY, J. M., MOREIRA, I., WANDERLEY, C., SIMÕES FILHO, F. F., MOZETO, A. A., An Alternative Method for the Determination of Excess ^{210}Pb in Sediments, Radiat. Prot. Dosim. **75** (1998) 111-115.

[10.18] UNITED STATES ENVIRONMENTAL PROTECTION AGENCY (EPA), Method 3050B: Acid Digestion of Sediments, Sludges, and Soils (2019).

[10.19] UNITED STATES ENVIRONMENTAL PROTECTION AGENCY (EPA), Method 6020b: Inductively Coupled Plasma-Mass Spectrometry (2019).

[10.20] UNITED STATES ENVIRONMENTAL PROTECTION AGENCY (EPA), W-846 Test Method 3550C: Ultrasonic Extraction, (2007).

[10.21] UNITED STATES ENVIRONMENTAL PROTECTION AGENCY (EPA), Method 8270D: Semivolatile Organic Compounds by Gas Chromatography/Mass Spectrometry (GC/MS), (2007).

[10.22] SANCHEZ-CABEZA, J. A., RUIZ-FERNÁNDEZ, A. C., ^{210}Pb sediment radiochronology: an integrated formulation and classification of dating models, Geochim. Cosmochim. Acta **82** (2012) 183-200.

[10.23] GODOY, J. M, OLIVEIRA, A. V., ALMEIDA, A. C., GOFOY, M. L. D. P., MOREIRA, I., WAGENER, A. R., FIGUEIREDO JUNIOR, A. G., Guanabara Bay Sedimentation Rates based on ^{210}Pb Dating: Reviewing the Existing Data and Adding New Data, J. Braz. Chem. Soc. **23** (2012) 1265-1273.

[10.24] LAVANDIER R, ARÊAS J, QUINETE N, DE MOURA JF, TANIGUCHI S, MONTONE R., SICILIANO, S., MOREIRA, I., PCB and PBDE levels in a highly threatened dolphin species from the Southeastern Brazilian coast, Environ Pollut. **208** (2016) 442–447.

[10.25] YOGUI G. T., TANIGUCHI S., DA SILVA J., MIRANDA D. A., MONTONE R. C. The legacy of man-made organic compounds in surface sediments of pina sound and suape estuary, northeastern Brazil, Braz J. Oceanogr. **66** (2018) 58–72.

[10.26] STOLL, G. C. "Desenvolvimento de Método Analítico para a Determinação de Resíduos de Bifenilas Policloradas (PCBs) em Águas Superficiais por Cromatografia Gasosa Acoplada à Espectrometria de Massas em Tandem." Ph.D.Thesis, Pontifícia Universidade Católica do Rio de Janeiro, Chemistry Department, Rio de Janeiro, Brasil, 2018.

[10.27] PENTEADO, J. C. P., VAZ, J. M. O., legado das bifenilas policloradas (PCBs), Química Nova **24** (2001) 390–398.

[10.28] SILVA, S. F. G. “Ocorrência de PBDEs e PCBs em mexilhões e peixes da Baía de Guanabara”, Ph.D.Thesis, Pontifícia Universidade Católica do Rio de Janeiro, Chemistry Department, Rio de Janeiro, Brasil, (2009).
[10.29] STOLL, G. C., CARREIRA, R. S., MASSONE, C. G., Polychlorinated biphenyls (PCBs) in water: Method development and application to river samples from a populated tropical urban area, Anal. Bioanal. Chem. **412** (2020) 2477–2486

11. TRACE ELEMENT ENRICHMENT TRENDS IN THE SOUTHERN GULF OF MEXICO AND ITS RELATIONSHIP WITH OIL INDUSTRY ACTIVITIES

A.C. RUIZ FERNANDEZ
Instituto de Ciencias del Mar y Limnología, Universidad Nacional Autónoma de México
Mazatlán, Mexico
Email: caro@ola.icmyl.unam.mx

A. GRACIA
Instituto de Ciencias del Mar y Limnología, Universidad Nacional Autónoma de México
Mexico City, Mexico

J.A. SANCHEZ-CABEZA
Instituto de Ciencias del Mar y Limnología, Universidad Nacional Autónoma de México
Mazatlán, México

M. RANGEL-GARCIA
Posgrado en Química. Universidad Nacional Autónoma de México.
Mazatlán, Mexico

L.H. PEREZ BERNAL
Instituto de Ciencias del Mar y Limnología, Universidad Nacional Autónoma de México
Mazatlán, México

P.G. LOPEZ-MENDOZA
Posgrado en Ciencias del Mar y Limnología. Universidad Nacional Autónoma de México
Mazatlán, México

M.L. MACHAIN-CASTILLO
Instituto de Ciencias del Mar y Limnología, Universidad Nacional Autónoma de México
Mexico City, Mexico

P. SCHWING
University of South Florida, College of Marine Science.
St. Petersburg, USA

D. HOLLANDER[†]
University of South Florida, College of Marine Science.
St. Petersburg, USA

F. PAEZ-OSUNA
Instituto de Ciencias del Mar y Limnología, Universidad Nacional Autónoma de México
Mazatlán, Mexico

J.G. CARDOSO-MOHEDANO
Instituto de Ciencias del Mar y Limnología, Universidad Nacional Autónoma de México
Ciudad del Carmen, Mexico

T.C. CUELLAR-MARTINEZ
Instituto del Mar del Perú
Lima, Perú
† Deceased on September 26th, 2020

Abstract

^{210}Pb sediment dating is useful to reconstruct historical contamination trends in aquatic environments. It is especially valuable to solve the lack of pre-anthropogenic baselines and long-term records, and allows assessment of both the impacts of industrial development and effectiveness of management measures for environmental protection. Temporal variations, within the past ~100 years, of potentially toxic trace element (As, Ba, Co, Cr, Cu, Hg, Ni, Pb, V and Zn) concentrations, enrichment and fluxes were assessed in ^{210}Pb-dated sediment cores from offshore and intertidal locations in the southern Gulf of Mexico, with the purpose to estimate the potential oil industry impacts in the region. In offshore cores, the concentrations of As, Ba, Cu, Pb, V, and Zn were usually higher, and of Co, Cr, Hg, and Ni were usually lower, than in intertidal cores. Based on cluster analysis results, trace element concentrations were mostly influenced by sediment grain size and terrigenous contribution through fluvial discharges. Cr, Cu and Ni concentrations are naturally elevated in the region, reaching levels of potential ecological concern. Enrichment factors indicated the absence of contamination for most elements in most offshore cores; whereas contamination was observed in cores collected from areas influenced by major river discharges and flushing from Terminos Lagoon. There was a recent decrease in trace element fluxes in most cores, which could be caused by environmental controls established since the 1980s. However, around the Campeche Sound area, trace metal inputs were mostly attributed to wastewater discharges and land erosion in the adjacent coastal areas, associated with population growth and land use changes, promoted by the Campeche Sound oil production activities. Results highlight that environmental controls are needed for the offshore oil industry, and for the control of watershed erosion and untreated wastewater discharges to reduce contaminant loads to the marine environment.

11.1. INTRODUCTION

The Gulf of Mexico (GoM) is a marginal sea, between the tropics and subtropics of the north-western Atlantic Ocean, bounded by the shorelines of southern United States of America and eastern Mexico. It encompasses ~1.5 million km^2 and a large coastal population (40,522,728 inhabitants in 2010; [11.1]). The major economic activities in the GoM include tourism, recreational and commercial fisheries, and petroleum exploitation. The coastal and marine resources in the southern GoM (sGoM) are under pressure from numerous anthropogenic stressors, including shoreline development and contamination from land-based (e.g. industrial and agricultural) and offshore (e.g. oil drilling and ship traffic) activities.

Naturally occurring petroleum seeps are widely distributed throughout the region [11.2], which is one of the most relevant regions for global offshore oil and gas production [11.3]. In Mexico, the oil industry is, by value, the fourth largest export commodity [11.4] and contributes 8% of the gross domestic product (GDP; [11.5]); the country produced 1,833 thousand barrels per day of crude oil and 4,847 million cubic feet per day of natural gas in 2018 [11.6].

Contamination resulting from the oil industry is a worldwide concern, as the discharge of industrial wastes and oil spills can have deleterious effects on aquatic environments. Metals and metalloids may naturally occur in crude oils (e.g. Al, As, Ba, Cd, Co, Cr, Cu, Fe, Ga, Hg,

Mg, Mn, Mo, Na, Ni, Pb, Si, Sn, Sb, Zn, Ti, U, V and Zn), and may also be enriched during the production, transport and storage phases [11.7, 11.8].

Sediments are the final repository for a numerous various contaminants, scavenged by sinking matter. Assuming negligible post-depositional mixing, sediment cores may be used to reconstruct long-term trends of oil-related contamination in the aquatic environment. ^{210}Pb-dated sediment cores, in particular, have been successfully employed to provide contamination trends of organic compounds and trace elements attributed to oil industry activities in coastal areas, such as Prince William Sound [11.9], Yangtze River Estuary [11.10], Gulf of Tehuantepec [11.11, 11.12] and Coatzacoalcos River [11.13–11.15].

The purpose of this study was to assess the recent temporal trends (<100 yr) of the transport and deposition of trace elements (As, Ba, Co, Cr, Cu, Hg, Ni, Pb, V and Zn) regularly associated with oil industry activities, based on the analysis of ^{210}Pb-dated sediment cores collected in offshore and intertidal areas at the sGoM. Most data have been already discussed [11.16, 11.17], although new data on concentrations and fluxes of Co in offshore cores, and As, Ba, Co, Cr, Cu, Pb, Ni, V and Zn in intertidal areas are included.

11.2. STUDY AREA

The sGoM ecoregion is surrounded by five coastal states of Mexico, between Tamaulipas and the northeastern end of the Yucatán Peninsula. It extends along the west and south, from a terrigenous and progressively narrow continental shelf (composed of reworked fluvial sediment, interspersed with coastal rocky outcrops) to a broad carbonate platform and carbonate sand beaches to the east (Fig. 11.1; [11.18]). The western and southern sGoM coasts receive abundant freshwater discharges from numerous rivers, including the four most important rivers in Mexico by surface runoff (Grijalva-Usumacinta, Papaloapan, Coatzacoalcos and Pánuco) [11.19], in contrast with the scarce fluvial discharges in the easternmost part (eastern shoreline of Yucatán Peninsula) where river discharges are practically non-existent [11.20].

Most of the oil production in Mexico takes place in the coastal plain and the continental shelf of the sGoM (~83% oil and 61% gas extraction of the country). By the end of 2018, the region accounted for 252 marine platforms, mostly located in the Campeche Sound, the main gas and oil offshore production area of Mexico since late 1970s, where the Ixtoc-1 blowout occurred in 1979. The sGoM fishing resources (271,971 ton in 2017) represent a minor share (~14%) of the total country production (most important fisheries: octopus, oysters, mojarra, shrimp, sea bass, horse mackerel, blue crab). However, almost all fisheries production in the sGoM is for direct human consumption [11.21], thus representing an important food resource for the region.

Términos Lagoon (TL), in the sGoM, is the largest coastal lagoon of Mexico (7,050 km^2 surface and 3.5 m average depth [11.22]) and it is considered a Wetland of International Importance (Ramsar Site 1356). It receives riverine discharges from the Candelaria, Chumpan and Palizada (a branch of the Usumacinta River, which accounts for half of the freshwater input to the lagoon) Rivers. It is separated from the GoM by a barrier island, although the openings at both extremes allow a permanent sea exchange [11.22]. Ciudad del Carmen (~248,300 inhabitants; [11.23]), located on the barrier island, is the main operation center of the Campeche Sound, which has had a three-fold population growth since 1970, when Campeche Sound offshore oil extraction activities started.

Chronic contamination of trace elements and hydrocarbons has been documented in sGoM coastal areas related to offshore and land-based sources [11.24–11.27]. TL is affected by urbanization, land and coastal erosion, waste water discharges, and especially by the petroleum industry. Oil pipes traversing the region coupled with exploration have promoted heavy marine traffic, resulting in oil spills that have reached TL [11.28–11.30].

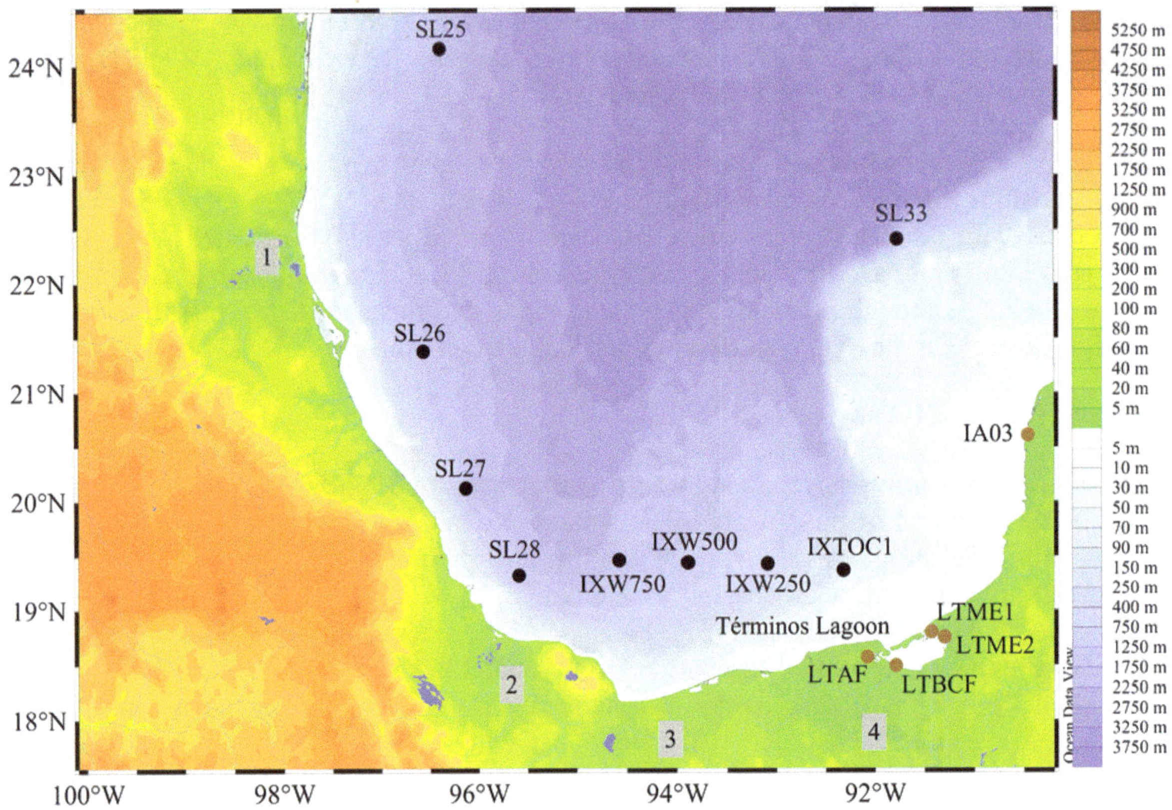

FIG. 11.1. Sampling sites in the southern Gulf of Mexico. Offshore cores are noted by black circles, and intertidal cores by brown circles. Numbers in grey squares indicate the rivers Pánuco (1), Papaloapan (2), Coatzacoalcos (3) and Grijalva-Usumacinta (4). The color bar indicates topographic elevation (white to orange) and bathymetric depth (white to purple).

11.3. METHODS

Most data included in this study have been previously used to discuss trace element contamination trends in offshore and intertidal areas of the sGoM [11.16, 11.17], where the methods are described in detail.

11.3.1. Sampling and sample preparation

Polyethylene liners (9.5 cm internal diameter) were used to collect sediment cores (multicorer system) in offshore areas on board of the R/V Justo Sierra of Universidad Nacional Autónoma de México in 2015, and push-cores in mangrove-surrounded intertidal areas between 2016 and 2018 (Table 11.1). Extrusion techniques were used to sub-samples all cores. For offshore cores, which were predicted to have lower sediment accumulation rates, the sectioning resolution (2 mm sections for the upper 100 mm, and 5 mm sections below) was lower [11.31] than for the intertidal cores (1 cm sections). After freeze-drying, samples that were not to be analysed for grain size were ground to powder with agate mortars and kept in polyethylene bags until analysis.

TABLE 11.1. SUPPORTING INFORMATION FOR SAMPLING SITES IN THE SOUTHERN GULF OF MEXICO AND CORRESPONDING ^{210}PB DATA (*Adapted from [11.17]*)

Core	Sample date	Lat. (N)	Long. (W)	Water depth (m)	Core length (cm)	$^{210}Pb_{sup}$ ($Bq\cdot kg^{-1}$)	SAR range ($cm\cdot yr^{-1}$)	MAR range ($g\cdot cm^{-2}\cdot yr^{-1}$)
Offshore cores								
SL25	09/08/2015	24°09.597'	96°23.655'	1603	26.0	53.0 – 61.0	0.01 – 0.17	0.01 – 0.03
SL26	31/07/2015	21°22.589'	96°34.473'	1499	40.0	33.0 – 57.0	0.03 – 0.43	0.02 – 0.20
SL27	08/08/2015	20°07.275'	96°07.811'	1522	37.0	32.0 – 44.0	0.02 – 0.14	0.01 – 0.07
SL28	07/08/2015	19°19.484'	95°35.462'	1564	28.0	28.0 – 32.0	0.02 – 0.17	0.01 – 0.05
SL33	01/08/2015	22°24.614'	91°46.682'	1326	34.0	20.0 – 38.0	0.005 – 0.06	0.003 – 0.03
IXW750	05/08/2015	19°27.598'	94°35.097'	1440	37.0	31.0 – 142	0.02 – 0.80	0.01 – 0.34
IXW500	06/08/2015	19°26.649'	93°53.323'	1010	25.0	37.0 – 179	0.02 – 1.50	0.01 – 0.80
IXW250	06/08/2015	19°25.841'	93°05.698'	583	25.5	35.0 – 134	0.07 – 0.74	0.02 – 0.37
IXTOC1	03/08/2015	19°22.205'	92°19.031'	60	31.0	19.3 – 25.8	0.02 – 0.66	0.01 – 0.12
Intertidal cores								
IA03	10/08/2016	20°36.940'	90°27.392'	0.5	23.5	11.3 – 32.2	0.01 – 0.56	0.01 – 0.11
LTAF	06/04/2016	18°34.297'	92°04.797'	0.5	56.0	15.3 – 47.1	0.18 – 1.17	0.09 – 1.49
LTBCF	05/04/2016	18°29.530'	91°47.418'	0.5	51.0	22.6 – 25.9	0.53 – 2.24	0.62 – 2.79
LTME1	09/04/2018	18°47.581'	91°27.827'	0.5	59.0	14.6 – 20.5	0.04 – 3.62	0.01 – 0.03
LTME2	09/04/2018	18°48.362'	91°27.455'	0.5	64.0	15.6 – 17.8	0.21 – 2.73	0.01– 0.37

$^{210}Pb_{sup}$ = supported ^{210}Pb. SAR = sediment accumulation rate. MAR = mass accumulation rate

11.3.2. Sediment core dating

^{210}Pb activities ($^{210}Pb_{tot}$) were determined through the analysis of ^{210}Po by alpha-particle spectrometry [11.32]. ^{226}Ra activities were used to assess the supported ^{210}Pb activities ($^{210}Pb_{sup}$) and ^{137}Cs activities to attempt ^{210}Pb dating corroboration. ^{226}Ra (through ^{214}Pb, 352 keV) and ^{137}Cs (661.5 keV) were determined by gamma-ray spectrometry [11.33]. Excess ^{210}Pb ($^{210}Pb_{xs}$) was estimated through the difference between $^{210}Pb_{tot}$ and supported ^{210}Pb ($^{210}Pb_{sup}$; assumed in equilibrium with ^{226}Ra activities). ^{210}Pb chronologies, mass accumulation rates (MAR) and sediment accumulation rates (SAR) were calculated with the constant flux (CF) model [11.34, 11.35], and the dating uncertainties were estimated by Monte Carlo simulation [11.36].

11.3.3. Sediment characterization and trace elements concentrations

Sediments were analysed for grain size distribution by laser diffraction using a Malvern Mastersizer 2000E; The loss on ignition method was used asfor organic matter (LOI_{550}) and carbonate (LOI_{950}) content, respectively [11.37]. Elemental composition was determined by X-ray fluorescence spectrometry (XRF, Spectrolab Xepos-3). Hg concentrations were analysed by cold vapor atomic absorption spectrophotometry (Varian SpectrAA 220) [11.38].

11.3.4. Analytical quality control

The analytical accuracy, determined by evaluating certified reference materials, provided results within the reported ranges: IAEA-300 for ^{210}Pb, ^{137}Cs and ^{226}Ra activities; QAS3002-Malvern™ for grain size percentages; and IAEA-158 and PACS-2 for elemental composition. Replicate analysis of a single sample (n = 6) resulted in variation coefficients <10% for ^{210}Pb, ^{226}Ra and ^{137}Cs activities, <5% for grain size distribution, <3% for elemental composition and <5% for mercury concentrations.

11.3.5. Data treatment

Analysis of variance (ANOVA) and Tukey post-hoc tests (95% confidence) were used to evaluate differences in values among the cores, and between offshore and intertidal cores. Cluster analysis was used to identify the association between geochemical variables and trace element concentrations in the cores. The screening values are threshold effect level (TEL, below which adverse biological effects are expected to occur rarely), and probable effect level (PEL, above which adverse effects are expected to occur frequently) [11.39], were compared to element concentrations to evaluate the likelihood of adverse effects to the benthic biota.

The degree and trends of trace element contamination were evaluated through a normalized enrichment factor (EF; [11.40]), i.e. the ratio of the element concentration at each core section to its pre-industrial concentrations. Each value was normalized by the corresponding concentrations of Ti, as reference element to compensate for variations of sediment grain size [11.38]. The pre-industrial values were calculated as the average element concentration in sections older than 1900s, according to the ^{210}Pb chronologies, and these values were confirmed through a calculated distribution function [11.41]. EF values were graded as follows [11.42]: EF<2 indicates no enrichment; 2<EF<3 minor enrichment; 3<EF<5 moderate enrichment; 5<EF<10 moderately severe enrichment; 10<EF<25 severe enrichment and 25<EF<50 very severe enrichment.

Temporal variations of trace element fluxes (the product of MAR and element concentrations, in $\mu g\ cm^{-2}\ year^{-1}$) were estimated by using the flux ratio (FR), i.e. the quotient of the element flux at each core section to its pre-anthropogenic flux [11.43], which is an unitless index useful to assess the temporal and spatial variability of element deposition, independent from variations in MAR [11.43].

11.4. RESULTS

11.4.1. Radionuclide activities and ^{210}Pb dating

^{210}Pb activities decreased steadily with depth in each core (Fig. 11.2), and therefore postdepositional mixing was considered negligible (Fig. 11.2). $^{210}Pb_{tot}$ and $^{210}Pb_{sup}$ (^{226}Ra) activities were significantly ($p<0.05$) lower in intertidal ($^{210}Pb_{tot}$: 14.8- 317, $^{210}Pb_{sup}$: 11.3-

47.1 Bq kg^{-1}) than in offshore ($^{210}Pb_{tot}$: 31.8- 1350, $^{210}Pb_{sup}$: 19.4 -143.1 Bq kg^{-1}) cores (Fig. 11.3; Table 11.1).

Most sediment records spanned more than 100 years [11.16, 11.17] except offshore SL26, and intertidal LTAF, LTBCF, LTME1 and LTME2 cores. In intertidal cores, the MAR range (0.01 - 1.1 g cm^{-2} yr^{-1}) and SAR (0.02 - 3.6 cm yr^{-1}) were broader than in the offshore cores (MAR: 0.01- 0.4 g cm^{-2} yr^{-1}, SAR: 0.01-0.8 cm yr^{-1}) (Table 11.1, Fig. 11.3).

^{137}Cs activities in all offshore cores and the intertidal core IA03 were below the minimum detectable activity; <5 Bq kg^{-1} in cores LTBCF, LTME1 and LTME2, and up to 157 Bq kg^{-1} in LTAF [11.16] (data not shown). None of the cores (where ^{137}Cs was detectable) showed a maximum attributable to the ^{137}Cs maximum fallout shortly before the signature of the Limited Test Ban Treaty [11.44], which could be used to corroborate the ^{210}Pb-derived chronologies [11.16, 11.17]. Detectable ^{137}Cs activities in the upper segment of these cores beginning in the 1950s verified that these sediments were deposited after the start of nuclear weapons tests (i.e. 1945, Trinity test, Alamogordo, USA) in agreement with ^{210}Pb derived age models.

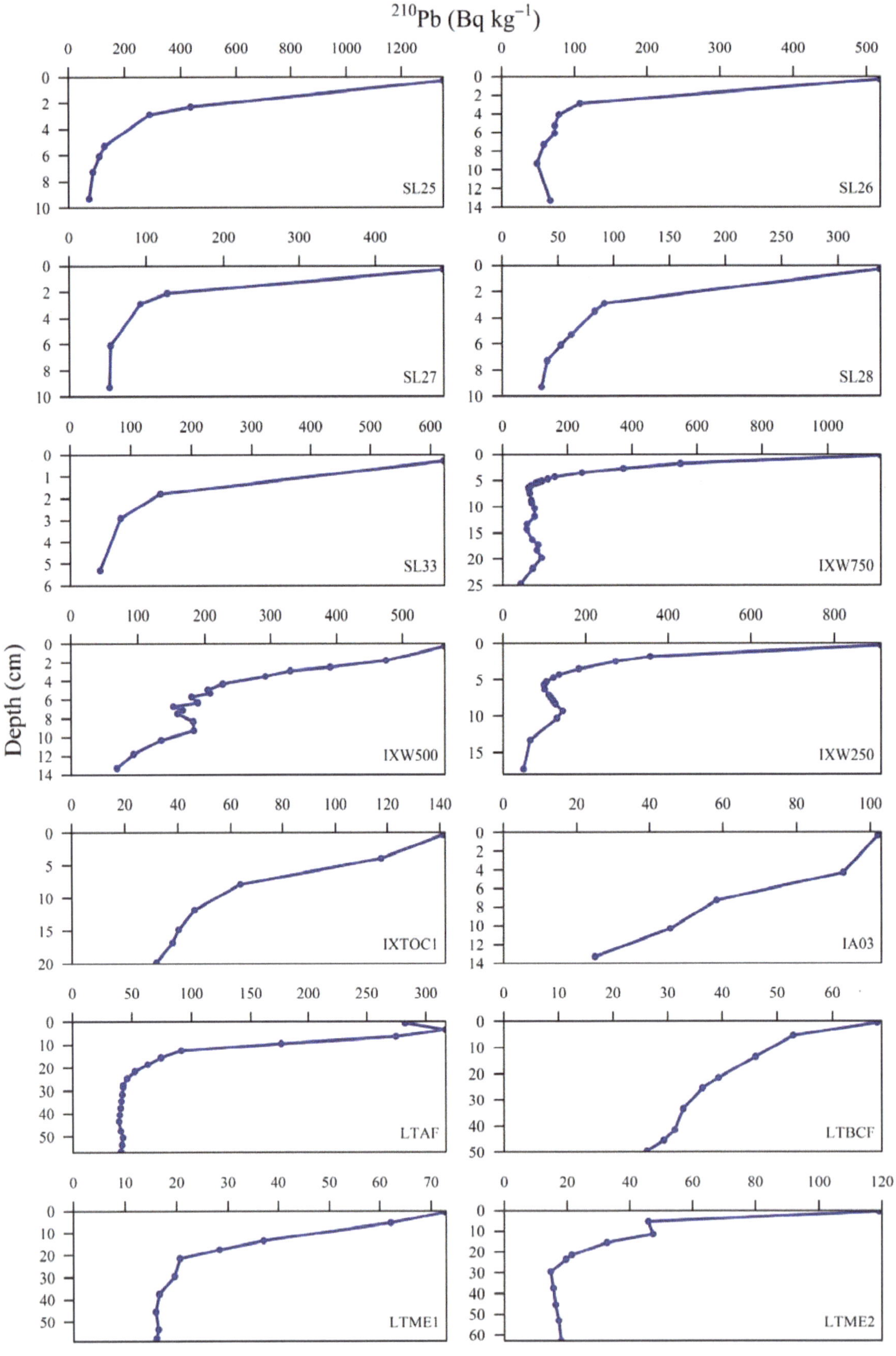

FIG. 11.2. Total ^{210}Pb activity depth profiles in sediment cores from offshore and intertidal areas in the southern Gulf of Mexico.

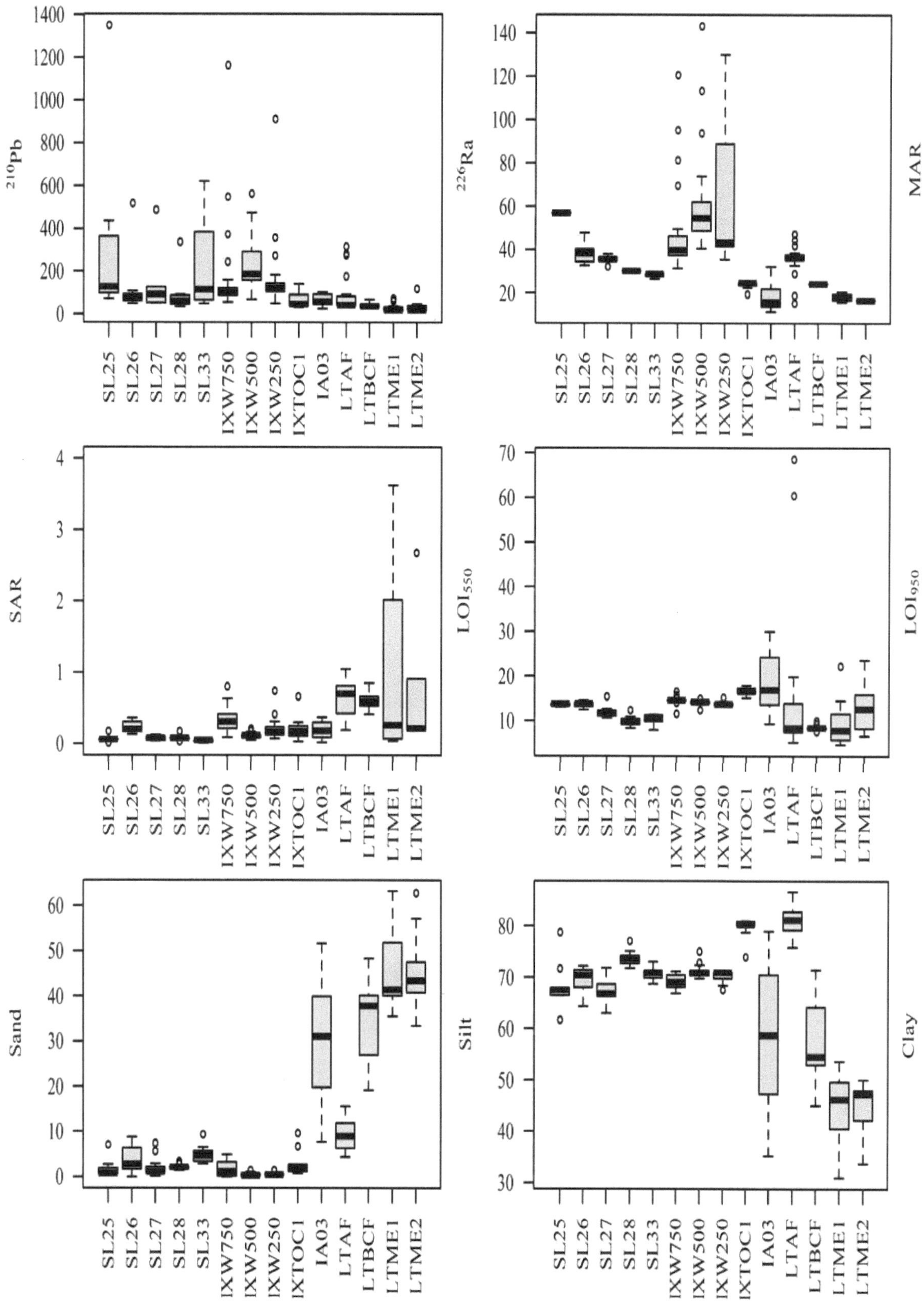

FIG. 11.3. Box plots of ^{210}Pb and ^{226}Ra (Bq kg^{-1}), mass accumulation (MAR, g cm^{-2} yr^{-1}) and sediment accumulation (SAR, cm yr^{-1}) rates, and geochemical characteristics (LOI$_{550}$, LOI$_{950}$, sand, silt, clay, in %) in sediment cores from southern Gulf of Mexico.

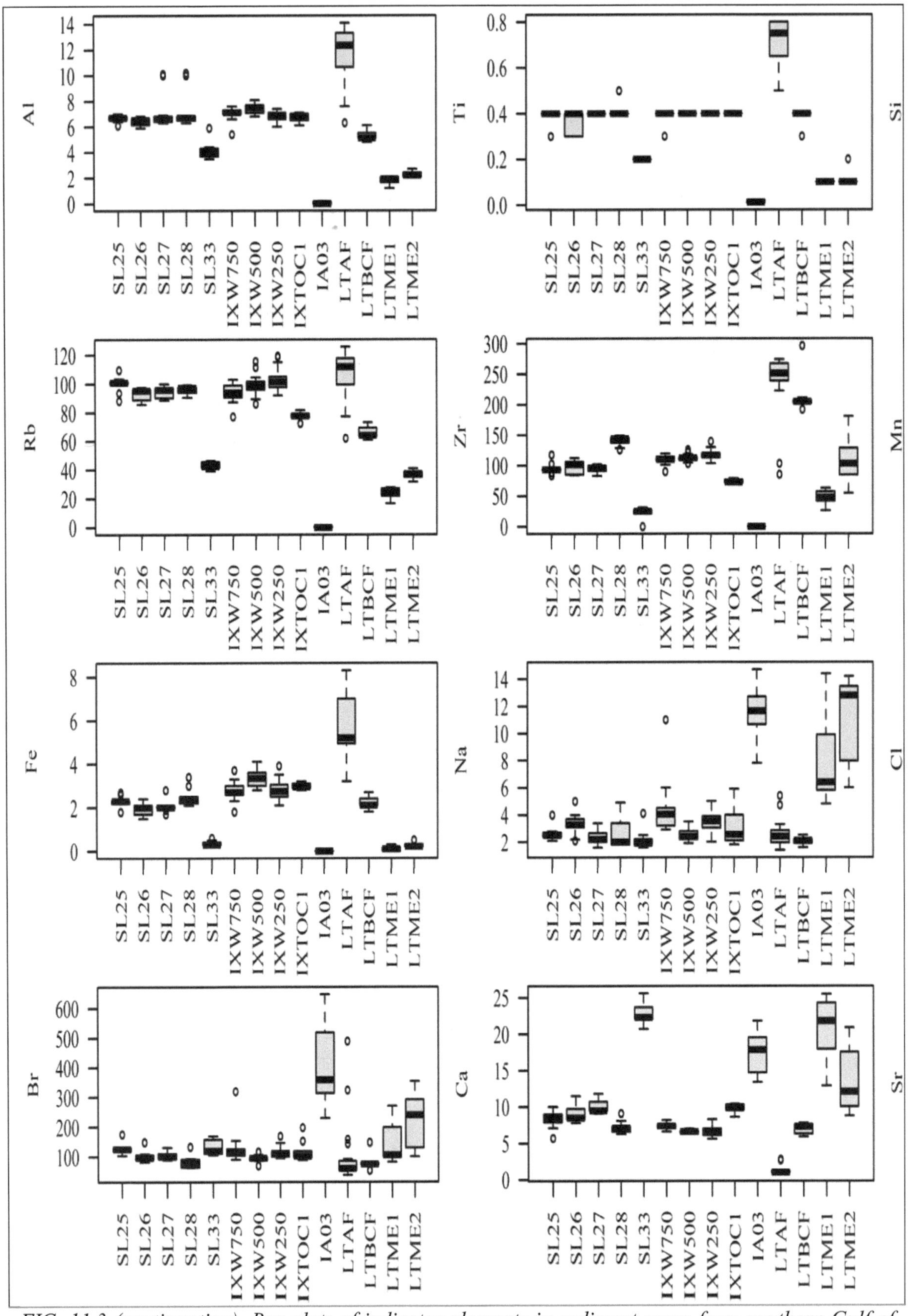

FIG. 11.3 (continuation). Box plots of indicator elements in sediment cores from southern Gulf of Mexico. Al, Ti, Si, Fe, Na, Cl, Br and Ca are in %; Rb, Zr, Mn and Sr in μg g^{-1}.

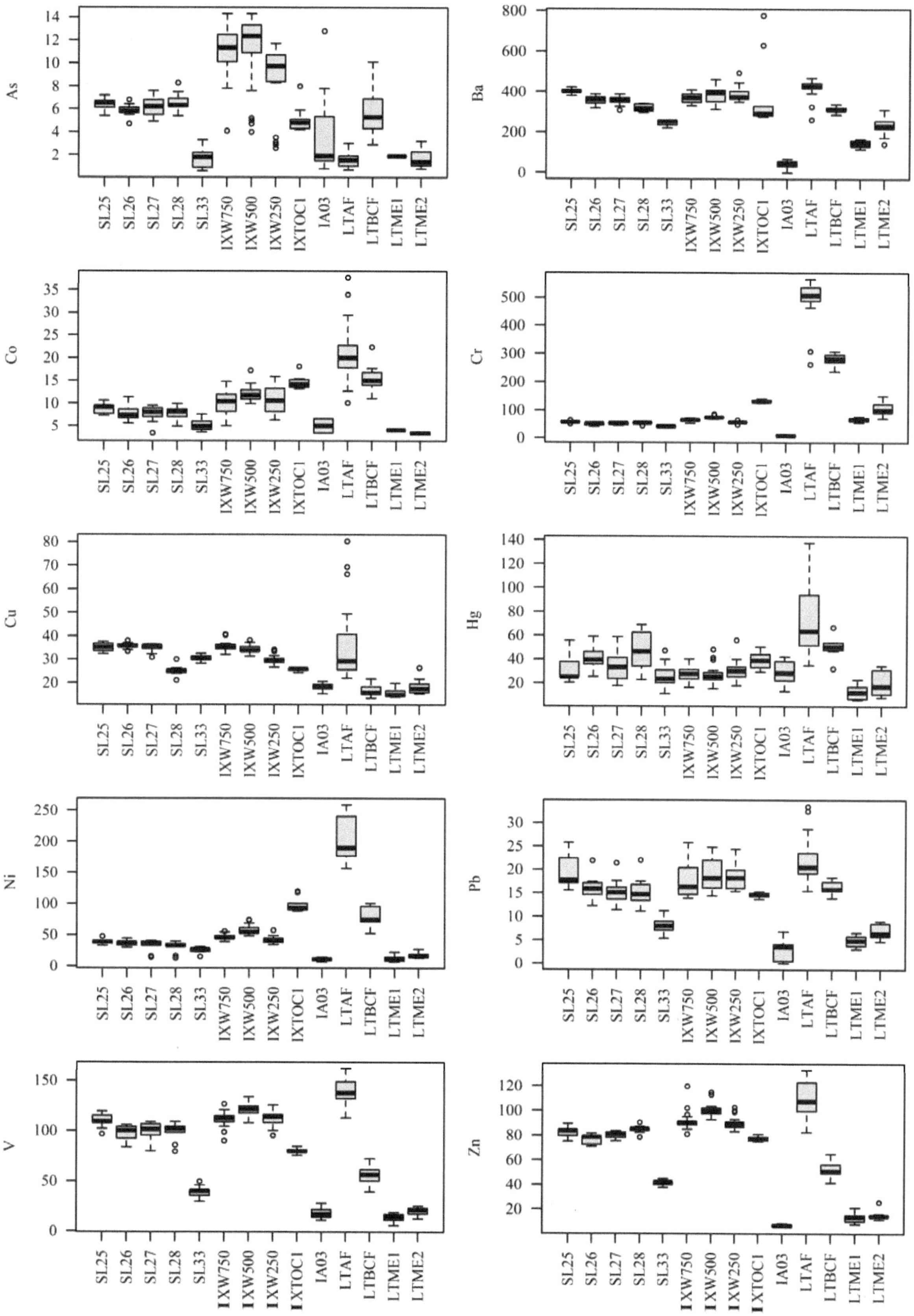

FIG. 11.4. Box plots of trace element concentrations (µg g^{-1}) in sediment cores from the southern Gulf of Mexico.

11.4.2. Geochemical characteristics, element composition and provenance

The intertidal sediments generally had significantly lower mean contents of clay, silt, Al and Mn, and higher mean contents of sand, LOI_{950}, Si and Zr than the offshore sediments, whereas both environments presented comparable values of LOI_{550}, Ti and Fe (Fig. 11.3). In general, intertidal cores showed significantly higher concentrations of Co, Cr, Hg, and Ni, and lower concentrations of As, Ba, Cu, V, Pb, and Zn, than the offshore cores (Fig. 11.4, Table 11.2). Trace element concentrations showed a high variability within and among the sampling areas (Fig. 11.5). For instance, Pb concentrations have steadily increased over time since early 1900s and reached the highest values at the most recent sections in most cores (except intertidal core IA03, where the Pb maxima was observed in 1960s). This is in contrast with Cr profiles, which showed subsurface maxima, but the minima were usually observed at the uppermost section of most cores (except intertidal cores LTME2 and LTBCF). In the offshore cores SL25, SL27, IXW750, IXW500 and IXW250, only Pb concentrations steadily rise, whereas in SL28 and SL33 the highest concentrations of As, Cu, Hg and Pb were observed in the most recent section. The cluster analysis showed three main groups (Fig. 11.6), consisting of: (1) clay, LOI_{550}, Mn, Br, Cl, Na, As and Cu; (2) silt, Al, Ti, Si, Rb, Zr, Fe, Ba, Co, Cr, Hg, Ni, Pb, Zn and V; and (3) sand, LOI_{950}, Sr and Ca.

11.4.3. Assessment of sediment quality and trace element enrichment

A preliminary screening of the ecological significance of trace element concentrations was made by comparing the values observed with the benchmarks threshold effect level (TEL), and probable effect level (PEL) ([11.39]; values in Table 11.2), except for Co and V, which are not available. In most cores Cr, Cu and Ni concentrations exceeded the TEL values (except the offshore cores IA03 for the four elements, and SL26 and SL33 for Cr; the pre-industrial sections of the intertidal cores LTBCF, LTME1 and LTME2 for Cu, and LTME1 and LTME2 for Ni). PEL values for Ni were also exceeded in cores IXW750, IXW500, IXW250, IXTOC1, LTAF and LTBCF; and for Cr in cores LTAF and LTBCF, even at pre-industrial sections.

Enrichment factors were low (EF<2; indicative of no enrichment) for most trace elements in most cores, including Pb concentrations which were consistently rising during the past 100 years (Fig. 11.7). EF >2 were observed in offshore cores IXW250, IXW500 and IXW750 (minor to moderate enrichment by As) and, for intertidal cores, IA03 had varying degrees of enrichment, including As (from minor to severe), Hg (minor), Pb (from severe to very severe), V (from minor to moderate) and Zn (minor); LTAF had minor to moderate enrichment of Cu and Hg since the 1970s; LTME1 from minor to moderate enrichment by Hg; and LTME2 from minor to moderate enrichment by Hg, Ni and Pb. Some cores had minor to moderate enrichment (2<EF<3) only in the topmost section: As in SL33 and LTME2; Ba in IXTOC1; Ba, Co, Ni in IA03; Hg in SL28 and SL33; Ni in LTME1; and As, Cu and Zn in LTME2 (data not shown). The cores with the highest EF values were (Fig. 11.8): IA03, from the eastern side of Yucatan Peninsula (for As, Ba, Pb, V and Zn), as well as LTME1 and LTME2, both from inside Términos Lagoon (for Hg). Pb enrichment in IA03 was EF=38 in 1960s, and even if this has decreased towards present, the topmost section of the core still indicates severe Pb enrichment (EF=19).

11.4.4. Trace element flux ratios

The flux ratios (FR; Fig. 11.9), ranged from <2 (all elements in the intertidal cores LTAF and LTBCF) to 84 (for As in intertidal core IA03), meaning that element fluxes were from less than 2 up to 84 times as large as the pre-industrial fluxes. The highest FR for all elements were observed in core IA03 (Fig. 11.9). Excluding this core, FR of most elements were significantly lower in the intertidal than the offshore cores, with IXTOC1 showing among the highest values. Most offshore cores (except IXTOC1) had lower FR in the last few decades in comparison with those observed by mid-20th century (Fig. 11.10).

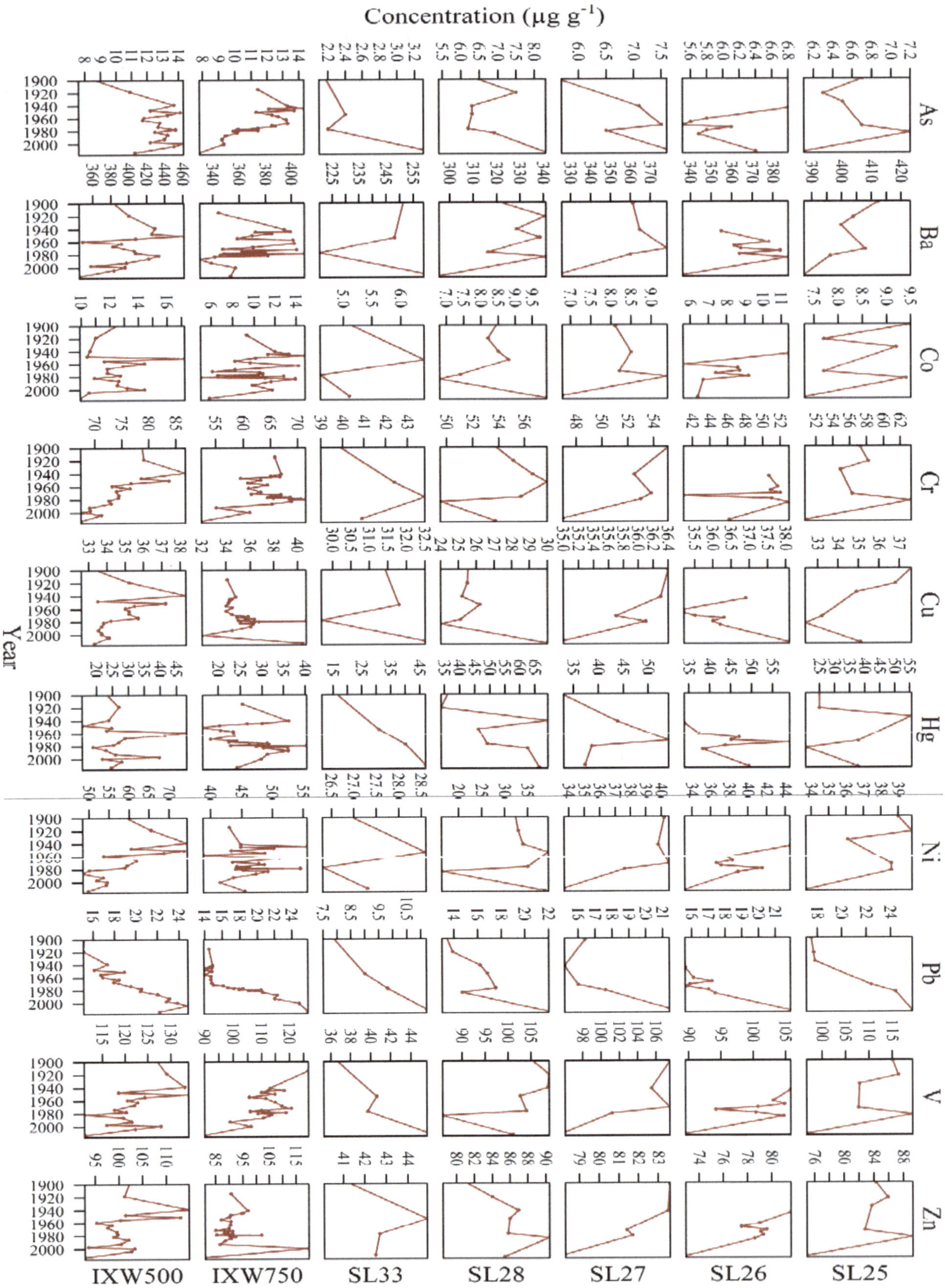

FIG. 11.5. Trace element concentration profiles in sediment cores from the southern Gulf of Mexico. Blank plots correspond to element concentrations below the detection limit (As < 0.5 μg g^{-1}, Co <3.0 μg g^{-1}).

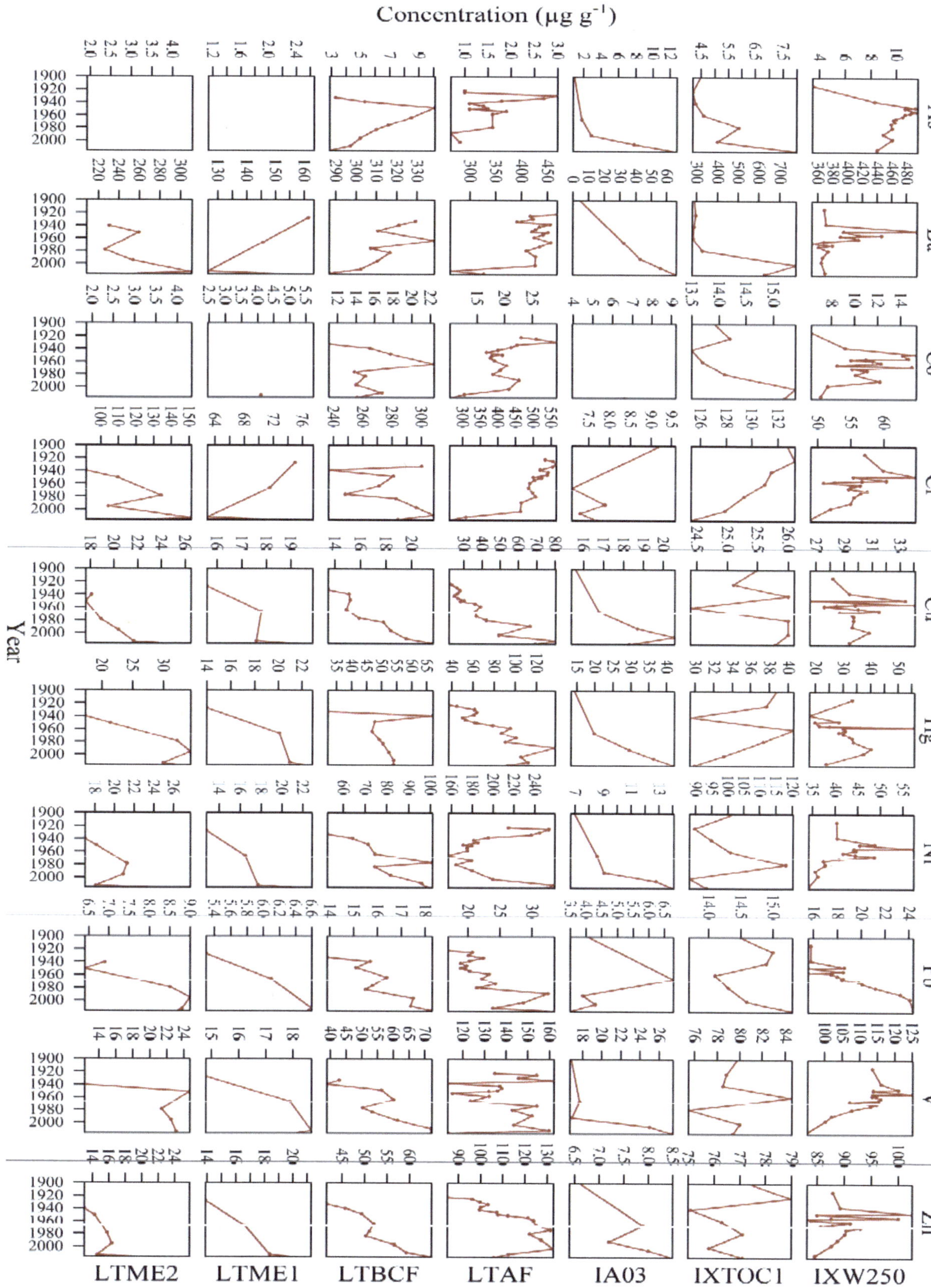

FIG. 11.5 (continuation). Trace element concentration profiles in sediment cores from the southern Gulf of Mexico. Blank plots correspond to element concentrations below the detection limit (As < 0.5 μg g^{-1}, Co <3.0 μg g^{-1}).

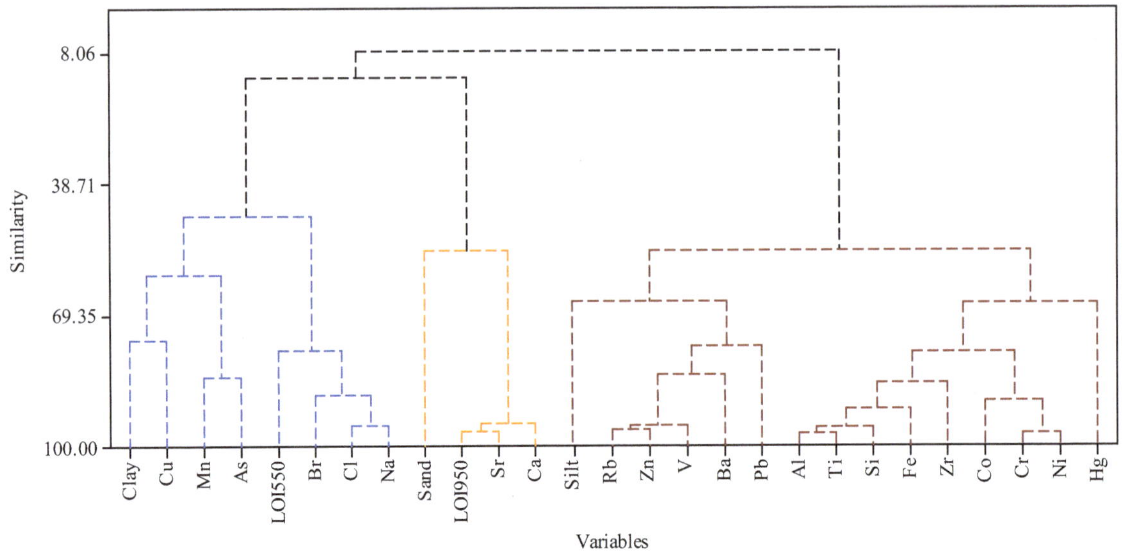

FIG. 11.6. Dendrogram of cluster analysis for geochemical characteristics, reference elements and trace element concentrations in sediment cores from the southern Gulf of Mexico. Groups: 1 (blue), 2 (yellow) and 3 (brown).

11.5. DISCUSSION

The high $^{210}Pb_{tot}$ and ^{226}Ra ($^{210}Pb_{sup}$) activity ranges in offshore cores are atypical in the region, belonging to a latitudinal band characterized by low ^{210}Pb fluxes [11.45], which can be attributed to a low continent land mass/ocean area ratio. These high values might be related to sampling depth, since most offshore cores were retrieved at depths > 1000 m. ^{226}Ra activities in deep sea sediments are commonly higher than in coastal areas, due to ^{226}Ra ingrowth from excess ^{230}Th, which is constantly scavenged from seawater to sediments [11.46]. In addition to the atmospheric flux, $^{210}Pb_{xs}$ activities in deep sediments can be enriched by i) water column scavenging of ^{226}Ra-derived ^{210}Pb [11.47], which is much more efficient by fine particles [11.48] that are more abundant in offshore than intertidal cores (Fig. 11.3), and ii) redistribution by horizontal transport [11.49]. Higher values of ^{226}Ra activities in cores IXW250, IXW500 and IXW750 (Fig. 11.3) were related to the occurrence of redox fronts [11.17], as indicated by the also high concentrations of Mn (Fig. 11.3 continuation), since manganese oxides have a strong ability to absorb radium [11.50].

The difficulties of using ^{137}Cs to corroborate ^{210}Pb chronologies have already been discussed [11.16, 11.17]. In summary, low ^{137}Cs activities result from low fallout in lower latitudes of the northern hemisphere, the time elapsed since ^{137}Cs fallout (only ~65% of the ^{137}Cs deposited in 1963 should still occur in the environment) and the high solubility of ^{137}Cs in seawater, which implies that a high amount of ^{137}Cs remains in solution and, owing to ocean stratification, ^{137}Cs is inefficiently transported to the bottom sediments. Furthermore, the riverine input of ^{137}Cs-bearing soils eroded from the catchment could blur the ^{137}Cs activity profiles [11.51]. In addition, ^{137}Cs may exhibit post-depositional migration after being desorbed from its carrying substrates (e.g. clay minerals, carbonates, iron-manganese oxy-hydroxides, and organic matter) owing to sediment diagenesis [11.52]. Therefore, other stratigraphic markers are needed to corroborate ^{210}Pb-derived chronologies at low latitude coastal and marine environments, and the ^{210}Pb chronologies presented in this study may be adjusted in the future upon additional available information.

Trace element concentrations in GoM sediment cores were similar to those reported as slightly contaminated sediments in the Gulf of Mexico and elsewhere (Table 11.3). The comparison of TEL and PEL benchmarks with observed concentrations suggested that exposure to these sediments might be of ecological concern, owing to high concentrations of As, Cr and Cu, and especially of Ni that is higher than PEL in some cores at both offshore and intertidal areas. However, elevated concentrations of Cr, Cu and Ni are recorded in the majority of sediment cores since pre-industrial times (Fig. 11.5). In fact, enrichment factors indicated negligible contamination by Ni, Cu, and Cr in most cores (except Ni in LTME2). Thus, natural concentrations of As, Cr, Ni, and Cu in the area might represent a hazard to the health of benthic biota, which may have been exposed to high metal concentrations over a long period of time.

TABLE 11.2. TRACE METAL CONCENTRATION RANGES ($\mu g\ g^{-1}$) IN SEDIMENT CORES FROM THE SOUTHERN GULF OF MEXICO (*Adapted from [11.17]*)

	Offshore cores									*Intertidal cores*				
	SL25	SL26	SL27	SL28	SL33	IXW 250	IXW 500	IXW 750	IXTOC 1	IA03	LTAF	LTBCF	LTME1	LTME2
n	19	23	20	18	16	40	37	40	32	10	20	9	11	11
As														
Min	5.4	4.7	4.9	5.4	<0.5	2.6	4.0	4.1	4.0	0.8	0.7	2.9	1.9	0.8
Max	7.2	6.8	7.6	8.3	3.3		14.3	14.3	8.0		3.0	10.1	1.9	3.2
Bckg	5.7	5.3	5.0	5.9	0.6	3.1	4.9	4.1	4.9	1.9	2.0	3.3	N.A	1.1
Mean value per zone: 8.0 ± 3.6										3.1 ± 2.9				
Ba														
Min	350	318	307	295	219	332	312	314	243	<2.0	262	287	117	142
Max	425	396	387	342	259	495	461	410	778	74.7	469	339	166	311
Bckg	393	337	387	324	235	3723	323	365	270	56.0	441	329	140	212
Mean value per zone: 364 ± 66										266 ± 140				
Co														
Min	7.3	5.6	3.4	4.9	3.6	6.3	9.9	5.0	13.2	3.4	10.1	11.1	<DL	<DL
Max	10.6	11.4	9.5	9.9	7.6	15.9	17.2	14.8	18.1	6.6	37.7	22.4	<DL	<DL

	Offshore cores									*Intertidal cores*				
	SL25	SL26	SL27	SL28	SL33	IXW 250	IXW 500	IXW 750	IXTOC 1	IA03	LTAF	LTBCF	LTME1	LTME2
Bckg	9.7	7.4	7.2	6.0	5.1	13.8	11.4	10.5	15.4	3.4	35.9	11.1	<DL	<DL
Mean value per zone: 18.1 ± 10.0										17.3 ± 8.0				
Cr														
Min	48.5	41.0	43.2	43.8	35.8	47.8	67.4	52.4	123	7.1	263	237	55.4	70.0
Max	63.6	55.2	55.1	57.7	45.2	64.9	86.7	71.4	142	15.1	565	308	77.5	151
Bckg	54.7	45.8	54.9	51.4	45.2	59.4	74.5	65.1	140	14.3	519	300	63.7	92.7
Mean value per zone: 141 ± 64.6										243 ± 202				
Cu														
Min	32.4	32.8	30.9	20.8	28.2	26.0	31.3	29.1	23.8	15.4	22.0	13.5	14.0	15.4
Max	37.6	38.1	35.3	30.0	32.5	31.1	38.3	40.7	27.5	20.7	80.4	21.7	19.9	26.5
Bckg	34.3	34.2	33.9	24.0	28.2	28.4	32.9	32.5	26.2	20.3	23.1	13.5	14.6	16.0
Mean value per zone: 32.4 ± 4.0										24.0 ± 13.7				
Hg														
Min	20.4	25.4	17.6	22.7	11.2	18.0	15.5	16.4	29.5	13.4	34.9	32.2	6.0	7.9
Max	55.7	59.1	58.7	69.2	47.0	56.1	48.1	40.3	50.3	42.3	137.7	67.1	22.9	34.4
Bckg	23.5	32.9	29.0	28.2	21.9	28.9	28.8	29.9	48.1	30.8	35.5	32.2	8.4	10.0
Mean value per zone: 69.2 ± 32.2										42.6 ± 30.8				
Ni														
Min	32.7	30.1	14.1	11.1	13.0	29.9	48.4	34.3	86.4	6.7	157	52.9	6.9	13.1

	Offshore cores									*Intertidal cores*				
	SL25	SL26	SL27	SL28	SL33	IXW 250	IXW 500	IXW 750	IXTOC 1	IA03	LTAF	LTBCF	LTME1	LTME2
Max	47.5	44.5	36.9	39.4	30.7	59.8	74.4	55.6	139	14.2	259	101	22.9	27.7
Bckg	36.9	33.3	32.5	31.2	30.7	37.4	52.1	38.4	93.9	13.5	240	52.9	9.3	13.9
Mean value per zone: 120 ± 46										88.4 ± 89.2				
Pb														
Min	15.6	12.3	11.4	10.6	5.4	15.4	14.4	13.9	13.7	0.1	15.5	13.9	3.1	4.7
Max	26.1	21.9	21.4	22.0	11.2	25.6	24.8	26.2	15.8	6.8	33.5	18.3	6.6	9.0
Bckg	16.6	13.2	14.9	13.1	8.5	16.2	14.6	15.0	14.3	0.1	16.3	13.9	4.1	5.7
Mean value per zone: 25.8 ± 16.7										125 ± 8.5				
V														
Min	94.9	82.4	76.3	76.1	29.3	95.3	108	89.9	75.5	10.7	113.0	39.0	5.5	12.3
Max	119	106	109	109	49.6	125	134	126.6	87.7	27.5	162.1	72.3	18.7	24.7
Bckg	104	88.5	109	98.7	39.6	116	116	117	84.2	12.9	84.6	41.6	11.2	14.0
Mean value per zone: 134 ± 102										64.1 ± 56.2				
Zn														
Min	75.0	69.1	75.3	78.5	37.6	83.2	92.8	81.1	75.0	5.5	82.7	41.6	8.0	11.7
Max	89.3	82.0	79.1	91.4	46.2	83.8	115	120	76.4	8.9	133	64.9	21.3	25.8
Bckg	80.1	70.6	80.9	82.8	42.8	92.5	97.8	95.4	79.7	5.6	138	42.9	11.3	18.1
Mean value per zone: 51.3 ± 45.1										120 ± 85				

n = sample size, Min = minimum concentration, Max = maximum concentration, Bckg = background concentration. <DL = below detection limit. Threshold effect level (TEL values in $\mu g\ g^{-1}$: Cr =52.3, As = 7.2, Ni = 15.9, Cu = 18.7, Pb = 30.2, Zn = 124 V = n.a.; [39], Ba = 2,218

[53]). Probable effect level (PEL values in µg g^{-1}, [39]): Ba = n.a., As = 41.6, Cr= 160, Ni = 42.8, Pb = 112, , Zn = 271, V = n.a., Cu = 108 [39].

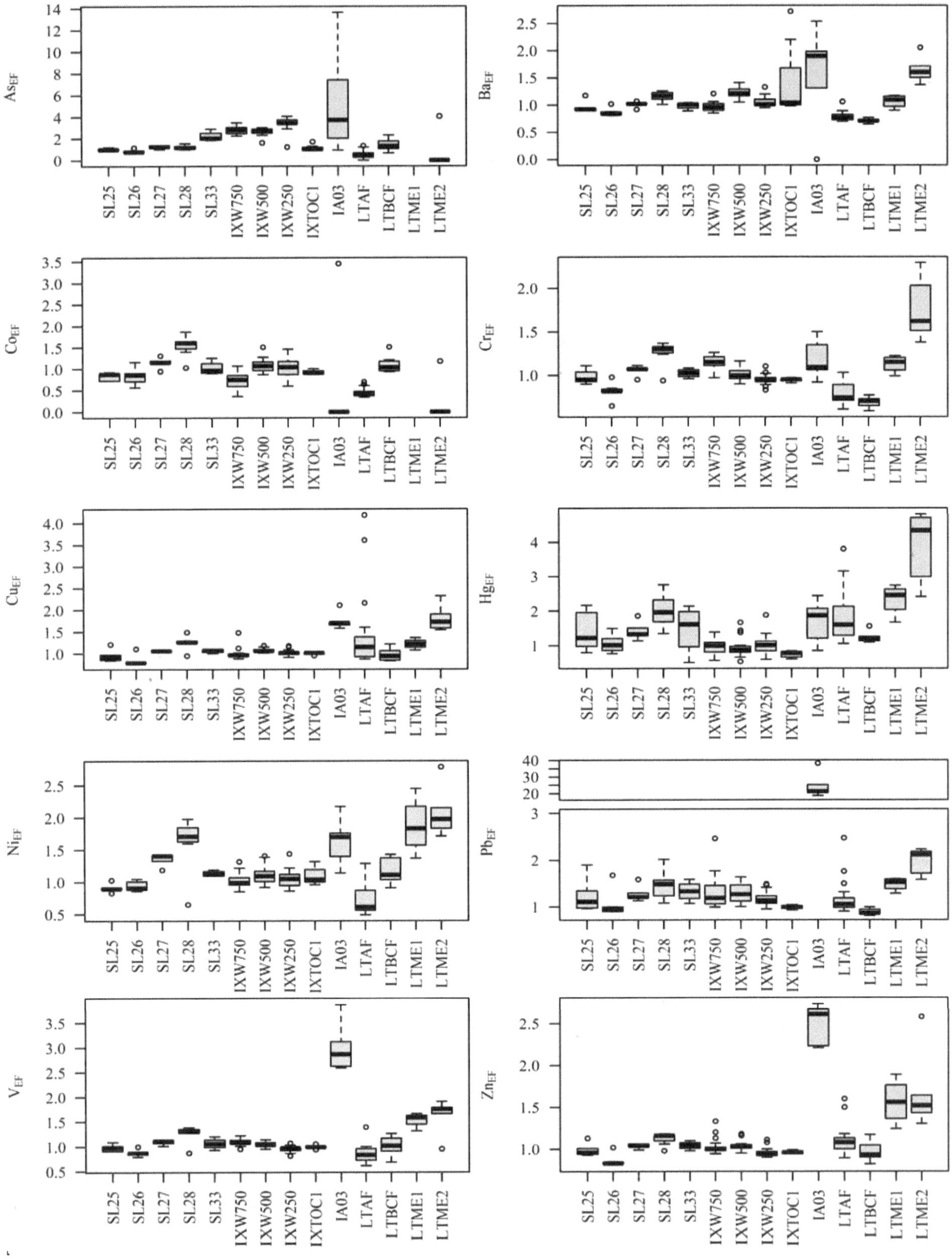

FIG. 11.7. Box plots of enrichment factors for trace elements in sediments from the southern Gulf of Mexico.

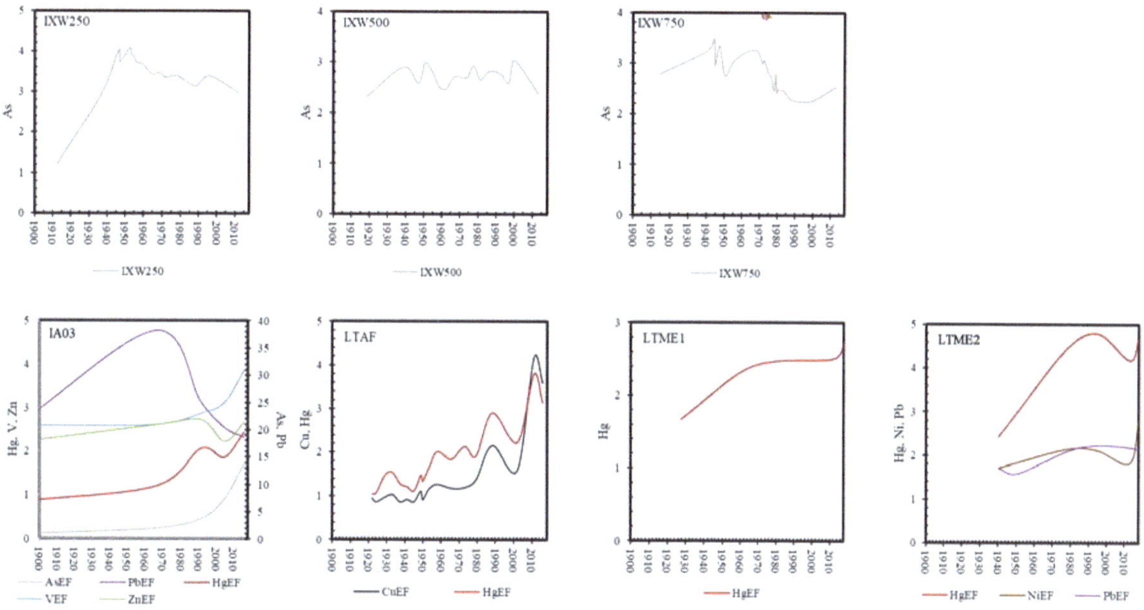

FIG. 11.8. Temporal profiles of the enrichment factor for trace elements in sediment cores from the southern Gulf of Mexico.

Trace element concentrations changes are associated with sediment grain size. As and Cu were mostly related with clay-sized particles with a prevalent marine origin, as suggested by Na and Cl present in group 1 (Fig. 11.6). Most trace elements grouped with lithophile elements (i.e. Al, Ti, Si, Rb and Zr) in group 3, indicating a dominant terrigenous source, with silt-sized particles as the main carrier phase. No trace elements were associated with sand-sized carbonate particles in group 2. This is because, in comparison to Fe and Mn oxides and clay minerals, carbonate particles are not significant trace element carriers to the ocean [11.54], and coarse particles have smaller surface to volume ratios and cation exchange capacities in comparison with fine particles [11.55]. Thus, results suggest that trace element supply (except As and Cu) is highly influenced by fluvial transport, either from natural and land-based contamination sources, or from offshore activities but carried by land-derived sediment particles.

According to the assessment of water quality in surface waters by the National Monitoring Network, that uses total suspended solid concentrations (TSS) as a contamination index (attributed to the input of waste waters and soil erosion [11.56]), some of the most contaminated sites (400 $\leq$ SST > 150 mg L^{-1}) in the sGoM are located in the coastal area adjacent to the collection sites IXW250, IXW500 and IXW750 (between the discharge areas of Coatzacoalcos and Grijalva rivers). A previous study [11.57] reported that the average discharge of suspended solids between Coatzacoalcos River and Términos Lagoon (TL) may vary between 1 and 78,150 t day^{-1}, with the highest values observed through Grijalva River and TL mouths. These sediments also carry large quantities of several metals (Al, Ba, Cd, Co, Cu, Cr, Fe, Hg, Ni, Pb, V, Zn) accounting, for instance, up to ~1.9 t day^{-1} of Ba and ~3 t day^{-1} for Cu through Grijalva River; and 0.04 t day^{-1} of Hg, >15 t day^{-1} of Ni, 2 t day^{-1} of Pb and 1.7 t day^{-1} of V through TL. Regarding As contamination, a previous study in the Coatzacoalcos River discharge area attributed increasing As concentrations to the input of clay minerals of terrestrial origin, most likely as lattice-bound components, resulting from erosion in the watershed owing to land use changes [11.14]. Thus, land-derived trace element input is an important factor contributing to the observed contamination in the study area. Studies elsewhere [11.58] have reported that ~70-75% of the global marine contamination is produced by human activities on land, and this is because >40% of the world's population lives within 100 km of the coast [11.59], and ~90% of the contamination produced in the countries in America Latina and the Caribbean, dominated by domestic and industrial waste waters, is transported by rivers to the sea.

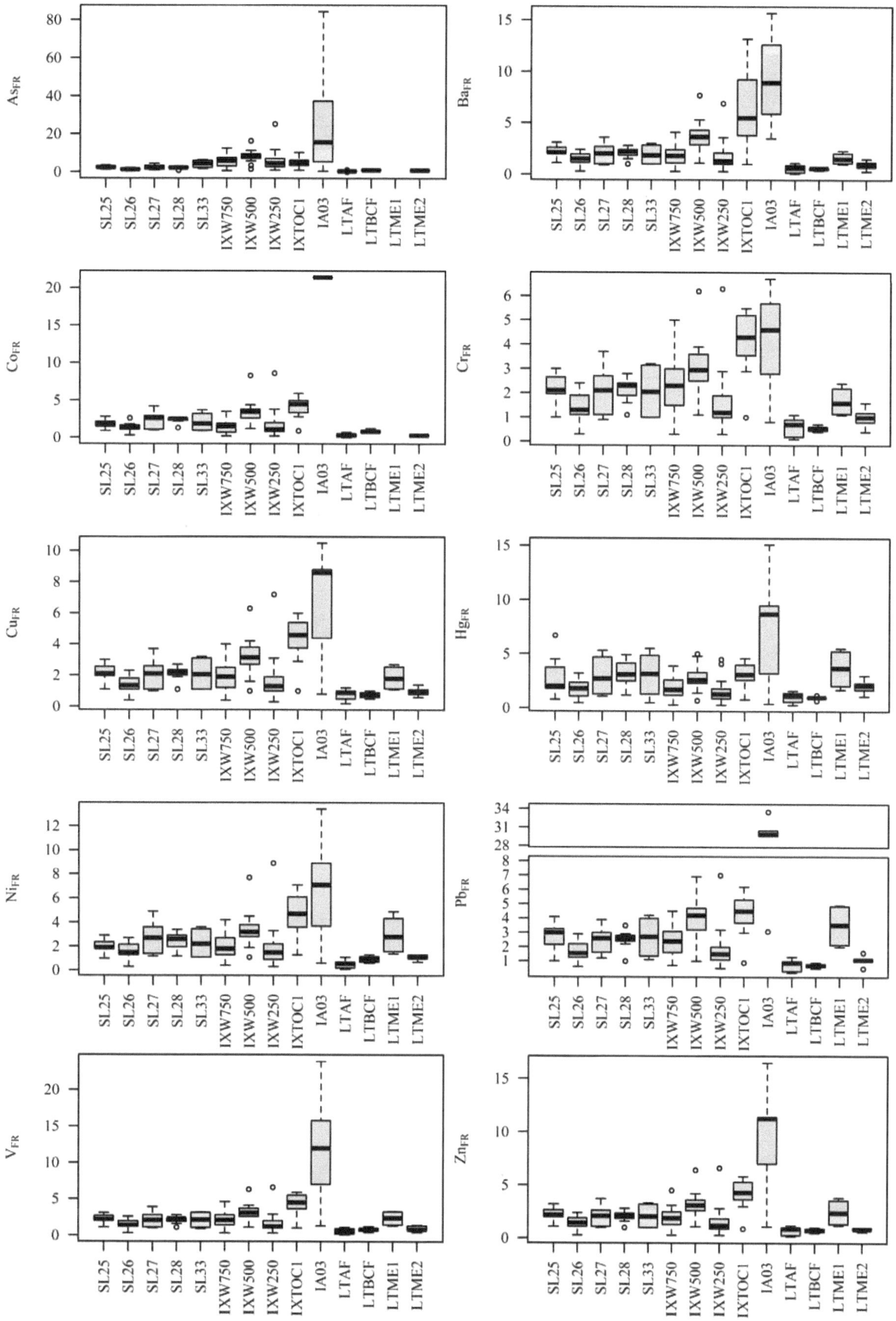

FIG. 11.9. Box plots of flux ratios for trace elements in sediment cores from the southern Gulf of Mexico.

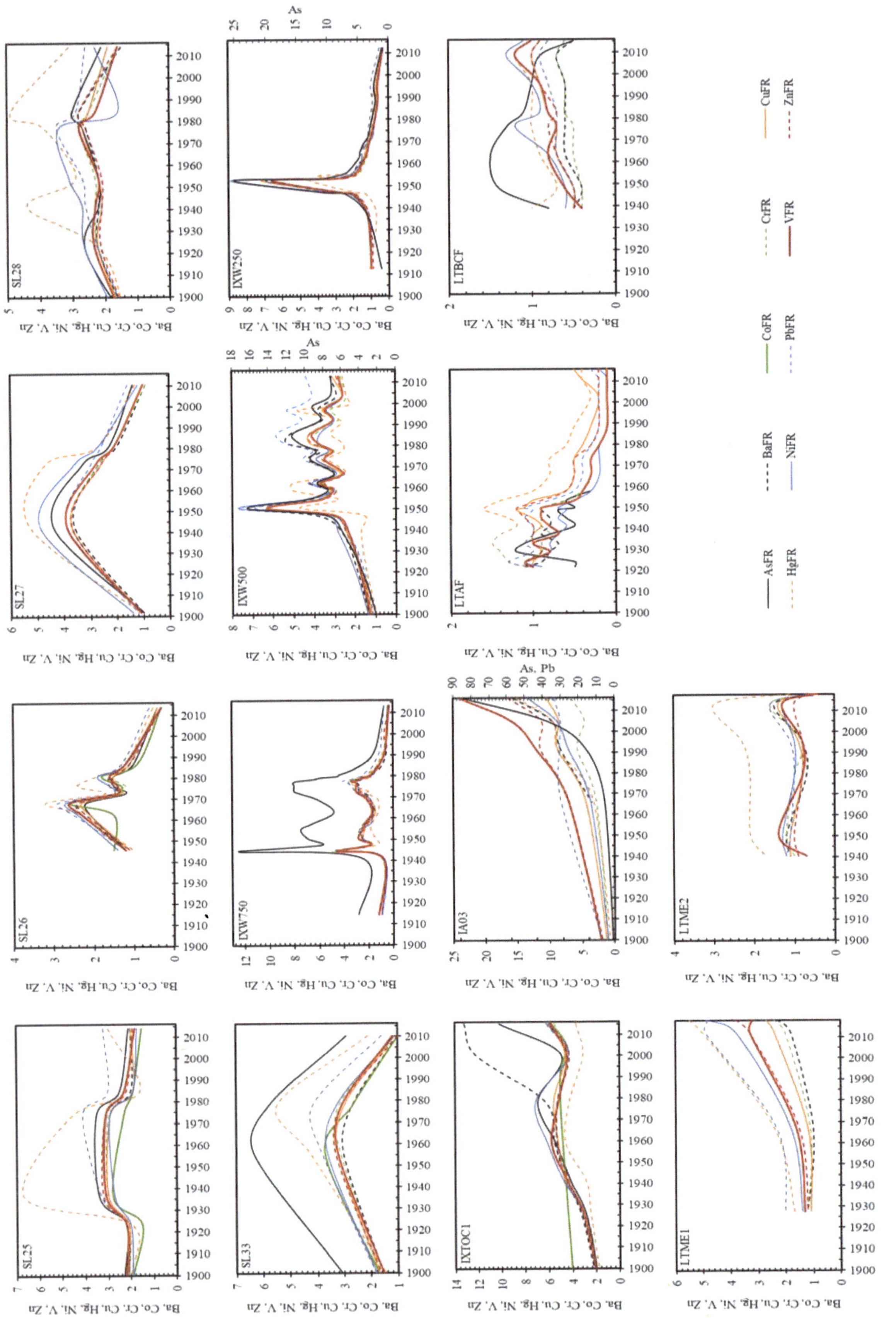

FIG. 11.10. Flux ratio temporal profiles for trace elements in sediment cores from offshore and intertidal areas in the southern Gulf of Mexico.

TABLE 11.3. TRACE ELEMENT CONCENTRATIONS IN SEDIMENTS FROM MARINE AREAS ASSOCIATED WITH GLOBAL OIL INDUSTRY ACTIVITIES (Hg IS IN ng g^{-1}, ALL OTHER VALUES AS μg g^{-1}) (*Adapted from [11.17]*)

Study site	Co	Cr	Cu	Ba	As	Zn	Pb	Hg	Ni	V	Status*
This study											
Gulf of Mexico (offshore)	3.4-18.1	35.8-186.7	20.8-40.7	219-778	<DL-14.3	37.6-119.7	5.4-26.2	11.2-69.2	11.1-139.3	29.3-133.8	C
Gulf of Mexico (intertidal)	<DL-37.7	5.4-27.9	15.4-26.4	<DL-74.7	<DL-13.1	4.0-8.9	0.1-7.5	6.0-137.7	6.7-28.7	2.9-40.5	C
Southern Gulf of Mexico											
Coatzacoalcos River Estuary [60]	9-29	44-100	14-44	na	na	47-131	38-49	na	22-50	n.a.	C
Coatzacoalcos River Estuary [14]	na	29-92	15-30	na	5-10	64-109	11-30	87-280	18-35	63-128	C
Tamaulipas shelf [61]	6-32	8-63	10-18	242-400	na	37-91	10-27	na	10-25	47-113	C
Campeche Sound [27]	0-20.5	11.8-204.5	15.4-43.3	0-3139	0-13.7	2.9-113.9	0-27.4	na	6-104.2	7.9-131	C
Northern Gulf of Mexico											
Texas oilfields [62]	na	48-155	na	763-17633	na	57-5033	na	20-150	na	na	C
Galvestone Bay [63]	na	68.32	13.94	798	na	107	26.61	80	na	na	NC
Tampa Bay [63]	na	20	9.1	32	na	30	15.7	60	na	na	NC
Mississippi River Delta [63]	na	71.62	21.31	871	na	144	27.41	139	na	na	NC
Mississippi Bight [64]	0.4-13.1	4-88	1-25	10-940	1-17	2-154	1-40	6-113	1-43	na	NC
Other areas in America											
California [65]	na	117-127	15-17	752-965	7.5-9.1	70-77	14-15	53-99	39-44	40-69	NC

Study site	Co	Cr	Cu	Ba	As	Zn	Pb	Hg	Ni	V	Status*
Brazil [66]	na	2.2-66.8	0.3-92.2	2.7-18968	na	2.28-53.8	0.3-40	na	0.3-33.9	2.2-114	C
Brazil [67]	3.6-10.6	35.7-79.4	12.6-26.9	135-4610	9.14-48.1	41.1-105	14.7-35.5	na	19.9-46.7	52.6-110	NC
Brazil [68]	na	10.0-11.7	1.50-2.30	112-176	1.2-2.0	6.50-9.50	3.60-4.60	na	na	14.0-15.7	NC
Guamaré, Brazil [69]	na	0.9-20.8	<0.1-28	<2- 831	na	<0.1-26.5	<0.1-10.5	na	<0.1-5.1	1.0-15.4	NC
Continental Shelf Brazil [69]	na	<0.1-31	<0.1-6.6	<2-700	na	<0.1-36	<0.1-19	na	<0.1-9.4	0.3-54	NC
Alaskan Arctic [70]	7.3	56.9	18.9	394	11.1	70.1	9.7	41	24.3	92.6	NC
Europe											
North Sea [71]	na	38.4-426	110-374	na	na	<0.25-6699	<0.25-4785	na	<0.25-137	na	C
North Sea [72]	na	22	46	484	na	927	195	na	na	na	C
Valhall, Norway [73]	na	na	na	68.5-6670	na	7.0-55.6	7.6-55.6	na	na	na	NC
Gyda, Norway [73]	na	na	na	65-3376.5	na	7.4-14	5.0-6.1	na	na	na	NC
Veslefrikk, Norway [73]	na	na	na	357-6725	na	9.0-118.2	4.0-98	na	na	na	NC
Asia											
Black Sea [74]	0.5-71.59	1-135	1.9-75.7	13-1164	na	1-178	0.5-45.8	na	1-144	1-120	C
Kuwait [75]	na	na	0.2-56.5	na	na	0.7-50.8	0.2-64	na	2.1-96	1.5-95.2	NC
Azerbaijan, Caspian Sea [76]	11.5-18.1	56.4-100	14.5-57.6	314-1080	8.87-22.6	51.1-110	12.2-28.6	50-450	34.5-68	73.9-136	C
Iran, Caspian Sea [76]	6.9-24.2	59.6-128	13.2-50.9	200-679	6.97-20.1	55.9-146	11.3-24.6	20-90	29.4-67.8	76.5-145	C
Kazakhstan, Caspian Sea [76]	0.7-12.1	1.9-103	1.2-49.5	75-1250	2.13-20.2	1.0-59.9	1.43-14.6	<DL-40	1.8-54.8	5.6-81.2	C

Study site	Co	Cr	Cu	Ba	As	Zn	Pb	Hg	Ni	V	Status*
Russia, Caspian Sea [76]	1.3-7.6	2.1-69.3	2.5-21.9	70-669	0.42-6.71	2.8-52.9	0.69-8.03	10-70	5.42-34.2	7.3-84.5	C
Caspian Sea [77]	10.9-15.8	na	na	na	na	67.6-97.9	na	na	24.2-36.9	na	NC
Qatar, Gulf of Oman [78]	0.1-2.2	11.5-40.8	1.22-8.17	na	1-6.3	na	0.43-3.88	0.7-16.7	0.74-20.8	2.7-24.7	NC
Gulf of Oman, UAE [78]	0.34-45.2	17.6-303	0.64-3.58	na	0.7-9.6	na	0.69-5.88	0.6-2.2	2.0-1010	4.5-35.5	NC
Gulf of Oman, Bahrain [78]	0.17-2.43	3.36-41.8	2.38-48.3	na	3.16-6.88	6.12-52.2	0.67-99	2.5-220.2	2.46-23.2	3.47-28.4	NC
Gulf of Oman, Oman [78]	0.13-6.92	6.46-133	0.60-6.76	na	0.74-5.01	1.57-11.4	0.25-1.8	1.9-11.2	1.84-77.8	4.7-44.1	NC
Gulf of Suez [79]	na	na	1.8-10.2	na	na	4.3-23.8	13.9-28.3	Na	na	na	C
Persian Gulf [80]	1-34	4-174	3-41	na	na	13-119	na	Na	6-198	3-120	C
Persian Gulf [81]	na	na	na	na	na	na	90.479	Na	64.897	52.003	C
Saudi Arabia [82]	na	3.4-53.0	na	na	na	4.2-22.6	0.6-4.2	2.9-3.6	20.5-64.6	na	NC

n = sample size. <DL = below detection limit (As < 0.5 μg g^{-1}, Ba <2.0 μg g^{-1}; Co <3.0 μg g^{-1}). na = not available. Status*= as reported in the original reference: C = contaminated, NC = not contaminated.

Offshore oil-associated activities may also contribute to the contamination in the marine environment. Trace elements in drilling muds, drill cuttings and produced waters may contain As, Ba, Cd, Cr, Cu, Hg, Ni, Pb, V and Zn at concentrations substantially higher than their concentrations in the natural marine environment [11.83, 11.84]. Drilling muds may be enriched by Ba, which is assed as barite ($BaSO_4$) as a weighting agent to prevent blowout; however, drill cuttings and produced waters should typically reflect the composition of the geological strata excavated and the oil extracted. Thus, it is difficult to distinguish the natural variability of trace element concentrations from those associated with oil extraction activities in most sGoM areas, and this is even more difficult considering that natural petroleum seepage occurs all around in the region. Nonetheless, previous studies demonstrated accumulation and sublethal effects of oil-related pollutants on sGoM biota (e.g. [11.85]). Considering that anthropogenic contamination may add to the already naturally high trace element inputs to the study area, and consequently increase the toxicological risk to biota (and humans, via fish consumption), further monitoring and site-specific research, including biological and chemical testing, is advisable.

The FR for most trace elements in offshore cores were usually lower in recent decades, in comparison with those observed by mid-20th century (Fig. 11.10). This decrease is most likely the result of the environmental regulations established in Mexico since the 1980s. Similar findings are reported for the northern GoM [11.86]. The exceptions are IXTOC1 and the intertidal cores IA03 and LTME1, collected in the surroundings of Campeche Sound, the most important offshore oil production zone in the sGoM. It is noticeable that Ba fluxes in the three cores significantly increased since the 1970s, upon the onset of offshore oil extraction activities in the area. However, it is not clear if the still increasing Ba fluxes are related to drilling activities or to the increased discharges of suspended solids (and associated trace elements) through fluvial input of Grijalva River and tidal flushing of Términos Lagoon. In any case, the coastal area adjacent to Campeche has undergone extensive population and economic growth as a result of oil industry activities in the region. For instance, the population of Ciudad del Carmen (the main operation center of Campeche Sound) doubled in only a decade (1970-1980) and land use changes, either directly associated with the oil industry (development of roads, pipelines, ports and other infrastructure to support offshore activities from land), with agriculture and cattle raising (to provide food to the growing population), and the generation of domestic and industrial waste waters, all constitute land-based contamination sources that are a collateral impact of the offshore oil-industry.

11.6. CONCLUSIONS

^{210}Pb-dated sediment cores from offshore and intertidal areas in the southern Gulf of Mexico were used to evaluate temporal variations of trace element concentrations, enrichment factors and fluxes, in relation with the oil industry activities in the region. Results show that As, Co, Cr, Cu, Ni, Pb, V and Zn concentrations were elevated during the past 100 years, even prior to the extensive development of the oil industry in the entire GoM, and even background concentrations of Cr, Cu and Ni are elevated enough to be a pose a health hazard to benthic biota. Enrichment in As, Cu, Hg, Ni, Pb, V and Zn was observed in only a few cores (collected in the region adjacent to the Coatzacoalcos River, Grijalva River and Términos Lagoon discharge areas), greatly influenced by the strong fluvial discharges, which carry high loads of suspended solids and land-derived trace elements from the catchment. The impact of offshore oil drilling activities are difficult to distinguish from the contribution of trace elements from fluvial and natural oil seepage in the sGoM; however, except for the cores collected around the Campeche Sound, trace element fluxes within the last 3 decades are lower than those observed by mid-20th century, most likely as a result of the environmental regulations established since the 1980s. Nonetheless, the continuously rising trace element fluxes observed in the cores collected between the Grijalva River and the eastern Yucatan Peninsula since the 1970s were attributed to trace metal inputs derived from wastewater discharges and land erosion in the adjacent coastal areas, associated with population growth and land use changes, as a result of the development of the Campeche Sound offshore oil production industry. These results highlight the need to manage continental erosion and improve wastewater treatment, in order to control trace element releases to the marine environment.

11.7. ACKNOWLEDGEMENTS

This research was supported by the grants C-IMAGE II No. SA 12-10 from the Gulf of Mexico Research Initiative (USA), CONACYT PDCAPN 2015-1/473 and the CRP IAEA K41016. Data are available through the Gulf of Mexico Research Initiative Information & Data Cooperative (GRIIDC) at https://data.gulfresearchinitiative.org (DOI:10.7266/n7-7bcg-yk08), and through Unidad de Informática Marina (UNINMAR-ICMyL-UNAM) at http://uninmar.icmyl.unam.mx. Authors are grateful to the crew and scientists aboard the R/V "Justo Sierra" in the southern Gulf of Mexico coring campaigns C-IMAGE 2015 and 2016, and to León Felipe Álvarez (data base curation), H. Bojórquez-Leyva and H. Alexander-Valdés, S. Rendón-Rodríguez, M. A. Gomez Ponce, A. Rodríguez-Ramírez, G. Ramírez Reséndiz, C. Suárez-Gutiérrez, J. A. Reda Deara and H. Álvarez Guillén (technical support). In memory of our dear colleague and friend David Hollander, whose generosity inspired and guided us through this investigation.

11.8. REFERENCES

[11.1] FANNING, L., MAHON, R., BALDWIN, K., DOUGLAS, S., Transboundary Waters Assessment Programme (TWAP) Assessment of Governance Arrangements for the Ocean, Volume 1: Transboundary Large Marine Ecosystems, Tech. Ser 119, IOC-UNESCO, France (2015).

[11.2] LOVE, M.S., BALDERA, A., YEUNG, C., ROBBINS, C., The Gulf of Mexico Ecosystem: A Coastal and Marine Atlas., Ocean Conservancy, Gulf Restoration Center, U.S.A. (2013).

[11.3] EIA, Today in Energy. Offshore Production Nearly 30% of Global Crude Oil Output in 2015, US Energy Information Administration (2016) ,https://www.eia.gov/todayinenergy/detail.php?Id=28492

[11.4] WITS, Mexico Trade Summary 2016 Data, World Integrated Trade Solution, World Bank (2018), https://wits.worldbank.org/

[11.5] OCDE, Estudios Económicos de la OCDE, México, Visión General, Organización para la Cooperación y el Desarrollo Económicos (2017).

[11.6] PEMEX, Anuario Estadístico, Petróleos Mexicanos, Mexico (2018).

[11.7] WILHELM, S.M., BLOOM, N., Mercury in petroleum, Fuel. Process. Technol. **63** (2000) 1–27.

[11.8] KHUHAWAR, M.Y., ASLAM MIRZA, M., JAHANGIR, T.M., “Determination of metal ions in crude oils”, Crude Oil Emulsions - Composition Stability and Characterization, IntechOpen (2012).

[11.9] PAGE, D.S., BOEHM, P.D., DOUGLAS, G.S., BENCE, A.E., BURNS, W.A., MANKIEWICZ, P.J., Pyrogenic polycyclic aromatic hydrocarbons in sediments record past human activity: A case study in Prince William Sound, Alaska, Mar. Pollut. Bull. **38** 4 (1999) 247–260.

[11.10]GUO, Z., LIN, T., ZHANG, G., ZHENG, M., ZHANG, Z., HAO, Y., FAN, M., The sedimentary fluxes of polycyclic aromatic hydrocarbons in the Yangtze River Estuary coastal sea for the past century, Sci. Total Environ. **386** 1–3 (2007) 33–41.

[11.11] RUIZ-FERNÁNDEZ, A.C., PÁEZ-OSUNA, F., MACHAIN-CASTILLO, M.L., ARELLANO-TORRES, E., ^{210}Pb geochronology and trace metal fluxes (Cd, Cu And Pb) in the Gulf of Tehuantepec, South Pacific of Mexico, J. Environ. Radioactiv. **76** 1–2 (2004) 161–175.

[11.12] RUIZ-FERNÁNDEZ, A.C., SPROVIERI, M., FRIGNANI, M., SANCHEZ-CABEZA, J.A., FEO, M.L., BELLUCCI, L.G., PÉREZ-BERNAL, L.H., PREDA, M., MACHAIN-CASTILLO, M.L., Reconstruction of hydrocarbons accumulation in sediments affected by the oil refinery industry: The case of Tehuantepec Gulf (Mexico), Environ. Earth Sci. **67** 3 (2012) 727–742.

[11.13] RUIZ-FERNÁNDEZ, A.C., BETANCOURT PORTELA, J.M., SERICANO, J.L., SANCHEZ-CABEZA, J.A., ESPINOSA, L.F., CARDOSO-MOHEDANO, J.G., PÉREZ-BERNAL, L.H., GARAY TINOCO, J.A., Coexisting sea-based and land-based sources of contamination by PAHs in the continental shelf sediments of Coatzacoalcos River discharge area (Gulf of Mexico), Chemosphere **144** (2016) 591–598.

[11.14] RUIZ-FERNÁNDEZ, A.C., SANCHEZ-CABEZA, J.A., ALONSO-HERNÁNDEZ, C., MARTÍNEZ-HERRERA, V., PÉREZ-BERNAL, L.H., PREDA, M., HILLAIRE-MARCEL, C., GASTAUD, J., QUEJIDO-CABEZAS, A.J., Effects of land use change and sediment mobilization on coastal contamination (Coatzacoalcos River, Mexico), Cont. Shelf Res. **37** (2012) 57–65.

[11.15] RUIZ-FERNÁNDEZ, A.C., SPROVIERI, M., PIAZZA, R., FRIGNANI, M., SANCHEZ-CABEZA, J.A., FEO, M.L., BELLUCCI, L.G., VECCHIATO, M., PÉREZ-BERNAL, L.H., PÁEZ-OSUNA, F., ^{210}Pb-derived history of PAH and PCB accumulation in sediments of a tropical inner lagoon (Las Matas, Gulf of Mexico) near a major oil refinery, Geochim. Cosmochim. Ac. **82** (2012) 136–153.

[11.16] RUIZ-FERNÁNDEZ, A.C., RANGEL-GARCÍA, M., PÉREZ-BERNAL, L.H., LÓPEZ-MENDOZA, P.G., GRACIA, A., SCHWING, P., HOLLANDER, D., PÁEZ-OSUNA, F., CARDOSO-MOHEDANO, J.G., CUELLAR-MARTINEZ, T., SANCHEZ-CABEZA, J.A., Mercury in sediment cores from the southern Gulf of Mexico: Preindustrial levels and temporal enrichment trends, Mar. Pollut. Bull. **149** (2019) 110498.

[11.17] RUIZ-FERNÁNDEZ, A.C., SANCHEZ-CABEZA, J.A., PÉREZ-BERNAL, L.H., GRACIA, A., Spatial and temporal distribution of heavy metal concentrations and enrichment in the southern Gulf of Mexico, Sci. Total Environ. **651** (2019) 3174–3186.

[11.18] MENDELSSOHN, I.A., BYRNES, M.R., KNEIB, R.T., VITTOR, B.A., Coastal Habitats of the Gulf of Mexico, Springer, U.S.A. (2017).

[11.19] CONAGUA, Estadísticas del Agua en México, Comisión Nacional del Agua, Mexico (2014).

[11.20] BAUER-GOTTWEIN, P., GONDWE, B.R.N., CHARVET, G., MARÍN, L.E., REBOLLEDO-VIEYRA, M., MEREDIZ-ALONSO, G., Review: The Yucatán Peninsula karst aquifer, Mexico, Hydrogeol. J. **19** 3 (2011) 507–524.

[11.21] SAGARPA, Anuario de Acuacultura y Pesca, Comisión Nacional de Acuacultura y Pesca, Mexico (2017).

[11.22] CONTRERAS RUIZ ESPARZA, A., DOUILLET, P., ZAVALA-HIDALGO, J., Tidal dynamics of the Terminos Lagoon, Mexico: Observations and 3D numerical modelling, Ocean Dynam. **64** 9 (2014) 1349–1371.

[11.23] INEGI, Instituto Nacional de Estadística y Geografía, Cuéntame. Información por Entidad, Campeche (2015),

http://cuentame.inegi.org.mx/monografias/informacion/camp/poblacion/
[11.24] PÁEZ-OSUNA, F., BOTELLO, A. V., VILLANUEVA, S., Heavy metals in Coatzacoalcos Estuary and Ostion Lagoon, Mexico, Mar. Pollut. Bull. **17** 11 (1986) 516–519.
[11.25] VÁZQUEZ-BOTELLO, A. VILLANUEVA-FRAGOSO, S. ROSALES-HOZ, L., "Metals distribution and contamination in the Gulf of Mexico", Environmental Analysis of the Gulf of Mexico, Harte Research Institute for Gulf of Mexico Studies, Mexico (2004).
[11.26] CELIS-HERNÁNDEZ, O., ROSALES-HOZ, L., CARRANZA-EDWARDS, A., Heavy metal enrichment in surface sediments from the SW Gulf of Mexico, Environ. Monit. Assess. **185** (2013) 8891–8907.
[11.27] MACHAIN-CASTILLO, M.L., RUIZ-FERNÁNDEZ, A.C., GRACIA, A., SANCHEZ-CABEZA, J.A., RODRÍGUEZ-RAMÍREZ, A., ALEXANDER-VALDÉS, H.M., PÉREZ-BERNAL, L.H., NAVA-FERNÁNDEZ, X.A., GÓMEZ-LIZÁRRAGA, L.E., ALMARAZ-RUIZ, L., SCHWING, P.T., HOLLANDER, D.J., Natural and anthropogenic oil impacts on benthic foraminifera in the southern Gulf of Mexico, Mar. Environ. Res. **149** (2019) 111–125.
[11.28] RSIS, Ramsar Sites Information Service, Área de Protección de Flora y Fauna Laguna de Términos (2006), https://rsis.ramsar.org/ris/1356
[11.29] REYES-GÓMEZ, H.G., VÁZQUEZ-LULE, A.D., "Caracterización del sitio de manglar Isla del Carmen", Sitios de Manglar con Relevancia Biológica y con Necesidades de Rehabilitación Ecológica, Comisión Nacional para el Conocimiento y Uso de la Biodiversidad (CONABIO), Mexico (2009).
[11.30] DÍAZ-GALLEGOS, J.R., ACOSTA-VELÁZQUEZ, J., RODRÍGUEZ-ZUÑIGA, M.T. CRUZ, I., VÁZQUEZ-LULE, A. TROCHE, C., URIBE, A., RESSI, R., JIMÉNEZ, R., "The mangrove forests of Mexico: Transformation, conservation and threats", Frontiers in Biodiversity Studies, I.K International Publishing House Pvt. Ltd, India (2011).
[11.31] SCHWING, P.T., ROMERO, I.C., LARSON, R.A., O'MALLEY, B.J., FRIDRIK, E.E., GODDARD, E.A., Brooks, G.R., HASTINGS, D.W., ROSENHEIM, B.E., HOLLANDER, D.J., GRANT, G., MULHOLLAN, J. Sediment core extrusion method at millimeter resolution using a calibrated, threaded-rod, J. Vis. Exp . **114** (2016) e54363.
[11.32] RUIZ-FERNÁNDEZ, A.C., HILLAIRE-MARCEL, C., ^{210}Pb-derived ages for the reconstruction of terrestrial contaminant history into the Mexican Pacific coast: Potential and limitations, Mar. Pollut. Bull. **59** 4–7 (2009) 134–145.
[11.33] RUIZ-FERNÁNDEZ, A.C., MAANAN, M., SANCHEZ-CABEZA, J.A., PÉREZ-BERNAL, L.H., LÓPEZ-MENDOZA, P., LIMOGES, A., Chronology of recent sedimentation and geochemical characteristics of sediments in Alvarado Lagoon, Veracruz (southwestern Gulf of Mexico), Cienc. Mar. **40** 4 (2014) 291–303.
[11.34] ROBBINS, J.A., "Geochemical and geophysical applications of radioactive lead isotopes", Biochemistry of Lead, Elsevier, Netherlands (1978).
[11.35] SANCHEZ-CABEZA, J.A., RUIZ-FERNÁNDEZ, A.C., ^{210}Pb sediment radiochronology: An integrated formulation and classification of dating models, Geochim. Cosmochim. Ac. **82** (2012) 183–200.
[11.36] SANCHEZ-CABEZA, J.A., RUIZ-FERNÁNDEZ, A.C., ONTIVEROS-CUADRAS, J.F., PÉREZ BERNAL, L.H., OLID, C., Monte Carlo uncertainty calculation of ^{210}Pb chronologies and accumulation rates of sediments and peat bogs, Quat. Geochronol. **23** (2014) 80–93.
[11.37] DEAN, W.E., Determination of carbonate and organic matter in calcareous sediments and sedimentary rocks by loss on ignition; comparison with other methods, J. Sediment.

Res. **44** 1 (1974) 242–248.

[11.38] LORING, D.H., RANTALA, R.T.T., Manual for the geochemical analyses of marine sediments and suspended particulate matter, Earth-Sci. Rev. **32** 4 (1992) 235–283.

[11.39] BUCHMAN, M.F., NOAA Screening Quick Reference Tables, U.S.A. (2008).

[11.40] BUAT-MENARD, P., CHESSELET, R., Variable influence of the atmospheric flux on the trace metal chemistry of oceanic suspended matter, Earth Planet. Sc. Lett. **42** (1979) 3 399–411.

[11.41] MATSCHULLAT, J., OTTENSTEIN, R., REIMANN, C., Geochemical background - can we calculate it? Environ. Geol. **39** (2000) 990-1000.

[11.42] BIRCH, G.F., DAVIES, K.I., “A Scheme for assessing human impact and sediment quality in coastal waterways”, Coastal GIS Conference (Proc. Conf. Wollongong, 2003), University of Wollongong, Australia (2003).

[11.43] HEYVAERT, A.C., REUTER, J.E., SLOTTON, D.G., GOLDMAN, C.R., Paleolimnological reconstruction of historical atmospheric lead and mercury deposition at Lake Tahoe, California–Nevada, Environ. Sci. Technol. **34** 17 (2000) 3588–3597.

[11.44] FAO/IAEA, Use of ^{137}Cs for Soil Erosion Assessment, Food and Agriculture Organization of the United Nations, Italy (2017).

[11.45] LIU, H., JACOB, D.J., BEY, I., YANTOSCA, R.M., Constraints from ^{210}Pb and ^{7}Be on wet deposition and transport in a global three-dimensional chemical tracer model driven by assimilated meteorological fields, J. Geophys. Res-Atmos. **106** D11 (2001) 109–128.

[11.46] KRISHNASWAMI, S., COCHRAN, J.K., U-Th Series Nuclides in Aquatic Systems, Vol. 13, Elsevier (2008).

[11.47] COCHRAN, J.K., MCKIBBIN-VAUGHAN, T., DORNBLASER, M.M., HIRSCHBERG, D., LIVINGSTON, H.D., BUESSELER, K.O., ^{210}Pb scavenging in the North Atlantic and North Pacific oceans, Earth Planet. Sc. Lett. **97** 3–4 (1990) 332–352.

[11.48] ONTIVEROS-CUADRAS, J.F., RUIZ-FERNÁNDEZ, A.C., SANCHEZ-CABEZA, J.A., WEE-KWONG, L.L., PÉREZ-BERNAL, L.H., Geochemical fractionation of ^{210}Pb in oxic estuarine sediments of Coatzacoalcos River, Gulf of Mexico, J. Radioanal. Nucl. Ch. **292** 3 (2012) 947–956.

[11.49] MASQUÉ, P., ISLA, E., SANCHEZ-CABEZA, J.A., PALANQUES, A., BRUACH, J.M., PUIG, P., GUILLÉN, J., Sediment accumulation rates and carbon fluxes to bottom sediments at the Western Bransfield Strait (Antarctica), Deep Sea Res. Part II Top. Stud. Oceanogr. **49** 4–5 (2002) 921–933.

[11.50] VALENTINE, R. L., SPANGLER, K. M., MEYER, J., Removing radium by adding preformed hydrous manganese oxides. J. Am. Water Works Ass. **82** 2 (1990) 66-71.

[11.51] RUIZ-FERNÁNDEZ, A.C., ALONSO-HERNÁNDEZ, C., ESPINOSA, L.F., DELANOY, R., SOLARES CORTEZ, N., LUCIENNA, E., CASTILLO, A.C., SIMPSON, S., PÉREZ-BERNAL, L.H., CABALLERO, Y., PEÑA-CASTRO, A., LÓPEZ-MONROY, F., QUEJIDO-CABEZAS, A.J., GARAY-TINOCO, J.A., DÍAZ-ASENCIO, M., GÓMEZ-BATISTA, M., PARRA LOZANO, J.P., SANCHEZ-CABEZA, J.A., ^{210}Pb-derived sediment accumulation rates across the Wider Caribbean Region. J. Environ. Radioactiv. **223–224** (2020) 106366.

[11.52] SHOLKOVITZ, E.R., MANN, D.R., The pore water chemistry of ^{239,240}Pu and ^{137}Cs in sediments of Buzzards Bay, Massachusetts, Geochim. Cosmochim. Ac. 48 5 (1984) 1107–1114.

[11.53] LEUNG, K. M. Y., GRAY J. S., LI, W. K., LUI G. C. S., WANG Y., LAM P. K. S., Deriving sediment quality guidelines from field-based species sensitivity distributions. Environ. Sci. Technol. **39** 14 (2005) 5148–5156.

[11.54]CHESTER, R., Marine Geochemistry, Blackwell Science Ltd., England (2000).
[11.55]HOROWITZ, A.J., A Primer on Trace Metal Sediment Chemistry, US Geol Survey, U.S.A. (1985).
[11.56]CONAGUA, Estadísticas del Agua en México, Comisión Nacional del Agua, Mexico (2017).
[11.57]GRACIA, A., F., VÁZQUEZ, G., ENCISO SÁNCHEZ, G. ALEXANDER VALDÉS, H.M., Composición y Volumen de Contaminantes de las Descargas Costeras al Golfo de México, UAC/UNAM-ICMYL/CINVESTAV, Mexico (2014).
[11.58]CEPAL, La Contaminación de los Ríos y sus Efectos en las Áreas Costeras y el Mar, Series No. 50, Comisión Económica para América Latina y el Caribe, Chile (2002).
[11.59]IOC/UNESCO, IMO, FAO, UNDP, A Blue Print for Ocean and Coastal Sustainability, IOC/UNESCO, Francia (2011).
[11.60]VÁZQUEZ-BOTELLO, A., PÁEZ-OSUNA, F., El Problema Crucial, la Contaminación, Serie Medio Ambiente en Coatzacoalcos, Centro de Ecodesarrollo, Mexico (1986).
[11.61]CELIS-HERNANDEZ, O., ROSALES-HOZ, L., CUNDY, A.B., CARRANZA-EDWARDS, A., CROUDACE, I.W., HERNANDEZ-HERNANDEZ, H., Historical trace element accumulation in marine sediments from the Tamaulipas Shelf, Gulf of Mexico: An assessment of natural vs anthropogenic inputs, Sci. Total Environ. **622–623** (2018) 325–36.
[11.62]CHAPMAN, P.M., POWER, E.A., DEXTER, R.N., ANDERSEN, H.B., Evaluation of effects associated with an oil platform, using the sediment quality triad, Environ. Toxicol. Chem. **10** 3 (1991) 407–424.
[11.63]SANTSCHI, P., PRESLEY, B., WADE, T., GARCIA-ROMERO, B., BASKARAN, M., Historical contamination of PAHs, PCBs, DDTs, and heavy metals in Mississippi River Delta, Galveston Bay and Tampa Bay sediment cores, Mar. Environ. Res. **52** 1 (2001) 51–79.
[11.64]PRESLEY, B.J., TAYLOR, R.J., BOOTHE, P.N., Trace metal concentrations in sediments of the Eastern Mississippi Bight, Mar. Environ. Res. **33** 4 (1992) 267–282.
[11.65]STEINHAUER, M., CRECELIUS, E., STEINHAUER, W., Temporal and spatial changes in the concentrations of hydrocarbons and trace metals in the vicinity of an offshore oil-production platform, Mar. Environ. Res. **37** 2 (1994) 129–163.
[11.66]PEREIRA DORE, M., FARIAS, C., HAMACHER, C., Offshore drilling effects in Brazilian SE marine sediments: A meta-analytical approach, Environ. Monit. Assess. **189** 44 (2017) 1–11.
[11.67]POZEBON, D., SANTOS, J.H.Z., PERALBA, M.C.R., MAIA, S.M., BARRIONUEVO, S., PIZZOLATO, T.M., Metals, arsenic and hydrocarbons monitoring in marine sediment during drilling activities using NAFs, Deep Sea Res. Part II Top. Stud. Oceanogr. **56** 1–2 (2009) 22–31.
[11.68]REZENDE, C., LACERDA, L., OVALLE, A.R., SOUZA, C.M., GOBO, A.A., SANTOS, D., The effect of an oil drilling operation on the trace metal concentrations in offshore bottom sediments of the Campos Basin oil field, SE Brazil, Mar. Pollut. Bull. **44** 7 (2002) 680–684.
[11.69]LACERDA, L.D., CAMPOS, R.C., SANTELLI, R.E., Metals in water, sediments, and biota of an offshore oil exploration area in the Potiguar Basin, Northeastern Brazil, Environ. Monit. Assess. **185** 5 (2013) 4427–4447.
[11.70]TREFRY, J.H., REMBER, R.D., TROCINE, R.P., BROWN, J.S., Trace Metals in sediments near offshore oil exploration and production sites in the Alaskan Arctic, Environ. Geol. **45** 2 (2003) 149–160.
[11.71]BREUER, E., STEVENSON, A., HOWE, J., CARROLL, J., SHIMMIELD, G., Drill

cutting accumulations in the Northern and Central North Sea: A review of environmental interactions and chemical fate, Mar. Pollut. Bull. **48** 1–2 (2004) 12–25.

[11.72] GRANT, A., BRIGGS, A.D., Toxicity of sediments from around a North Sea oil platform: Are metals or hydrocarbons responsible for ecological impacts?, Mar. Environ. Res. **53** 1 (2002) 95–116.

[11.73] OLSGARD, F., GRAY, J., A comprehensive analysis of the effects of offshore oil and gas exploration and production on the benthic communities of the Norwegian continental shelf, Mar. Ecol. Prog. Ser. **122** (1995) 277–306.

[11.74] SECRIERU, D., SECRIERU, A., Heavy metal enrichment of man-made origin of superficial sediment on the continental shelf of the North-Western Black Sea, Estuar. Coast. Shelf S. **54** 3 (2002) 513–526.

[11.75] AL-ABDALI, F., MASSOUD, M.S., AL-GHADBAN, A.N., Bottom sediments of the Arabian Gulf—III. Trace metal contents as indicators of pollution and implications for the effect and fate of the Kuwait oil slick, Environ. Pollut. **93** 3 (1996) 285–301.

[11.76] DE MORA, S., SHEIKHOLESLAMI, M.R., WYSE, E., AZEMARD, S., CASSI, R., An assessment of metal contamination in coastal sediments of the Caspian Sea, Mar. Pollut. Bull. **48** 1–2 (2004) 61–77.

[11.77] KARBASSI, A.R., AMIRNEZHAD, R., Geochemistry of heavy metals and sedimentation rate in a bay adjacent to the Caspian Sea, Int. J. Environ. Sci. Technol. **1** 3 (2004) 191–198.

[11.78] DE MORA, S., FOWLER, S.W., WYSE, E., AZEMARD, S., Distribution of heavy metals in marine bivalves, fish and coastal sediments in the Gulf and Gulf of Oman, Mar. Pollut. Bull. **49** 5–6 (2004) 410–424.

[11.79] EL-MOSELHY, K.M., GABAL, M.N., Trace metals in water, sediments and marine organisms from the northern part of the Gulf of Suez, Red Sea, J. Marine Syst. **46** 1–4 (2004) 39–46.

[11.80] BASAHAM, A.S., AL-LIHAIBI, S.S., Trace elements in sediments of the Western Gulf, Mar. Pollut. Bull. **27** (1993) 103–107.

[11.81] POURANG, N., NIKOUYAN, A., DENNIS, J.H., Trace element concentrations in fish, surficial sediments and water from northern part of the Persian Gulf, Environ. Monit. Assess. **109** 1–3 (2005) 293–316.

[11.82] SADIQ, M., ZAIDI, T.H., Metal concentrations in the sediments from the Arabian Gulf coast of Saudi Arabia, B. Environ. Contam. Tox. **34** 1 (1985) 565–571.

[11.83] NEFF, J., LEE, K., DEBOIS, E.M., "Produced water: Overview of composition, fates and effects", Produced Water, Environmental Risks and Advances in Mitigation Technologies, Springer (2011).

[11.84] SIDDIQUE, S., KWOFFIE, L., ADDAE-AFOAKWA, K., NJUGUNA, YATES, J., "Oil based drilling fluid waste: An overview on highly reported but less explored sink for environmentally persistent pollutants", 3rd International Conference on Structural Nano Composites (IOP Conf. Series: Materials Science and Engineering, United Kingdom, 2016) IOP Publishing Ltd (2017).

[11.85] GOLD-BOUCHOT, G., RUBIO-PIÑA, J., MONTERO-MUÑOZ, J., RAMIREZ-MISS, N., ECHEVERRÍA-GARCÍA, A., PATIÑO-SUAREZ, V., PUCH-HAU, C.A., ZAPATA-PÉREZ, O., Pollutants and biomarker responses in two reef fish species (*Haemulon aurolineatum* and *Ocyurus chrysurus*) in the Southern Gulf of Mexico, Mar. Pollut. Bull. **116** 1–2 (2017) 249–257.

[11.86] WARD, C.H., BYRNES, M.R., DAVIS, R.A., KENNICUTT, M.C., KNEIB, R.T., MENDELSSOHN, I.A., ROWE, G.T., TUNNELL, J.W., VITTOR, B.A., Habitats and Biota of the Gulf of Mexico: Before the Deepwater Horizon Oil Spill: Volume 1: Water Quality, Sediments, Sediment Contaminants, Oil and Gas Seeps, Coastal Habitats,

Onshore Plankton and Benthos, and Shellfish, Springer (2017).

12. CONCLUSIONS

This compilation integrates the experience of specialists from different regions of the world, on the study of sources, distribution and trends of various types of contaminants in different marine environments, through the analysis of diverse environmental compartments (biota, water and sediments), including the reconstruction of long-term environmental changes using the ^{210}Pb method as a dating tool.

The topics covered by the studies included the identification of sources, contamination trends and inventories of plutonium isotopes by using biota and surface sediments from the southern Baltic Sea; the assessment of ^{137}Cs concentrations and inventories in sediments from coastal areas of Japan in connection with the Fukushima Daiichi Nuclear Power Plant accident occurred in 2011; the use of ^{210}Pb dating for the reconstruction of sediment accumulation rates and contamination trends in a coastal lagoon in Morocco, coastal areas in the Gulf of Aqaba (Jordan) and the Gulf of Mexico (Mexico), estuarine areas in Penang (Malaysia), the bays of Ha Long (Vietnam) and of Tvären (Sweden), and a coastal upwelling system in Namibia; and the use of contaminant profiles as an alternative for validation of ^{210}Pb chronologies in Brazil.

Regardless of the region of the world, most of the studies showed the widespread anthropogenic impact on seas globally. The knowledge of the concentrations levels, transport and fate of contaminants is vital for evaluating the effects of human activities on the coastal environments, as well as the enactment of environmental policies. All studies presented provided evidence-based information that might serve as the basis to sustain integrated management programs or to evaluate the success of interventions implemented to address environmental problems. Despite its limitations, the ^{210}Pb dating method is one of the most powerful tools for interpreting the depositional history of contaminants in the coastal areas. The sedimentary record is complex and often, not easily interpretable, for which, the knowledge of the study site's geologic and anthropogenic history are key for a proper reconstruction of the contamination trends. The studies in this publication also show alternative methods for looking at historical trends through time when ^{210}Pb dating is not feasible or needs additional validation.

Scientific knowledge is indispensable for informed decision making. The studies in this publication will hopefully inspire readers to continue and even expand monitoring efforts to assess the status and trends of pollution in order to protect the oceans, and to continue working for the sustainability of marine and coastal environments around the world.

ABBREVIATIONS

BCLME	Benguela Current Large Marine Ecosystems
BUS	Benguela Upwelling System
CF:CS	Constant Flux-Constant Sedimentation
CIC	Constant Initial Concentration
CRM	Certified Reference Materials
CRS	Constant Rate of Supply
CRP	Coordinated Research Project
DL	Detection Limit
EDXRF	Energy Dispersive X-ray Fluorescence Spectrometer
EF	Enrichment Factor
EPA	Environmental Protection Agency
FDNPP	Fukushima Dai-ichi Nuclear Power Plant
FR	Flux Ratios
HPGe	High Purity Germanium
ICP-MS	Inductively Couples Plasma Mass Spectrometry
ILC	Interlaboratory Comparisons
IAEA	International Atomic Energy Agency
LME	Large Marine Ecosystem
LOI	Loss on Ignition
MAR	Mass Accumulation Rate
MDA	Minimum Detectable Activity
NAA	Neutron Activation Analysis
NORM	Naturally Occurring Radioactive Material
PCB	Polychlorinated Biphenyl
PEL	Probable Effect Level
PLB	Phosphate Loading Berth

SAR	Sediment Accumulation Rate
SD	Standard Deviation
SSM	Swedish Radiation Safety Authority
TECDOC	Technical Document
TEL	Threshold Effect Level
TEPCO	Tokyo Electronic Power Company Holdings
TOC	Total Organic Carbon
TR	Términos Lagoon
XRF	X-ray Fluorescence

CONTRIBUTORS TO REVIEW

Abril-Hernández, J. M.	University of Seville, Spain
Al-Rousan, S. A.	The University of Jordan, Jordan
Barsanti, M.	Italian Agency for New Technologies, Energy and Sustainable Economic Development, Italy
Eriksson, M.	Linköping University, Sweden
García-Tenorio, R.	University of Seville, Spain
Godoy, J. M.	Pontifical Catholic University of Rio de Janeiro, Brazil
Heijnis, H.	Australian Nuclear Science and Technology Organization, Australia
Laissaoui, A.	National Center for Nuclear Energy, Science and Technology, Morocco
Reguigui, N.	National Center of Nuclear Sciences and Technologies, Tunisia
Ruiz-Fernández, A. C.	National Autonomous University of Mexico, Mexico
Sanchez-Cabeza, J. A.	National Autonomous University of Mexico, Mexico
Strumińska-Parulska, D.	University of Gdańsk, Poland
Telfeyan, K.	International Atomic Energy Agency, Monaco
Törnquist, P.	Linköping University, Sweden
Uddin, S.	Kuwait Institute for Scientific Research, Kuwait
Yii, M. W.	Malaysian Nuclear Agency, Malaysia

IAEA
International Atomic Energy Agency

CONTACT IAEA PUBLISHING

Feedback on IAEA publications may be given via the on-line form available at:
www.iaea.org/publications/feedback

This form may also be used to report safety issues or environmental queries concerning IAEA publications.

Alternatively, contact IAEA Publishing:

Publishing Section
International Atomic Energy Agency
Vienna International Centre, PO Box 100, 1400 Vienna, Austria
Telephone: +43 1 2600 22529 or 22530
Email: sales.publications@iaea.org
www.iaea.org/publications

Priced and unpriced IAEA publications may be ordered directly from the IAEA.

ORDERING LOCALLY

Priced IAEA publications may be purchased from regional distributors and from major local booksellers.

24-03449E

Zeitfracht Medien GmbH
Ferdinand-Jühlke-Straße 7
99095 Erfurt, Deutschland
produktsicherheit@kolibri360.de